COMMERCIAL STEEL ESTIMATING

Commercial STEEL Estimating

A Comprehensive Guide to Mastering the Basics

Kerri S. Olsen

Industrial Press

A full catalog record for this book will be available
from the Library of Congress.

ISBN: 978-0-8311-3439-6

Industrial Press, Inc.
1 Chestnut Street, South Norwalk, CT 06854
Tel: 203-956-5593, Toll-Free: 888-528-7852
Email: info@industrialpress.com

Sponsoring Editor: John Carleo
Copyeditor: Robert Weinstein
Interior Text and Cover Design: Janet Romano

books.industrialpress.com
ebooks.industrialpress.com

Table of Contents

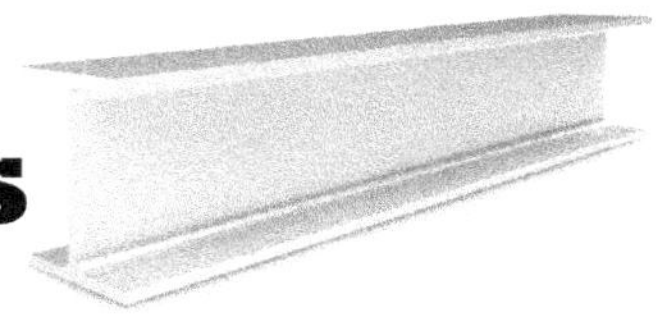

Acknowledgments

This book is dedicated to all who have encouraged its creation. To my colleagues and friends, Jerry Heltne, Harlan Meier, Tami Hoover, Peggy Breen, Jeff Morris, Jennifer Eschweiler, and Dan Wyland, I thank you especially. My development of this guide has been due to your continued support.

The following have kindly granted me permission to use many of the plans, sections, and details found herein:

- Lewis T. Griffith, P.E., with the City of Tacoma,Washington
- Scott Kuebler, with KPFF Consulting Engineers and KPG, Inc. of Tacoma, Washington

Introduction

Steel estimating is an intriguing career that requires a strong constitution, a gambler's heart, and an insatiable thirst for knowledge. The steel industry is in constant movement, with ever-changing conditions and processes that both simplify and complicate the ability to understand all there is to know about the trade.

Commercial Steel Estimating is designed as a comprehensive guide for anyone wanting to master the basics and know more about the steel fabrication trade. The book includes the following twelve chapters and additional notes and references:

Chapter 1, The Steel Estimator, describes in general the job requirements and the estimating process. Information is given on how work to bid is found and how an estimate is created.

Chapter 2, The Pricing Breakdown, lists the specific categories of that pricing to create a completed quote. The bid items listed there can be used in every project; they are intended to help you keep from missing any aspect of that quote. Each item is explained in detail so that you may understand why they are being used. This format may be utilized in some capacity for every quote, and serves as a good reminder for items that may otherwise not be considered.

Chapter 3, The Bid Letter, provides an example of a standard bid letter used to define the specific scope of work to be presented to your potential customer. The Bid Letter is considered a living and legal document; it is the guide for everything the quote contains—from the day the job is bid to the final closeout billing.

Each qualification, inclusion, and exclusion is listed for a specific reason, These reasons are explained in detail, some of which are further defined within subsequent chapters. Items for which you find only partial details and information in the bid documents need to be specifically established, identified, and described in the bid letter. At times you may need to provide quantities and sizes to define ambiguities in the bid set information. (The subsequent chapter *Reading Architectural Drawings* will show you how.)

Chapter 4, The Bid Documents, explains these documents, including how to interpret and apply the information they contain. These documents identify the scope of work intended for Steel Fabrication. Samples of a Division 5—Metals specification are shown, followed by an explanation of every point of condition that is listed together with what each one means to the steel estimator.

Chapter 5, The Contract Drawings, gives examples of some architectural drawings along with an explanation of how to read them. Detailed information on how to view and understand these drawings is provided through samples of plan views, elevation views, and section cuts, presented in a step-by-step format.

Chapter 6, Steel Materials, provides an introduction to the most common steel material sizes and shapes. Information is given regarding material grades, production, and common uses for the particular material types. This chapter will help you to become familiar with what the steel shapes look like while learning to read the contract drawings.

Chapter 7, Reading Structural Drawings, provides specific information in how to read these drawings using sample contract drawings and details. In reviewing the section cuts and detail views from the point where they are called out at the drawing plan views, the steel estimator will be exploring the nuances of reading and using structural steel contract drawings.

Chapter 8, Structural Steel Material Listing, uses the same contract drawing samples that were presented in Chapter 7, and explains how the information for materials is derived from the structural contract set and is then transformed into steel items to be fabricated. Information on how to calculate the lengths and sizes of parts and pieces is provided in detail. Sample take-off sheets listings are shown at the end of each example.

Chapter 9, Reading Architectural Drawings, explains the process of reading and interpreting the information presented in these drawings in such a way that the reader can create a listing of steel materials. Samples for improvising steel fabrication information that is not specifically shown at the architectural drawings are given for both the bid letter and the take-off listing.

Chapter 10, Fabrication Labor, provides commonly-used theoretical labor time applications as well as the thought process in applying those labor factors. Information is included on welding and weld symbols, and how to read them. Sample calculations with extensions demonstrating the application of shop labor factors are shown.

Chapter 11, Paint and Painting, reviews sections of a typical specification on Painting and Coatings Section 099000. It explains how the information required for the steel fabricator is derived from this document. Information on applying paint labor and theoretical paint coverage is presented along with samples of how that information is utilized.

Chapter 12, Shipping and Handling, offers basic information with regard to obtaining pricing for trucking—it calculates the steel to be shipped based on the take-off listing. Information and application of handling theory and labor is provided.

This guide provides the beginner and those less experienced in steel estimating with a strong foundation in the trade. All of the suggested practices within these pages are designed to help you stay on the track to success. Remember: the information provided must be adapted to suit the fabrication company for whom the steel estimator is working in order to achieve a successful project.

1

The Steel Estimator

The Job of Steel Estimators

Steel estimators identify all the steel items to be fabricated from the set of bid or contract documents. They list and price these items. They then provide a bid price to the potential customer, prior to their required bid date and time, in the format requested.

Learning to become a steel estimator takes time, practice, and repetition. It is a profession that is learned by doing. There is no set format for doing the job—only unwritten practices that have been handed down by predecessors to eager apprentices. Each estimator's skill is developed over time and trial. What works or does not work to gain success is learned quickly.

All the information needed to prepare a steel estimate with regard to take-off and pricing is indicated in the bid and contract documents. A complete price must include all the labor hours to fabricate the work; all costs for steel detailing, steel materials, fasteners, buyouts, forming, and specialty items; all paint and related labor; shipping; and last; all markup and profit extensions.

It is the steel estimator's goal to include everything needed for the completed scope of the work and still be the low bidder against the competition. Although this task sounds impossible, it is not because all the other steel estimators bidding on the projects have the same set of information. The utilization and interpretation of that bid set information, together with the knowledge and experience of the steel estimator, are the trick to being successful.

It is highly unlikely that any two projects in the commercial marketplace will ever be the same. Because all jobs are different, it stands to reason that the methods for steel estimating vary from project to project. Nevertheless, the basics applications, as defined in this manual, remain consistent.

Consistency in the work structure helps to avoid confusion, both for steel estimators and for others who need to use the information created within the steel estimate. Developing a consistent format for the take off and estimating practices will enable others to quickly understand the information provided. Because the specifics of the steel fabrication company's requirements will vary, the estimating format needs to be shaped individually.

The basic information contained in this book will get you started and move you along in the right direction. Your work activity will help you learn more. Each project may reveal something new and different in the plans and specifications. Be flexible in dealing with project inconsistencies. Learn to work with potential customers in order to resolve any problems that may arise from those issues.

Steel Estimating is a little like gambling—you throw down your dollar and place your bet. You never really know how other people arrive at their bid numbers. What is truly important is that you know how you arrived at your price. Don't spend a whole lot of time trying to figure out how your competition arrived at theirs—the unknowns will leave you with many unanswered questions. In the end, the project sells for what the market will bear. Stay on top of what the current market values are by obtaining post bid information whenever possible. Having this information will help you know where your finished pricing has to be in order to create a successful quote.

Mentoring with someone who has a successful background and years of steel estimating experience is a great way to get started. Build a strong foundation for yourself by working with those who are successfully established in the industry, and stick with the winners.

The ability to read the contract drawings and the shop fabrication drawings is essential to steel estimators. Acquiring a working knowledge of these drawings is important. By understanding contract drawings—what the steel fabricated items look like and how they are used—you will know exactly what to look for and where.

After they are familiar with the drawings, steel estimators must then learn to list the steel materials, also referred to as "performing a take off." Listing the materials from the contract drawings becomes easy once you know what you want to find. After the take off is complete, the pricing is installed and the shop labor is applied. All the rest is just math steps—with the extensions of weights, pricing, and markups—to arrive at the proposed sale amount.

It is essential that steel professionals know everything there is to know about the steel items being quoted. They must know the plan for detailing the steel; purchasing the materials, goods, and services; and the required time to schedule for the shop, as well as the anticipated final shipping date for the steel. As the steel estimator, you must create a good map to the project by way of your take off, using adequate and current pricing and labor hours; you should demonstrate the expediency and efficiency toward which the project managers and shop managers will need to perform.

Create a complete plan with everything itemized and listed, including contract drawing details and drawing references. Those who refer to the plan should be able to see the necessary steps for fabrication, using the guides provided for them to follow. This plan created by the steel estimators will be referenced during the entire course of the project. From bid day through to the final billing, all the information you put in your quote will be very valuable.

All information provided by steel estimators must be clear and concise. Any items that are not clear and concise will be the ones that will come into question later—guaranteed! When questions come up during the fabrication process, the steel estimate will always be referenced for answers as to what was bid.

Every job has situations that are unclear. Therefore, even the ambiguous items must be well documented along with what was priced to cover that work. The materials list together with the bid letter precisely documenting all of the information goes a long way towards a successful project.

Choosing a Job to Bid

Many shops assign their estimating department to their sales department. In many cases, the estimators double as the ones who will also choose the projects to quote. The estimator as salesperson is common practice with smaller fabrication shops; estimators have the working knowledge of the project they quoted or will be quoting, enabling them to work more efficiently with the customer.

Initially, choosing a job sounds like a simple task. One would think that if a project requires fabricated steel, well, you would then quote on providing that steel. However, when the bid documents are analyzed based on the items to be fabricated, the process of choosing or not choosing a particular job to quote becomes a bit more involved.

Selecting projects to quote for the shop is a responsibility that involves a realistic view of a combination of items. Use Table 1.1, The Elements of Project Evaluation, to help with your choices. Know where you are with these items listed before you choose to bid on a job.

Consider these items carefully. It is important to understand both your own estimating capacity and your work backlog. Don't overload yourself and create a quote in a rush if you can avoid it. Overselling the shop has negative implications to project schedules and cash flow requirements.

Steel estimators must understand the types of projects that are a successful fit for the company. If the company is interested in expanding or diversifying, then looking for small projects with fabrications that are different than what the shop normally performs might be prudent.

Table 1.1 The Elements of Project Evaluation

1. Your own strengths and weaknesses as an estimator
   - The type of project that fits your expertise
   - The time available in your current estimating work load
   - The magnitude of the scope of work

2. The type of work on which your company thrives
   - Commercial buildings, smaller or larger projects
   - Specialty fabrication (nonferrous)
   - Private fabrication

3. The type of work your company may want to do
   - Common steel fabrication
   - Specialty steel fabrication
   - Nonferrous fabrication

4. The size of your shop
   - Manpower—how many hours a week does your shop have to apply to the work
   - Physical characteristics of the shop
   - Square footage under roof and yard fabrication space
   - Handling capacity, overhead lifting, and rolling stock
   - Equipment capacity and versatility

5. Current work load
   - How this potential project will nest with the current shop work load
   - Project completion requirements-duration of schedule
   - Future work to help create a backlog

Be clear on the handling capacity of the shop equipment and what kinds of tools are available to use. Understand the specifics of the available shop equipment, such as:

- the throat allowance on the band saw
- the ironworker tonnage capacity for bending and cutting
- throat allowances for the different materials
- types and functions.

Know what type of welding equipment is available. Make a list of all the types, sizes, and capacity of

- overhead moving equipment
- forklifts
- hand trucks and dollies
- yard rolling stock
- small tools (like drills, grinders, beveling equipment)
- flat layout areas
- work tables
- saw horses

so that you are aware of the shop's resources as you consider your estimate for a project.

Understand the strengths and weakness of the shop with regard to personnel, tools and equipment, handling capacity, and floor space. Shop size has to be considered, not only the square footage, but also the height limits of the building, the allowance for door openings, and space for yard storage. Would special equipment have to be rented or purchased to work the job you are bidding? Are the fabricated members in the contract drawings extra large and heavy? Allowance for such conditions need to be made in the pricing, either by additional labor or a buy out of certain labor functions, including, for example, special cutting of extra heavy or oversize materials.

Rolling, forming, bending of materials, bevel cutting of plate, special shapes cut to size—choices about these items need to be made prior to quoting the projects. Will they be provided as a labor function of the shop or as a buy out item from a vendor?

Target the markets that contain the elements that the fabrication shop is best at building. In the quest to fill the shop, don't waste estimating time and dollars to quote on a project that has unrealistic fabrication and delivery requirements. If the shop labor staff is good at fabricating beams and columns, but lacks the expertise for fabricating stainless steel, then steer clear of bidding anything with stainless steel work. Avoid projects that may be too small or too large for the shop.

Check project specifications; make sure that there is nothing there that will disqualify your company. Watch for special requirements that you would not be able to meet—like being an AISC-certified shop or a union shop if you are not. Check on the project schedule that is indicated in the general conditions section of the specifications or may be advised by the customer; make sure that this schedule fits within current work load requirements.

Research past projects, especially if you are new to the shop where you are working. Get to know what jobs were successful and why. Also, get to know the jobs that were not successful and why not. Specific opinions on the good and the bad may or may not be the actual reason a project resulted in a loss for the company. There will be clues that will lead to the true answers when comparing the working file to the original quote.

Most companies now have computer programs for their job costing. Review the historical job cost files for projects to research. Once a project is identified as interesting, get the hard copy job file, then locate and review the original steel estimate. Review the entire quote for the total labor hours and all cost information that should be there.

If the project chosen to review was a successful project, then all too often it is automatically assumed that every element was completed under budget. However, even if the project as a whole was completed under budget, some specific fabrications may have had labor hours and costs that ran over budget, while others were under budget. Find out why. What did the job look like? Are the contract drawings still available for review? What about a set of shop drawings? Reviewing all of this information will enable you see how and why your company was successful with that work.

The goal is to find what works best for your company and focus on those elements. Find the work that suits your shop the best and you will be more successful as an estimator for them. Once you are clear on Table 1.1's The Elements of Project Evaluation relative to the fabrication shop, you will have better defined your direction. You will be less likely to waste time quoting projects that are not going to be beneficial for your company.

Plan Centers

Actually travelling to a plan center or a general contractor's plan room has become a thing of the past. Most plan centers and general contractor plan rooms are online; you can review drawings right from your desktop and print out your own drawings. Using online plan rooms takes much less time and effort than borrowing bid documents. Now you can buy a set or get one on deposit—even at night or on weekends. Online plan centers vary in costs and services. Some offer free support for the subcontractors to the general contractors that are members.

Construction Newspapers and Bid Advertisements

Bid advertisements can be found in construction newspapers. They often can come over the fax or via email. Public works projects may also be advertised in local newspapers or posted on government web sites. These are the most common ways to find open bid construction projects.

You may get occasional phone calls from general contractors or other customers requesting a quote on a specific project. In addition, someone from your company may drop a set of drawings on your desk. Phone calls to your favorite general contractor or past customers with an inquiry on any up and coming projects is a helpful way both to both get your name out there and to find out about any attractive opportunities.

Most of the time, you can tell by the project's title what type of project it is. You may also be able to tell just from the name of the job whether it is new construction or a remodel. For example (these are not actual names) "Joy Rider Middle School" might be a new school. But "Joy Rider Middle School Modernization" would be a remodel. Another example is "Big City WWTP." This name would be a Waste Water Treatment Plant. You can never really tell exactly what a project is until you look at the drawings, but the name of the project can be a pretty good indicator of what you might find there.

No matter what the resource, approach every job like it may be the right opportunity. Then consider The Elements of Project Evaluation while making your decision. If it passes all five, then the project may be a good opportunity for the company.

Compliance Requirements

Never knowingly bid work that is going to get your company into trouble! Positioning your company to take a job with a specific compliance your company cannot meet or specific scheduling requirements that are unrealistic is bad business practice.

If the project specifications indicate that structural steel fabrication is to be performed by an AISC-certified shop (see references) and pricing is provided by a shop that does not have this certification, this action may invite the opportunity for litigation should you be low bidder. Stay away from such projects. If the project schedule calls for steel delivery in six weeks and it would take your shop twelve weeks to complete, it may be best to pass on this opportunity.

Errors and Omissions

If you are aware that you may have missed some item in your quote that has made your number too low, make this known to the manager or owner as soon as it is discovered. Moving forward with a project knowingly without disclosing such errors is a serious disservice to your company. The negative ramifications of such action will never go away.

Exposing errors and omissions upon discovery will allow the company and the customer to process them. Errors and omissions eventually become apparent all on their own over the course of the work; they will make a bad situation worse. Steel estimators are only as good as their last job, after all. Thus, so much work is needed to cover all bases.

Bid Results

Within a few days of bidding on a project, call the company where you submitted your bid and ask for bid results. You want to learn how your number stacked up against your competition. Knowing the pricing helps steel estimators become aware of what the current market is doing, guiding them towards efficiency in pricing. Get the names of the other companies that competed in the bidding as well as their bid price, if you can. If you can't get exact numbers, ask for a percentage of the spread between the prices. After a few times of gathering this information, you may discover a consistency with the pricing and the companies that submitted them.

After gathering and reviewing the historical quote information, steel estimators may be able to surmise what the market will bear with the next quote. This accumulative history of projects quoted, together with what was won and by whom, provides clues as to which shops may be getting backlogged and which ones may become more aggressive in quoting projects.

Historical bid information helps steel estimators determine how aggressive they need to be, and whether their companies would benefit from expanding or switching markets. Check with the managers who make these types of business decisions to determine your company's sales direction.

The industry standard for the bids won vs. jobs quoted indicates that good steel estimators usually get one job in ten. If the steel estimators find themselves consistently or unexpectedly doing better than that, there may be concern for an error in the pricing extensions or for some other judgment error.

The remaining chapters in this book will break down all the information needed to create a complete steel estimate. The guidance and suggestions will be specific with regards to the materials application of your estimating and, in general, the applied shop labor, painting labor, and handling time.

By the time that steel estimators have finished listing all the materials, extended all the labor hours, installed the pricing, and completed the bid letter, they should have a working knowledge of the contract drawings well enough to "see" the entire job. Mentally the project is completed—all the steel beams, columns, and miscellaneous fabrications have been fabricated in the shop; loaded on trucks; and sent to be framed together in the field.

2

The Pricing Breakdown

We begin with the steel estimator's goal, which is to create a complete and comprehensive fabrication pricing of structural steel, miscellaneous steel, architectural metals, and any specialty fabrications as derived from a set of bid document drawings and specifications. To achieve this goal, you must perform a complete listing or "take off" of materials. You then add accurate pricing for all labor, materials, goods and services, painting, handling, and shipping necessary to perform the work.

Explaining this end goal in detail makes it easier for beginning steel estimators to understand the process of creating the steel estimate. For this reason, the pricing breakdown is presented first. Using simple elements, this summary itemizes the categories that comprise a steel estimate and creates the foundation for every quote.

Components of a Steel Estimate

Most projects have a common list of required components that make up the individual line items. We look to this list to assure that we do not to miss anything that may be needed in our pricing. Table 2.1 summarizes this list.

Table 2.1 Components of a Steel Estimate

Bid Summary	*Hours Weight*	*Cost lb/hr*	*Cost Extensions*	*Sale Amount*
Detailing	0		$ -	$ -
Labor	0		$ -	$ -
Materials	0		$ -	$ -
Hardware			$ -	$ -
Buyouts			$ -	$ -
Paint Material	0		$ -	$ -
Paint Labor	0		$ -	$ -
Galvanizing	0		$ -	$ -
Handling	0		$ -	$ -
Shipping	0		$ -	$ -
			$ -	$ -

Summary of Components

Steel Detailing	Cost to provide shop detail drawings
Labor Hours	The time for shop labor to fabricate the steel and furnish a competed project
Materials	All stock steel necessary to complete the project
Hardware	Anchor bolts, bolts steel to steel, headed anchor studs
Buyouts	Anything that is not fabricated by your shop, but must be pur chased from an outside source and used as part of your shop fabrication, such as grating, expanded metals, and brackets for railing
Paint	Shop primer and finish paint if required
Paint Labor	Labor to apply paint
Galvanizing	Cost to galvanize materials (add 10% more weight to the total weight of fabricated steel to cover for the additional weight of the galvanized material itself)
Handling	Staging raw materials in the shop work stations and moving finished fabrications to shipping area
Shipping	Trucking materials to and from sub-contractors and suppliers, as well as to final destination

Steel Detailing

Steel detailing is the very complicated and exact process of extracting all the steel items shown at the contract drawings, then accurately drawing or "detailing" these steel items to be shop fabricated in such a way as to be easily understood. Steel detailers will create shop fabrication drawings from the very contract drawings with which you are creating your quote.

The contract drawings show the steel support framing of the building as the finished product. The shop fabrication drawings will show the very detailed and specific information required to allow the steel shop welders to actually make all those parts and pieces intended for the framework. The steel detailers provide the steel erection drawings that locate each and every part used in the framing, including all of the miscellaneous steel fabricated items.

There are industry standards for the steel detailing process that we will not address in this book. As you are building your library of publications that are part of your trade, it is a good idea to get a copy of a steel detailing manual (see notes and references) to study and reference. This will not only help you to understand more of the specifics of what steel detailers do, but will also help you learn to read shop drawings if you are not already familiar with them.

It is important to understand that steel detailers take the estimated work to a finished product level. Steel estimators can get away with a plus or minus in inches or even feet in doing the "take off" or material listing, but steel detailers must be accurate to the nut with what they do. Every part and piece that they draw must be such that the steel will fit up in the field 100%. If the steel does not fit as it should, it will cost the fabricator (as well as the general contractor and all the other trade work that follows the steel) much in the way of time and money.

The steel detail drawings are typically a buyout item for most fabricators. Some steel fabricators do have in-house detailers, although the practice is not as common as it once was. Contract drawings used for bidding tend to be ambiguous in many areas. It is the steel detailer's job, together with the project manager and the steel estimator, to iron out all the issues as they move along with the drawings. Good project managers along with the steel detailers will perform an initial review of the contract drawings to identify early on some of the project's most glaring issues.

Requests for Information (better known as RFIs) are sent to customers to cover any questions that the steel detailers have regarding the job. The project manager is responsible for tracking them and making sure that the questions are answered in a timely way so that the steel detailers can continue with their work.

Creating and submitting RFIs is the job of the project manager and the steel detailer together. Steel estimators need to understand this process because they are the first to see these issues while creating the quote for a project. The need for RFIs becomes apparent to steel estimators during the take-off process.

Details will be found that don't look right, or are missing in relationship to the information that is shown in the drawing's plan views. This book includes sample conditions for possible RFIs within Chapters 7, 8, and 9. These sample conditions help estimators begin to understand what to look for.

Once you have decided to quote a specific project, it is best to alert your chosen steel detailers early on of your intent to quote. Ask them to provide you with pricing to prepare the shop detail drawings on the job. By contacting them early, you enable them to get your bid request in line with their own work load. They are more likely to get a price to you in a timely manner. Giving them a copy of your bid letter as a guide will help them create a quote more quickly. If you have any break out items in your final bid, be sure to let your detailer know so that they will break out their pricing accordingly.

Labor Hours

Labor hours reflect the amount of estimated time for the shop to fabricate the steel for the entire job. This labor time can be added to your quote items in several different ways—it should be specific to the way your shop needs the labor applied. Some shops labor every item individually; others labor in groups. In some cases, the individual part cuts, miters, cut to shape plates, holes, and welds might be timed, whereas others simply indicate a certain number of hours or minutes per types of parts.

However the individual labor times are applied, the total hours are reflected in this line item. The cost per labor hour will vary from shop to shop, by region, and by project requirements (prevailing wage, Davis Bacon wage rates, or others). The cost per hour is installed next to the labor hours for that extended pricing.

Materials

Materials will be the total cost of steel on the job. This cost should be the actual purchase amount of the steel that is needed to build the project, including waste. Material purchases will be based on available lengths from either the warehouse or the mill.

Material suppliers often have steel reference guides they provide for their customers to use. These reference guides provide specific data including material weights and stock lengths, designed to assist their customers with making a purchase.

Hardware

The hardware slot is where you enter all your bolts, nuts, and washers. Having this category itemized at the bid stage makes budgeting easier after you get the job; you will already be aware of how much of your budget is for hardware without having to go back and break that out.

For the hardware line item, we generally consider bolts for the steel fabricator's work only. Therefore, we include here anchor bolts for columns and embeds, steel-to-steel bolts for columns to beams, and beam-to-beam connections.

Buyouts

Buyouts can include grating, expanded metals, and railing brackets—anything that is necessary to complete the fabrication of the project. Some companies may require further breakdown of these items individually. If so, just make another line item for each one.

Large buyouts, like steel grating or heavy plate cutting, may be listed separately, including the weight of the item. This breakout will make it easier for other departments to see the applied budget. Supportive notes regarding lead time in scheduling and the timing of the financial commitments are helpful.

Paint

Because many shops do their own prime painting, it is good to have the cost of paint materials broken out for budgeting. This category is where you enter your estimated paint material cost. The company may also require costs for expendables associated with the paint, like cleaners, filters, or small tools (e.g., grinding wheels). Such expendable costs might be reduced to an average percentage for amortization purposes; this percentage would then be added to the actual cost of the paint.

Paint Labor

Structural and miscellaneous steels are generally prime painted unless shown as fireproofed or galvanized. Some shops do their own prime and finish painting; others ship the steel to a shop that will paint for them. This category is the place to enter either the labor hours to paint the coating or the cost of a supplier to perform that service. If painting is outsourced, costs may also have to be added at the line item for shipping and handling.

Galvanizing

Galvanizing is a buyout item that is often utilized for steel fabrications that will be exposed to extreme weather or caustic conditions. These items are broken out by weight, plus the additional (average) 10% weight factor for the galvanizing itself in order to get accurate pricing. Some shops consider galvanizing and some don't. Be certain. Ask if the 10% (or other factor) needs to be added to the base weight to verify that the correct pricing is being generated.

Galvanizing pricing is particular to the types of fabrications requiring galvanizing; therefore, a galvanizing shop will need to be contacted to get this information. A list of items needing to be galvanized, along with the weight and size for them, will need to be compiled to obtain a correct price. It is a good idea to compile and send this information to the galvanizer as early as possible prior to bid date, allowing them enough time to price it accurately.

The line item for galvanizing provides an alert to the budget. It points out scheduling and financial concerns that other departments will have regarding the items to be galvanized. Remember to consider the handling and shipping costs involved as well so they may be added at the proper line item.

Handling

Some shops use this line item as a way to identify the "extra" labor hours it takes to work through a job; others do not. For example, some especially large fabrications may need special staging or moving around during fabrication. The time required for these activities may not be actual fabrication labor; this category is where those hours would go.

Shipping

Some shops ship their own steel, and some shops hire trucks. Depending on the size of your project, even if you ship your own steel, the project may warrant outsourcing. If so, call your selected trucking outfit or two and get a price.

To give you a reliable price that matches your needs, trucking contractors will need to know the weight, the destination, and approximate ship date. Advice about any oversize or overweight items will need to be provided. See Chapter 12 for more information on shipping.

Completing the Estimate

After estimators calculate the individual category totals, these prices are entered into their respective line items. The result will produce a total cost for the fabrication project. From here, the extensions for pricing markups will be added. The items in this presentation may be manipulated to accommodate the

Table 2.2 Completed Bid Summary

Bid Summary	Hours Weight	Cost lb/hr	Cost Extensions	Sale Amount
Detailing	160	55	$ 8,800	$ 10,120
Labor	267	75	$ 20,025	$ 20,025
Materials	100000	0.63	63000	$ 72,450
Hardware			780	$ 897
Buyouts			280	$ 322
Paint Material	68	35	$ 2,380	$ 2,737
Paint Labor	102	75	$ 7,650	$ 7,650
Galvanizing	10000	0.57	$ 5,700	$ 6,555
Handling	8	75	$ 600	$ 690
Shipping	3	1250	$ 3,750	$ 4,313
			$ 112,965	$ 125,759

client's format, but the basics outlined in this chapter will remain a constant. Table 2.2 shows a completed sample.

Summary

This pricing breakdown will keep you from any omissions while preparing the final number. This list of basic components identifies the typical bid categories. Further explanation regarding their use and significance has been provided. As steel estimators work to compile all the costs associated with most projects, line items may be added for special purchases that don't fall into these categories.

3

The Bid Letter

Steel estimators must produce bid letters that properly convey the work proposed and the conditions by which the intended work is to be performed. The bid letter (or scope letter) to the customer should identify:

1. A complete list of items that are to be included and excluded as the condition of the pricing
2. A list of qualifications that indicates compliance with the project documents
3. The specific payment terms and conditions

Inclusions, Exclusions, and Qualifications

During the initial review of the bid documents and contract drawings for the project being estimated, steel estimators should list all the bid items to be included, excluded, and qualified within the project scope of work.

In the project specifications, the section for steel subcontractors is called Division 5, which is explained in Chapter 4, The Bid Documents. For commercial construction, customers expect all items indicated in the Division 5 section of the specifications—which is everything that is metal—to be included in the scope of work to be provided. The steel fabricator may or may not provide everything indicated at Division 5; there may be items listed in the specifications that will not apply to the project scope of work indicated at the contract drawings.

Specialty metals such as stainless steel or aluminum are most often provided by a specialty fabricator. You must make your customer aware of your intent to provide (or not to provide) such items by listing them specifically in the inclusions or exclusions part of your bid letter.

Structural steel, miscellaneous metals, steel joist, steel deck, ornamental metals, and cold formed metal framing are all items that are listed at their specific sections within Division 5 of the Contract Specifications. Read through the specifications carefully for items that you would provide; include these in your inclusions. All items that your company will not provide should be listed in your list of exclusions.

Composing and Sending the Bid Letter

Once you have satisfied yourself that you have listed everything on your project in your inclusions, exclusions. and qualifications, send your bid letter to your selected detailer for pricing. Your letter will be the guide for this job for all who are involved in the process, for example, detailing the items to be delivered and billed by the company shipping and accounting departments. It will also be the key for negotiations with the potential customer.

The take off should be listed in the same order as the inclusion/exclusion numbers on the drawings from which you are working. Items can then be easily located in some logical order. The ability to cross reference between the bid letter and the take off listing will save much in time and leaves little to question.

Furthermore, the bid letter should be sent to the potential customer a few days ahead of bid day, without the pricing. This advance notice will make the customer aware of what you have and don't have in your scope of work. Some readers tend only to lightly review these letters. Prior to the bid date, make a bid item review phone call to the potential customer referencing this letter. This communication would be helpful for everyone.

Bid documents often leave out specifics with regard to the steel fabrication. A conversation with your potential customer prior to bid day to discuss any items that are unclear or ambiguous in the drawings will go far to support a competitive price.

Sample Bid Letter

This Bid Letter is fundamental and comprehensive to any quote. One cannot afford to be complacent when preparing a bid document for pricing contract work.

Using this bid letter template will help you cover all the inclusions, exclusions, and qualifications that may be applicable to many different types of

The Company Name
1234 Address
City, Sate, 12345
Phone (000) 000-0000
Fax: (000) 000-0000
Email address

Date

Customer Name
Address
City, State and Zip Code

Attn: Estimator

Re: Project Name
Section 05120 Structural Steel,
Section 05500 Miscellaneous Steel

We are pleased to offer our quotation for the supply and fabrication of structural and miscellaneous steel on this project as follows:

Qualifications:

1. This proposal is based on drawings *(list all the drawings here that you reviewed by which you created your quote)* dated *(indicate latest revision date shown on the drawings)* as prepared by *(specifically write in the name of the architect and engineer, with the city and state they reside)*, together with Addendums *(list addendum numbers, or indicate that there are no addendums incorporated)*
2. All invoices are net 30 days with **no retention**. 1.5% interest on all overdue invoices. Detailing to be billed as a separate line item. A 50% deposit is required upon receipt of order to guarantee mill rolled materials, generate and purchase engineering, and shop detail drawings.
3. This quotation is **valid for 15 days** and is based upon labor and material cost as of the date indicated on this quotation. Awards subsequent to the 15 day period may be subject to mill price increase and / or scrap surcharges at time of material ordered.
4. This quote, if accepted, will form the basis of any purchase order awarded and will be included as an attachment to said contract as pertaining to specific inclusions and exclusions.
5. This quotation is based on availability of steel from the warehouse or mill at time of award.
6. Detailing and fabrication in this quote is based on release for construction documents being per “AISC, Code of Standard Practice” latest edition. **(Steel Fabrication is not performed by an AISC-Certified shop; all welders are WABO certified.)**
7. Delivery of steel is based on the shop schedule, with the first delivery in the time frame of 2-to-4 weeks after approved shop drawings.
8. Changes to submitted design by the contractor/owner or engineer on items designed by the fabricator will be incorporated as an added cost.

Figure 3.1 shows a sample of a typical bid letter.

(figure continues on following page)

9. This proposal is based on Prevailing Wage rates (if applicable) that are current as of the date of this bid document.

Inclusions:

1. Steel columns, beams, bracing; and girt framing; all materials A992 FY50 or A36 as required, inclusive of all connections and gussets; galvanized finish per revision A
2. Steel Stairs with railing to mezzanine and catwalk areas, as well as 250 ft of catwalk with railing, galvanized finish per revision A
3. Anchor bolts for columns, and A490 fasteners steel-to-steel framing for members provided by this fabricator
4. Design engineering performed by a state certified engineer
5. Detail shop fabrication drawings with erection plans
6. Total weight of project is 000 tons, freight allowed to destination with offloading and storage by others

Structural and Miscellaneous Steel Supply
Lump Sum <u>$0,000,000.00</u>

Exclusions:

1. All taxes (including all sales taxes), bonding, and **liquidated damages, builder's risk insurance**
2. Holes to and for other work
3. All light gauge (including 1/8" thick) and/or perforated plate, unless noted above
4. All penetrations frames through deck and roof for mechanical opening
5. Field erection and all field welding
6. Verification of all contract, missing elevations, dimensions, and field verifications
7. All engineering and stamped drawings, unless noted above
8. All costs for testing or special inspections, both shop or field
9. All metal decking and studs through deck
10. All stair nosings
11. Joist, joist girders, joist bridging, and bracing
12. Grout and concrete fill
13. Steel shims, backing plates, and sleeves
14. Finish paint, painting of galvanized items, field touch-up of paint or galvanizing, clear wax finish, blackened on steel parts, and prime paint
15. Unattached fasteners and bolts to or through wood and concrete (including lag screws, pins, and Hilti anchors)
16. All catch basins, manholes and manhole ladders, and trench drains
17. Downspout, gutters, hangers, and all appurtenances, roof drains.
18. All nonferrous metals not specifically noted at inclusions
19. Rebar and rebar not shop welded to our steel
20. Any framing for roof penetrations not listed specifically in our inclusions
21. All light gauge steel stud framing and appurtenances
22. Insulated panels
23. Bridge cranes and supports
24. Templates for anchor bolts

Please don't hesitate to call if we can be of any further assistance.

Sincerely,
signature,

typed name and title

projects. You can then add or delete items as appropriate. A letter of this style, used to define your proposed scope of work, will clarify that which you are providing, which you are not providing, and the contractual conditions by which you plan to complete your work.

Each one of the inclusions, exclusions, and qualifications has a specific purpose. The bid letter supports and protects both the fabricator and the customer from ambiguities and assumptions. The bid documents use some information that is general and other that is specific in application to the project, leaving the trade professionals to identify the difference. Any item left unattended and unaddressed by the bid letter may result in the steel fabricator bearing the financial responsibility.

Understand that the bid letter may be viewed as a contractual and binding agreement set forth by the steel estimator at the time the project is quoted. The terms and conditions stated within the letter enable the fabricator to produce the best and final low number. From the time the project is awarded through the last payment from the customer, this letter (together with any cited contract documents) is the guide by which the work will be performed.

During project negotiations, the Bid Letter often provides the only clarification of the steel fabricator's role. The letter specifies to the customer the steel fabricator's responsibilities in performing the work.

It is best to send the bid letter and the pricing prior to bid day. The customer then has ample opportunity to review these points of qualifications, inclusions, and exclusions, ensuring a complete scope of work. Most potential customers appreciate the clarity and completeness of scope as well as the opportunity to prevent future possible ramifications.

Qualifications

This section lists and explains the specific points within the Qualifications portion of the bid letter:

1. **This proposal is based on drawings** (list all the drawings here that you reviewed by which you created your quote) dated (indicate latest revision date shown on the drawings) **as prepared by** (specifically write in the name of the architect and engineer, with the city and state they reside), **together with Addendums** (list addendum numbers, or indicate that there are no addendums incorporated).

List all of the drawings that were used in creating your estimate. It is very important that this list be complete, including the revision numbers and issue dates, to provide verification of the documents used to create the steel estimate.

Contract drawings issued for construction may have changes that were not noted by addendum or any other means. When a set of drawings is issued with the intent for producing shop detail drawings, it is essential that the drawing list is checked—drawing by drawing and detail by detail—to verify that the drawings are the same as what was used in creating the quote.

If you forego this opportunity to perform a diligent drawing review, it is very likely that any changes affecting the scope of work won't be discovered until the project is in process. The window of opportunity to alert the customer of such changes and charge for that extra work will most likely be lost.

Document all changes, making certain that costs are covered with regard to any exposure to the steel fabrication. The differences in the drawings between the bid set and the construction issue set will come up eventually, and the steel estimate will be called to question. It is always that which is left undocumented or undone that will raise questions and create problems later.

2. All invoices are net 30 days with **no retainage**. 1.5% interest on all overdue invoices. Detailing to be billed as a separate line item. A 50% deposit is required upon receipt of order to guarantee mill rolled materials, generate and purchase engineering, and shop detail drawings.

Management makes the judgment call whether to accept or reject retention as part of the purchase order agreement. The reasoning behind retention being held on the billings is in case errors arise in the product or services being provided. If the steel fabricator is strictly a supplier, retention isn't as necessary because the steel fabrication is usually completed and installed before the end of the billing cycle.

If a fabricated item doesn't fit, then the "right to remedy" is executed and the steel is fixed or replaced right away. Once the fabricated steel has been delivered, and the steel fits, there is no reason to retain funds.

The 1.5% on overdue invoices is a given, a policy made by the company's accounting department.

The detailing is billed as a separate line item on the schedule of values for the job. Billing prior to providing the fabricated steel will ensure that the steel detailer gets paid in a timely manner. The steel detailer is on a 30-day billing cycle, like all the other subcontractors or suppliers.

The deposit requirement provides cash flow for steel and other materials on the project. Requiring a deposit is a good business practice; it secures the material delivery and enables steel fabricators to better control their schedule when the material resources are secure.

Financial markets require a quick turn on cash; deposits will secure the materials before they are sold out. Material availability impacts delivery schedules. The time span of 30–60 days may occur between award of a project and the shop drawing approvals. Often, materials are not purchased until the steel shop drawings have been approved. Customers need to be made aware that deposits are necessary to the project's success because every job has a quick turn schedule.

Most projects are quoted based on optimum project management conditions. Having ready cash flow for your suppliers is one of them. Financing costs to pay vendors, suppliers, and payroll cannot be bid into the project for you to be low bidder. The argument for deposits provides a good negotiating tool if you choose to use it. However, if this is not the case, just eliminate deposits from your list of qualifications.

3. This quotation is **valid for 15 days** and is based upon labor and material cost as of the date indicated on this quotation. Awards subsequent to the 15 day period may be subject to mill price increase and / or scrap surcharges at time of material ordered.

Validating the quotation is essential because the steel market is very volatile. If your customer waits too long to award a project, you may lose your low price on the steel detailing, the materials, and potentially all remaining buy out items.

Labor availability in the shop may change; you may need to allow for overtime costs once a job has been awarded to you. All of these factors change over the course of time from bid date to project award. Therefore, a validation time frame is important.

4. This quote, if accepted, will form the basis of any purchase order awarded and will be included as an attachment to said contract as pertaining to specific inclusions and exclusions.

It is very important that the bid letter tie into whatever contract or purchase order that your customer gives you. When you submit a bid on a project, it is contractually binding. Although the bid documents are the law for the project, your bid letter outlines the rules by which you will perform the work based on your pricing.

The contract specifications are usually generic in their presentation. Your work is specific to each job. Therefore, it has specific conditions by which it must be performed at the price you have submitted. For this reason, the bid letter must be included and must override any contract conditions that may conflict.

5. This quotation is based on availability of steel from the warehouse or mill at time of award.

Material availability changes all the time. The first step any estimator or project manager should take upon getting a new job is to check the steel availability. Steel mill rolling dates are undependable and materials move out of the warehouses on a daily basis. Materials need to be secured as soon as possible to insure availability for project fabrication.

Your estimate has a set budget for materials. It is very important that you maintain that budget. Unavailable materials or delayed deliveries due to the timing of mill rolling or shipments that come in from obscure locations may force you to acquire materials at a higher price, affecting both the budget and fabrication schedule.

Substitution requests can be made by petitioning the engineer for a different size material that is readily available in lieu of a size that is not. The paperwork involved in managing such a request may affect your schedule. Additional engineering costs may be incurred.

6. Detailing and fabrication in this quote is based on release for construction documents being per “AISC, Code of Standard Practice” latest edition. **(Steel Fabrication is not performed by an AISC-Certified shop; all welders are WABO certified.)**

Make certain of the AISC status of your company. Be advised that if you are not an AISC-certified shop, you may not bid on work that requires it. *Complying* with the AISC Code of Standard Practice is an industry standard; most shops comply.

Make certain that the company's specific status with regard to AISC membership or union involvement is clear to your customer—for their own safety as well as yours.

7. Delivery of steel is based on the shop schedule, with the first delivery in the time frame of 2-to-4 weeks after approved shop drawings.

Any changes in delivery for any reason are not viewed well by the customer—unless they are improvements! The steel package is the "skeleton" of the building; as such, scheduling delays will have a domino effect on all the other trades down the line.

Steel delivery to the job site is set prior to your even quoting the job. Your shop fabrication schedule isn't determined until you have approved shop drawings returned from the general contractor. Project scheduling information is usually available in the general conditions portion of the specifications. This delivery statement is intended to help protect you from delays caused by circumstances beyond your control, with approval of shop detail drawings being one of them and a potential major cause for holding up the project.

8. Changes to submitted design by the contractor/owner or engineer on items designed by the fabricator will be incorporated as an added cost.

Changes happen all the time during shop drawing approvals and even after approvals. This statement is a continuation of that which is already in the AISC code of standard practice. If the architect or engineer makes a change that will cost your company extra money to create, then you have the option to put a cost to those changes and submit them to your customer.

Require written approval on the funding from your customer for the extra work *before* you start that work. If work proceeds without agreements in place, the steel fabricator runs the risk of being liable for all costs incurred.

9. This proposal is based on Prevailing Wage rates (if applicable) that are current as of the date of this bid document.

Watch the wage rates because they do change periodically. Prevailing wage or Davis Bacon requirements may put your shop fabricators at a higher rate than they normally earn and create additional paperwork for your accounting department. The Davis Bacon Act was created to force trade pay scales to be equal to union scale requirements.

Inclusions

This section lists the inclusion items in the bid letter, with an explanation of each numbered item.

Inclusion #1

In this section, list the main framing members. Include the material grades as well as information regarding the connections and the finings of the steel. This sample letter style has the entire structural package all included in one bid item. A more specific breakdown is preferred, with a line item for each component of the building. The breakdown would also list the plan drawings with the applicable details, and drawing references with quantities as required. Indicate the type of finish to be applied to the material.

Inclusion #2

This bid item talks only about the steel stairs together with the catwalk and railing as a complete system. Note that the total linear feet of railing is indicated with the finish as being galvanized.

Inclusion #3

In this listing, estimators are specific about the fasteners being provided. This information is important because steel fabricators should not be responsible for supplying fasteners to be used for connection to other trade materials.

Inclusion #4

Estimators indicate that their quote includes the design engineering for the fabrication per some requirement in the documents. This information is important to note specifically because design engineering is an expensive item. For the most part, steel fabricators tend to omit this listing because the pricing can vary so much from engineer to engineer.

Inclusion #5

In this inclusion, estimators affirm that shop detail drawings will be provided inclusive of erection drawings. Some projects have the steel detail drawings already provided to them.

Inclusion #6

Advising potential customers of the total weight of the project allows them a frame of reference to better compare your quote with the other prices they get from the competition. Customers need to know about the shipping and offload point; it allows customers to better evaluate your quote with what your competition may or may not have.

Note: Additional inclusion items may need to be listed according to the items requiring fabrication, as shown at the contract drawings.

Exclusions

1. It is important to show clearly the exclusion of any additional costs associated with doing business. Taxes, sales taxes, bonding, and liquidated damages are all costs that are referenced in the general conditions. This exclusion eliminates any assumptions that might be made about their provision

2. This exclusion prevents any assumption that the steel fabricator will provide holes in the steel for other trades to use. Such work by the fabricator is not shown in the contract drawings and cannot be assumed as part of the labor for your portion of the work. Excluding holes for other trades is a standard exclusion for every job.

3. All light gauge materials less than 1/8" are a standard exclusion; the light weight materials are in a different section than the Division 5 (see explanation of Division 5 requirements in Chapter 4, *The Bid Documents*) that we are working with. Light gauge materials are often used for flashing, roofing, and cold formed metal framing.

4. Framing for openings are usually shown as a typical detail. If the contract drawings specifically show locations and details

for floor and roof openings, then it is clear the steel fabricator would include them. Typical roof opening details are a gray area because there usually are no references to them in the floor and roof plans. For this reason, they are excluded. Unit prices on fabricated items may be offered as an alternate.

5. Field erection and all field welding are standard exclusions if the steel fabricator is not installing this finished product.

6. Verification of missing dimensions is a standard exclusion designed to protect the steel fabricator from exposure to field work.

7. This exclusion lets the contractor know that the steel fabricator is not including any costs for engineering.

8. Costs for special inspection are picked up by the owner as a rule. However, because the specifications put the inspection requirement in the Division 5 section (see Chapter 4), it could be mistaken that the steel fabricator is responsible for paying for them.

9. The section for metal floor deck and roof deck is in Division 5. Therefore, it is to be excluded if the steel fabricator is not providing this material.

10. The nosings for stairs are a buyout item installed in the field at the time the concrete is poured. Customers do not need to have the steel fabricator supply something they can buy themselves, saving themselves the additional profit markup.

11. Steel joists are also in the Division 5 section. Be clear whether or not they are included in your pricing.

12. Both the grout for leveling the columns and the concrete fill for stair pans and landings are handled by others. They are standard exclusions.

13. Steel shim material and backing plates are usually light gauge steel. The referenced sleeves are not sized at the specification; they could be light gauge as well, or any size for that matter. Steel fabricators have to be clear whether or not they are providing these items.

14. Steel fabricators need to reinforce that they are not providing any special finishes to the steel.

15. This category clarifies the provision of fasteners, which are a standard exclusion.

16. Catch basins, manholes and manhole ladders, and trench drains are typically shown in civil drawings. The mistake is often made, however, that because the items are steel, the fabricator would have them. In most cases, these are buyout items that the general contractor provides.

17. Downspouts and gutters are in the roofing section, although on occasion either they are designed with special attachments or the downspouts themselves are made from pipe that would be designated in Division 5. The mistaken assumption could be made that the steel fabricator will provide them. Thus, it is best to exclude them.

18. Stainless steel, aluminum, and brass are all nonferrous materials. These items are considered metal by the designers and are included in Division 5 of the specifications. Not all steel fabricators get involved with providing the items that are not carbon

steel. If steel fabricators are not providing these items, then this exclusion must be clearly stated.

19. Rebar is an item that is included with the contractors providing the concrete and will be found in Contract Specifications Division 3, as explained in Chapter 4. This exclusion is standard, provided for clarity.

20. Roof penetration framing is sometimes shown as a typical detail in the contract drawings. Typical details force assumptions to be made by the estimator. This exclusion then notifies the customer to look to the inclusions for advice on this provision.

21. There are specialty companies that supply steel stud framing. There is a separate section at the project specifications that allows for provision of these items. This listing has been provided as a standard exclusion to be clear that the framing is not the fabricator's/estimator's responsibility.

22. Insulated panels are provided by specialty companies that bid directly to the contractors. This is a standard exclusion that is provided for clarity

23. Bridge cranes and supports are sometimes included in the building framing drawings. The provision of them is listed in the Division 5 specifications. Bridge cranes and their supports need to be specifically designed and engineered for their specific use and location. Due to the complex nature of this provision, bridge cranes are often purchased directly from the vendor by the contractor.
24. Templates for anchors are extra work. Anchor bolt placement is

shown at the shop fabrication drawings. If templates are included in the steel fabricator's price this does not help you come in with a low bid. If the customer decides they want them, a separate price may be offered.

Note: Additional exclusions that need to be listed will be determined by the actual bid documents of the job being quoted.

As steel estimators create the bid letter, it is easiest to list the items as they are found on the drawings during the initial review. Starting the letter while going through the drawings will lead to a more thorough guide for take off and prevent duplication.

Creating the bid letter first is an activity that will also prevent double listing items between the architectural and structural drawings. By listing the individual details that apply to locating specific items, the estimators can more quickly understand the project. Additionally, the estimators may become aware of missing or unclear information early on in the estimating process, providing time to send out an inquiry to the architect or engineer for clarification prior to bid time.

Summary

This chapter has reviewed the significance of providing complete and accurate bid letters to the customer. A copy of a standard bid letter was presented, followed by a detailed description of each item with the explanation of why each one was being used. The bid letter is an important part of the estimating process; it should be utilized for every job quoted.

4

The Bid Documents

What are bid documents? Also referred to as *contract documents*, bid documents are the set of plans and specifications, as well as any addendums, that are released to the public for bid or quoting by the building owner's representative. The plans are all the drawings that show how the building is built; they are intended to contain all the information necessary for each trade to bid its portion of the work. The specification manual that accompanies these drawings describes all the requirements by which the contract is to be executed as well as the scope of work for all the trades that are to be involved.

The owner's representatives for building projects are the design team hired (by the owner) to create the building. This team generally consists of the project architect, structural engineer, civil engineer, mechanical engineer, and electrical engineer, with the architect as the head of the group. With the expertise of this team, all building codes and requirements are met. In addition, all the information necessary to convey those requirements is compiled to create a complete set of plans and specifications for the project.

The Project Specifications

The project specifications, generally assembled by the architect, are the guidebook for executing the work. This document is intended to govern the general contractor as well as the sub-trades bidding the job. The project specifications contain the instructions for all the trades involved in creating the building. They include all the necessary information from contract issuance through contract close out.

Reading the information in a book of specifications is as exciting as watching paint dry. Yet it is very important that you proceed slowly, making a concentrated effort to understand all the information and the way it is presented. Be especially careful at interpretation. Language and how it is used is everything here. Interpretations can and do vary according to need (for example, the steel fabricator versus the customer and document writer). If you have questions about items to be supplied, get the correct information in writing from the architect or engineer instead of making your own assumptions.

The General Conditions section will specify the contract and delivery expectations. The supply of steel or metals will be covered in what is called Division 5—the chapter of the specifications that will contain subchapters (or Sections) for everything considered metal by the designers. Structural steel, miscellaneous steel, and ornamental metals are examples of some of the titles for these sections. The Paint and Painting provisions are in Section 09900, which is a subchapter of Division 9, *Painting*. This section will be found near the middle of the general conditions, and will provide the complete information regarding the exterior finish of the steel items to be fabricated.

Read through these bid documents before you start developing the list of materials. There may be items that either do not apply or have a specific impact on the items that steel fabricators are required to supply.

It is important to list both the items that do apply and those that do not apply to a specific project. The reason is that all too often information is just copied from one job to the next without much thought to the specific conditions of the newer job. You must refer specifically to what exist in the contract drawings. It is best to find out for certain what applies and what does not; otherwise the job could be overbid or underbid.

After you have been through a few specification manuals for various jobs, you will begin to understand the difference between typical information and project specific information. Often both will be used in the creation of the documents. You will then be able to identify the general format and construction of the specification manual. In turn, you will be able to target those issues that have the most impact on your work.

The project specifications begin with Division 00—Procurement and Contracting Requirements. This division indicates the qualifications that must be met by the contractors bidding the job.

The rules that govern the completion of the work are in the General Requirements or General Conditions, which explain everything with regard to the information of the architect and engineers of record, the project location, the bid invitation, the instruction of bidders, the execution of the contract (including first billing information through to the final terms), and conditions of contract close out.

Following the general conditions, the specification book is broken down by construction divisions relative to the specific trades. Sections within these

Table 4.1 Divisions of Sample Specifications

Division 1—General Conditions
Division 2—Site Work
Division 3—Concrete
Division 5—Metals
Division 6—Wood and Plastics
Division 7—Thermal and Moisture Protection
Division 8—Doors and Windows
Division 9—Finishes
Division 10—Specialties
Division 11—Equipment
Division 12—Furnishings
Division 13—Pre-Engineered Structures
Division 15—Mechanical
Division 16—Electrical

Note that there are no Divisions 4 and 14; these are held for cases where special reference might be needed. Generally the rest of the numbered divisions remain the same.

divisions identify the different types of work to be performed by each individual trade.

Sample Project Specifications

The steel fabricator's scope of work is always defined within the project specifications as Division 05—Metals. This division contains all the information for those construction items related to steel. Sections for structural steel, miscellaneous steel, steel joist, steel deck, architectural metals, steel stairs, grating, and railing are found in Division 5. Generally anything that is metal will be included in this division.

The sections in Division 5 are numbered. For example, Section 051200 Structural Steel and Section 055000 Miscellaneous Steel list some of the major components that the designer intends you (steel estimator) to supply as part of your package. Check any other sections within Division 5 for which your company is supplying materials.

Be aware that designers are not consistent about how they construct the sections for Division 5 items. For example, steel stairs and rails generally are included in the 055000 Miscellaneous Metals, but sometimes they may have their own designated sections. The decision to provide a separate subsection for certain fabricated items tends to be at the whim of the designers; it may be

Section 051200-Structural Steel Framing

Part 1—GENERAL

1.01—RELATED WORK SPECIFIED ELSEWHERE

A. Drawings and general provisions of the Contract. Including General and Supplementary Conditions and Division 01 specifications sections, apply to this section.

This section is telling you that you need to comply with the provisions in bid document drawings and the particulars in the general conditions as they apply to you.

1.02—SUMMARY

A. Section includes structural steel and grout

The structural steel and grout statement is a standard provision. It is particular to fabricators who are erecting the steel structure needed to grout in columns and embedded steel items that are supplied by the fabricator.

1.03—DEFINITIONS

A. Structural Steel: Elements of Structural Steel frame, as classified by AISC 303, "Code of Standard Practice for Steel Buildings and Bridges"

This statement is a reference to the ISC Code of Standard Practice, the guide by which the engineer designs the steel framing in order to comply with the design requirements. It is important that you have a copy of this manual for easy reference to material sizes for use in your estimating. (See section on references for more information.)

1.04—SUBMITTALS

This section contains a standard list of submittal requirements for the project manager to give to the general contractor upon your project award, and continuing through the project as the need arises.

1.05—QUALITY ASSURANCE

This section contains a list of the quality assurance requirements that fabricators must comply within their performance of the work. If the shop must be " qualified fabricator that participates in the merican Institute of Steel Construction (ISC) Quality certification Program and is designates as an ISC-Certified Plant," then the project cannot be quoted by a non- ISC-certified shop.

ISC certification requirements are important to look for first if you are looking for jobs to quote. If your shop is not ISC Certified, then all the structural steel found on the 'S' drawings for the project is off limits. You may, however, still quote the miscellaneous steel items found in the architectural drawings. Not all projects have this requirement,

Figure 4.1 Sample of Section 051200—Structural Steel.

Section 055000—Metal Fabrications

Part 1—GENERAL

1.01—RELATED SECTIONS

This section lists the other sections that are understood as part of the Metal Fabrications work. Specialty fabricators who work solely with stainless steel, aluminum, and bronze will often pick up items in this section and quote them directly to the general contractor. If you are working for a fabricator that does not do nonferrous work, then be sure to list that in your exclusions in your bid letter. Otherwise, this list of related sections will alert you to the items that you need to include.

1.02 REFERENCE STANDARDS

The reference standards section refers to the material requirements for both the steel and the fasteners used in Metal Fabrications.

1.03 QUALITY ASSURANCE

This section gives advice on the experience requirements of the fabricator and the expected quality of the finished products.

1.04—SHOP DRAWINGS

This section gives the requirements for shop detail drawing presentation.

Part 2—PRODUCTS

2.01—MATERIALS

This section lists the material products to be used in fabrication, as well as cleaning and finish requirements.

2.02—FABRICATION AND METAL FABRICATIONS

This section gives the specific finished product preferences.

2.03—ANCHOR BOLTS

This section lists the type of anchor bolts to be supplied with the Metal Fabrications items

PART 3—EXECUTION

Figure 4.2 Sample of Section 055000—Metal Fabrications.

based on the quantity of the individual items to be fabricated. For example, if there is not much to be fabricated in the way of steel stairs and railing, then the guide for fabricating them would be included in the section for miscellaneous metals. If there is a significant amount of this work to be performed, however, then the designers would create a separate subsection for stairs and another one for handrail.

Read through the specifications carefully. Be sure you comprehend the information as the writer intended. When in doubt, ask questions—whether of a work colleague, the customer, or the general contractor. You can also call the architect or engineer to be sure you get a correct interpretation. It is always good to obtain these clarifications in writing.

Figure 4.1 shows a typical description of the items found in Section 051200—Structural Steel. Figure 4.2 then shows a sample of Section 055000—Metal Fabrications. Typical headings are shown as examples, followed by explanation of the items in italics.

Additional Comments

References to other trades often occur within Division 5. An example is the common reference to Section 099000 for Paint and Painting. The reference to Section 099000 for Paint and Painting may be necessary to direct your attention to the specific painting or galvanizing requirement that will apply to the fabricated steel. When you find these references, read them carefully so that you will be aware of any labor or material cost impacts to your pricing.

You must be aware of any references to other trades and make them clear to others in your bid letter's provisions.

If the steel fabricator fails to specifically address whether these sections are included or excluded in the bid letter, the contractors bidding this project will assume by your omission that you are including all items in your Division. Therefore, the steel fabricator could end up being required to provide or supply items that were not priced in the quote.

These sections are written as guides; much of the information may be borrowed from other similar projects. Sometimes the information given may not apply to the job you are bidding. Know that the specifications will list items both generally and specifically. Therefore, you must ferret out from the set of contract drawings all of the structural and miscellaneous steel items to be fabricated. Then be clear and concise about your provisions in the scope letter with the quoted price.

While working on the take off, steel estimators become aware of what is and is not being used in the specifications. Items that do not apply to their scope of work are listed at the exclusions in the bid letter to the customer. Prior to the bid date, it is good to provide a scope letter to the general contractor prior for their review. You can then bring to their attention any issues that you see within the scope of work and come to an early resolution of those issues. Unanswered questions may cause you to add extra money to the bid price to cover the unknown. To assure you are providing a competitive bid, clear up these issues early.

If the project is smaller, the specifications may not be broken down by divisions. Still, there will be some information provided in the project specifications as to what the designer expects you to provide.

If no information is provided for the steel fabricator in the way of specifications, any standard set of specifications may be used as a guide to help with what should be included or excluded with the quote.

If you have difficulty finding information on the contract drawings and specifications, you can always call the architect or engineer who created the documents. Sometimes details are not shown where they should be, or more clarification is needed for a certain condition in the drawings.

Discovery of missing information or mistakes within the bidding specifications and drawings that the architect and engineer have released will create an Addendum—a revision to the contract set. Questions that the designers are asked to clarify ambiguous conditions, and the answers to those questions, are clarified in an Addendum to the bid documents. Addendums are intended to help everyone understand how to assimilate the information correctly.

Summary

This chapter should help steel estimators understand how the project specifications affect the creation of a steel estimate. It reviewed the specific sections of Division 5—Structural and Miscellaneous Steel, showing what the different elements within the specifications mean for steel fabricators and how these items affect pricing. Particular attention has been given to key items within the bid documents that may need to be added at the inclusions, exclusions, or qualifications in the bid letter.

5

The Contract Drawings

Contract drawings are the group of drawings that are created by the architect, structural engineer, civil engineer, mechanical engineer, and electrical engineer.

Numbering Systems

These drawings are categorized by Civil, Landscape, Architectural, Structural, Mechanical and Electrical sheets. The Civil drawing number system will begin with a *C,* landscape drawings with an *L*, architectural drawings with an *A,* structural drawings with an *S,* mechanical drawings with an *M,* and electrical drawings with an *E.* Most numbering systems for contract drawings work this way, but there are the exceptions. Therefore, you will need to temper this information with what you are given for your prospective project.

Steel for the project is mostly shown in the *A* and *S* drawings. On occasion there may be steel items that are shown only on the Civil, Landscape (e.g., pipe bollards, site handrail, trash enclosures), and Mechanical drawings (support brackets for mechanical). These would need to be provided by the steel fabricator. Therefore, all drawings will need to be checked for steel items in the initial review prior to take off.

The numbering system that is used most often for each group begins with the letter to designate the type of work. After the letter, the drawings are placed numerically, although there appears to be no standardized system. Most designers use a numbering system that will let them insert drawings later, providing for more details. The need for this flexibility is individual to the job. Therefore, numbering systems vary, although generally the lettering indicators remain the same.

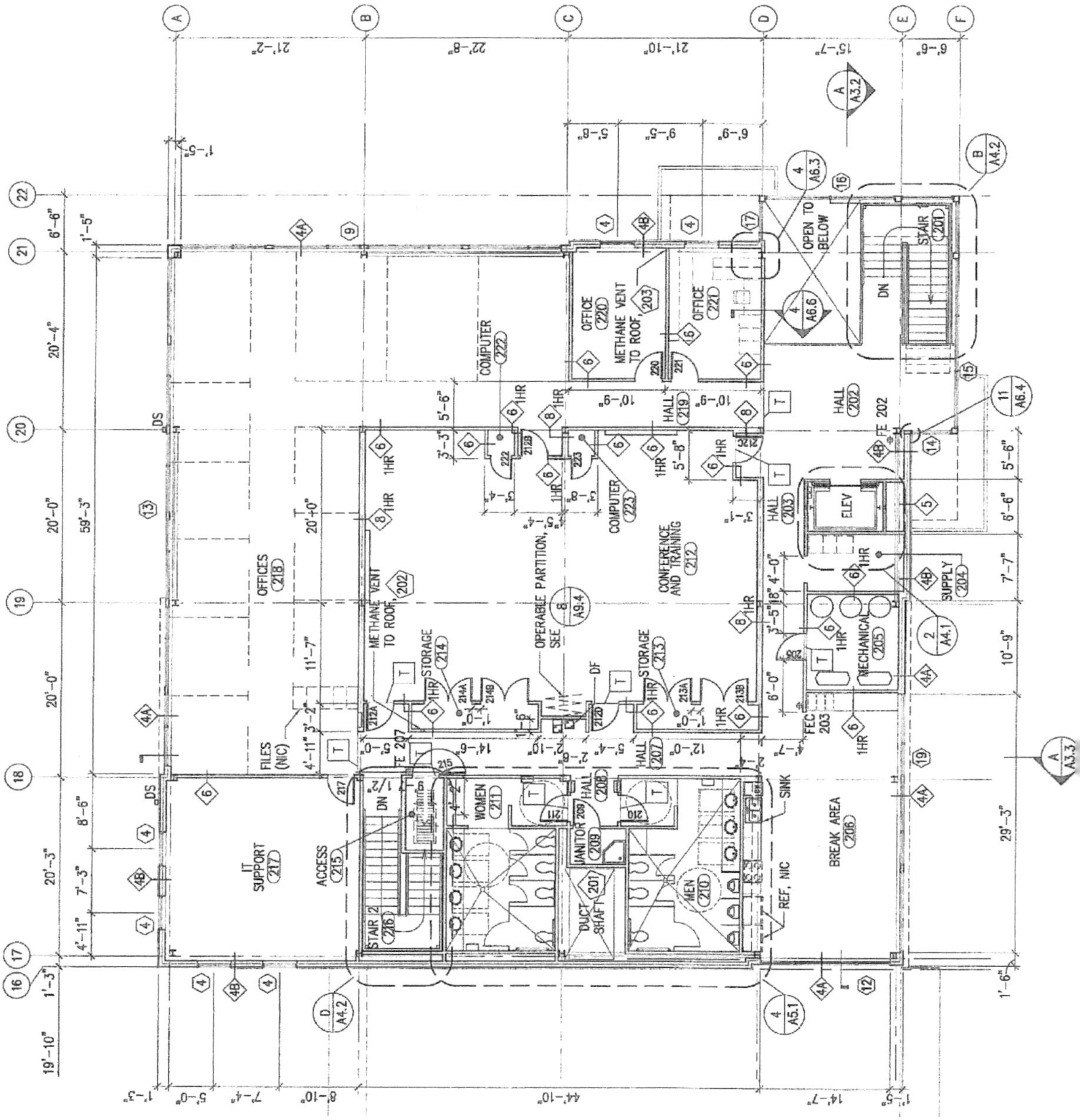

Figure 5.1 Plan view of a building's foundation

Reading Contract Drawings

Although there are industry standards regarding the creation of drawings, different design teams have their own *style of presentation*. As with any spoken language, inflections and accents come out in speech; the same happens in bid document drawings. When you begin to recognize these design professionals' traits, the drawings become easier for you to read and understand.

Reading contract drawings is a skill you learn by doing. Designers have individual styles for presenting the information in their drawings. It's your job to understand their language. If you don't understand what they are trying to show, find out. Call the architect or engineer who prepared the drawings; ask them to help you. Another option is to contact one of the trusted general contractors for assistance. Ultimately, the best help for interpreting drawings will come from the designers who created them; after all, they should know what they were trying to accomplish.

Sometimes the detail cuts are in the wrong place or have not been made where they are needed. Catching such errors or omissions can be important not only for the steel estimators, but also for other contractors who use the drawings.

A *Plan View* drawing is a map or a floor plan where you are looking down on the building. It is standard that the drawing is laid out so that the nominal north of the building is at the top of the sheet, south at the bottom, and thus west at the left and east at your right. (*Note:* The drawing will also indicate "true" north, which may not be the same as the directional north on the sheet.) This method, which makes the drawings easier to read, ensures that everyone is on the same page.

To explain how to read these drawings, we will look at a typical floor plan view and an elevation section, then review the subsequent details given to help us see the designer's intent.

Although every project is different, the general application of the plans, section views, and detail views remain the same. This is true for the architectural drawings and structural drawings alike. Look for clues like grid line locations, which will remain consistent from the plan view to the elevation views, to verify how you view the picture. Elevation cuts and detail section arrows point the way that you need to look when reviewing the sections and details.

Keep in mind the size of the elements you are viewing. Buildings can be hundreds of feet long; the grid lines section the buildings into 20-, 30-, and sometimes even 40- or 50-foot increments. Each section and detail gives you closer views of the building, much like a microscope gives you a closer look.

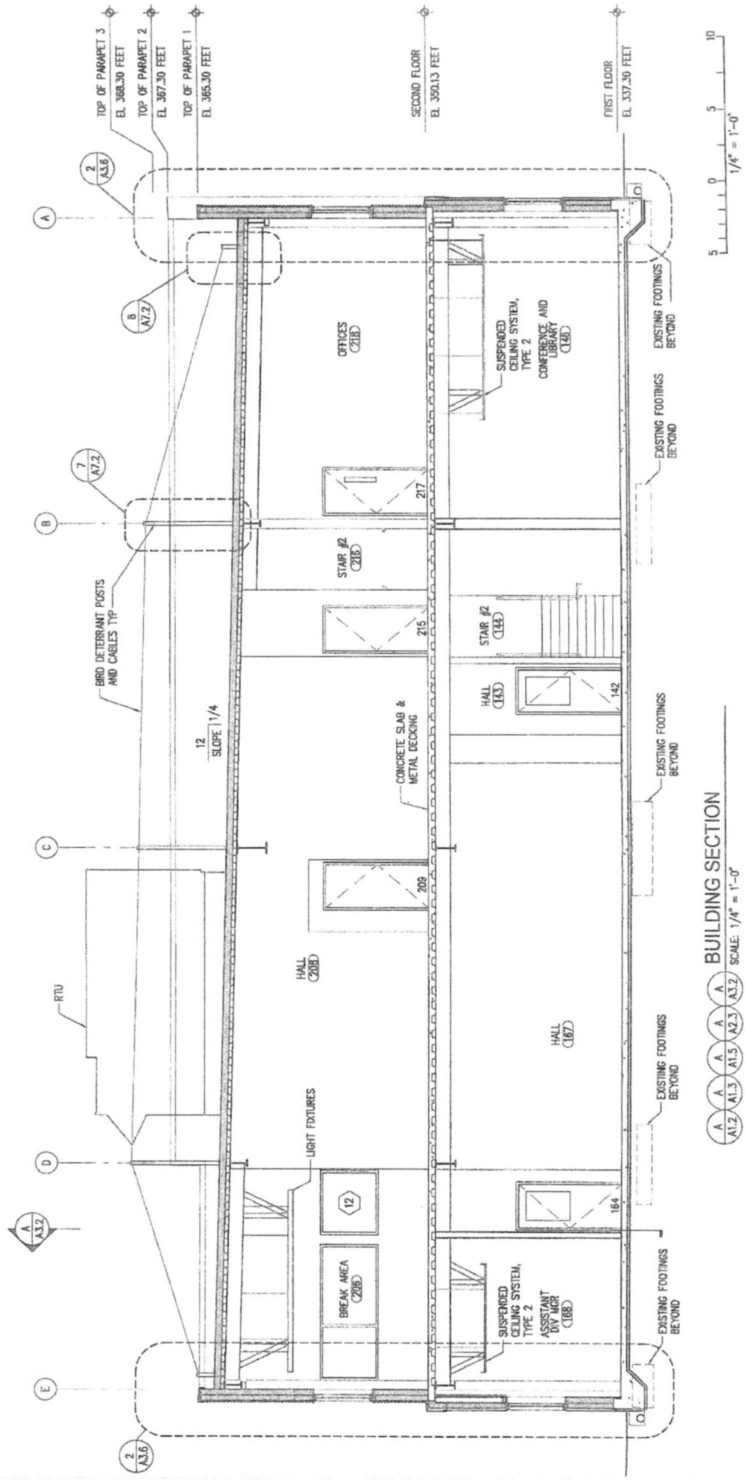

Figure 5.2 Building elevation views

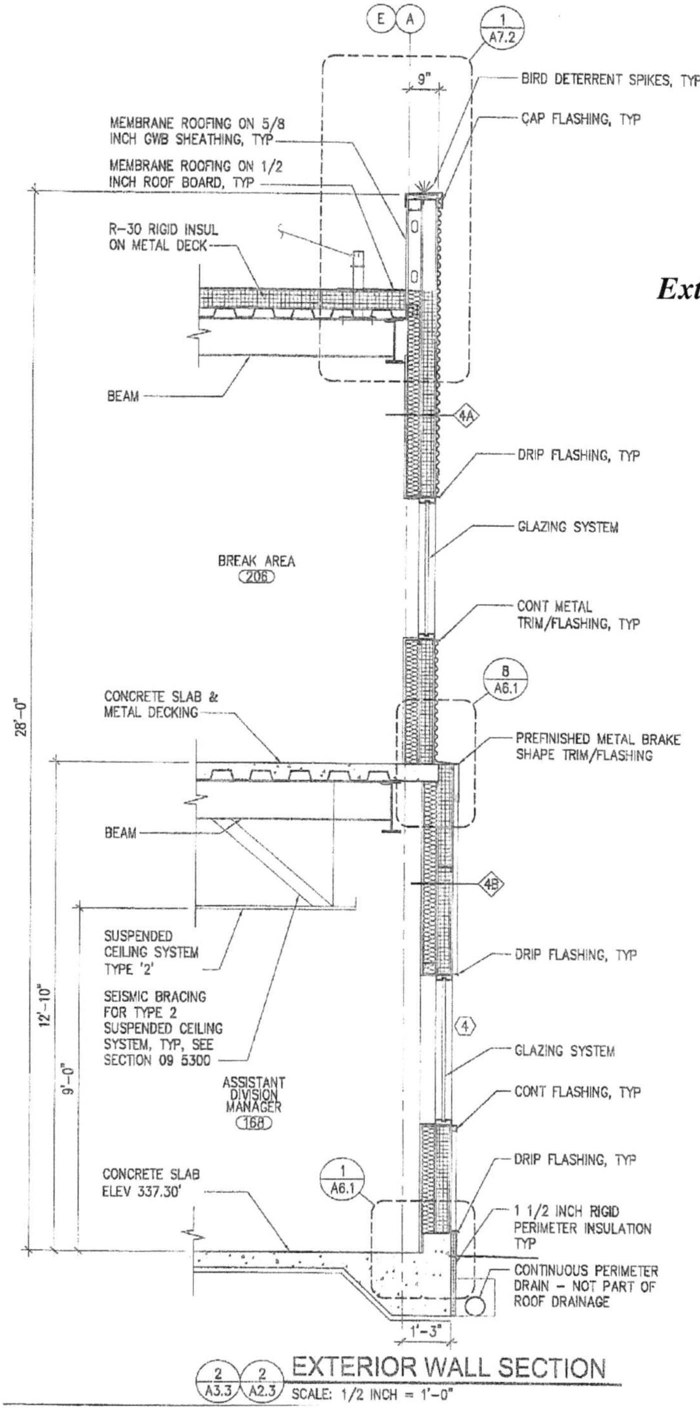

Figure 5.3
Exterior wall section

First, we will examine a sample *plan* view from an architectural sheet of a building's foundation (see Figure 5.1). The plan view has a section view indicator (A/A3.3) at the south end of this building, continuing all the way through the building until it ends at the north or top of this picture. This indicator tells us there is an elevation view through the building—a sideways view, usually beginning at the nearest grid line location.

The grid lines identify the quadrants of the building, indicated by a line through specific locations, usually dimensioned; a number or letter identifies each particular grid. These will be located at the top and to the left or right of the page in the plan views. In Figure 5.1, we see grids 16 through 22 at the top and grids A through F on the right of the drawing.

A *plan view* looks down on a building from above; a horizontal cut has been made through the building, removing the roof or floor above. The effect is like looking down on a box with the top cut off. The inside *elevation view* cuts through a building vertically from ceiling to floor—like looking at the inside walls and what is on those walls. (Another example is when a child opens a dollhouse and you can see the height of the upright walls, the windows, doors, stairs, etc.)

To see the inside elevation view, place a box on a flat surface so that you are looking directly into the box from the top—its plan view. For purposes of this example, mark the bottom or plan surface with a *P*, and mark the left or west side surface of the box as *A/A3.3*. Now, cut through the corners of the box so that its sides lie down flat on the table. It will form a cross, with the part *P* in the middle and the one marked *A/A3.3* on your left. This is the elevation that the cut intends for you to see.

Thus, Elevation Section *A/A3.3* looks to the left of the plan view. The grid lines are shown on the elevation so that grid line E is to the left and grid line *A* is to the right in Figure 5.2. Compare these locations of the grids to the plan view in Figure 5.1 to understand their relationship to the building.

In the elevation view in Figure 5.2 we can see the grid line locations *E* through *A* at the top. To the left of its title, this building section lists all of the drawings from which this section is cut. The building elevations are indicated at the right side of this drawing. Another section cut, *2/A3.6,* is also named; this section cut will give us a close up of that wall, allowing us to see more information about how this building is put together. These listings provide detail layering. For example, in this section *2/A3.6,* we will find more details called out, providing us even closer views of particular areas of the structure.

Refer back to the floor plan in Figure 5.1. The building section *A/A3.3* is located at the bottom of this plan view, just to the right of grid line 18. Section

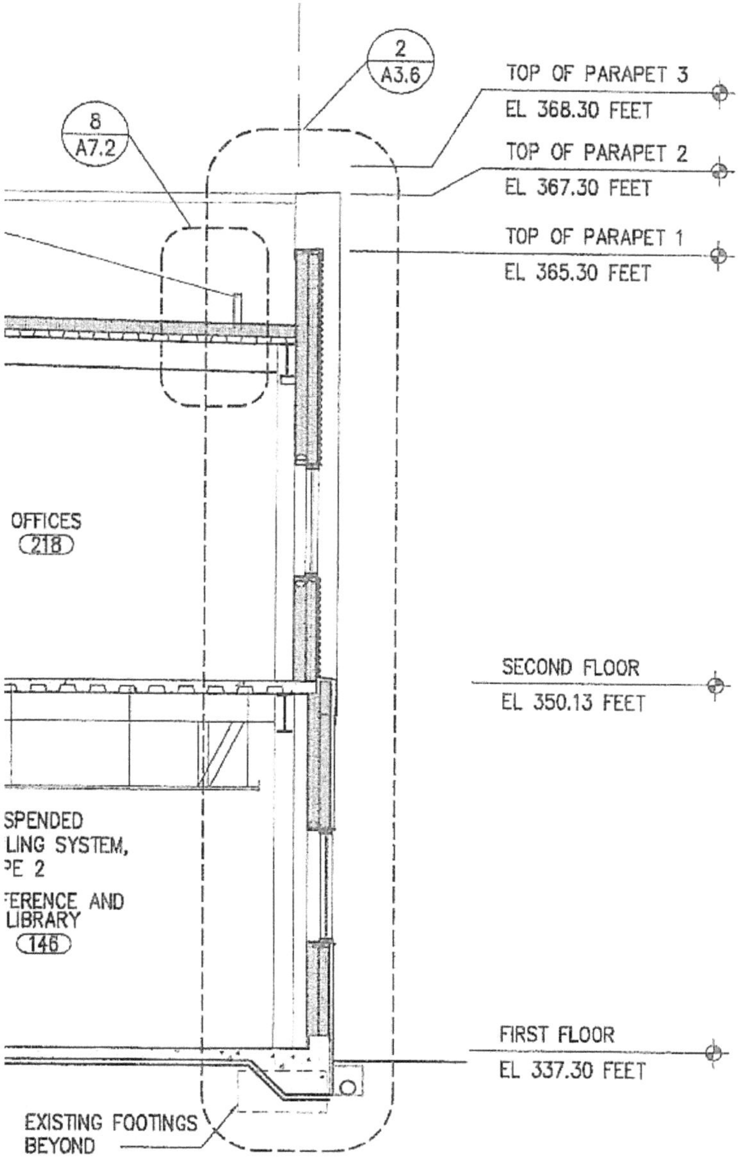

Figure 5.4 Elevations and parapets

A/A3.3 is cut through the entire building, from below grid line F up to above grid line A. The circle or 'detail bubble' around the call out *A/A3.3* has a darkened triangle underneath the left side half of the circle. This triangle is pointing to the left (or west) of the page, indicating that the user should look to the left (or west) side of the building to see what the designer is trying to show. Note also that the grid line *A* is at the top right hand side of the plan.

In Figure 5.2, this same grid line *A* is shown at the top right of the page. The proximity of these grid lines helps you see how the section is cut to make sure you are viewing it correctly. For example, if the section were cut at the plan view looking to the right or east end of the building, the grid line *A* at the elevation view would then be at the top left of the page.

In Figure 5.3—the elevation cut through the wall—we can see the glazing (windows) that interrupt the vertical plane of the wall, set at about mid-span between the first and second floors, and then again above the second floor, below the roof. We can see the suspended ceiling below the 2nd floor. We can see the steel beams just under the 2nd floor and also just under the roof, set just to the left of the wall.

The elevations are the dimensions that are taken "above sea level" (see Figure 5.4). These elevations can be used to calculate the dimension between the first and second floor, and from second floor to the top of parapet. The parapet is the section of the wall that extends above the roof level. In looking at the parapet, we see additional elevations. From this observation, we know that the parapet elevations vary in this building.

In Figure 5.4, detail *2/A3.6* circles the wall itself using a broken line to indicate that there is a larger view of this area of the wall. In looking at the wall section *2/A3.6*, more detail areas are cut at the specific floor and roof levels.

From Figure 5.3, we can see that more information will be shown at the foundation level (*1/A6.1*, Figure 5.5), floor level (*8/A6.1*, Figure 5.6), and roof level (*1/A7.2*, Figure 5.7). This information enables the architect to tell us in more detail how this building is built.

Starting at the bottom and working our way up, Figure 5.5 shows us the foundation detail *1/A6.1*, the concrete curb on which the wall sits. Notice the insulation that is outside the curb (the 1 1/2" thick rigid perimeter insulation area is indicated by the little squares), and also the siding and insulation (indicated by the radius lines) sitting on top of the wall. The indicators 4A, 4B, and 4C in the diamonds joined by the horizontal line that touches the wall refer to drawing notes provided elsewhere in the drawing set. We have not shown those sheets here as they all refer to items that are not steel. We talk about them here to help you understand what the detail is intended to show the reader.

The bent sheet metal pieces that support and enclose the siding and the insulation—the cont (continuous) sealant, the 4" rubber case, the continuous resilient channel, the continuous prefinished metal drip flashing, the neoprene profile closure—all these items make up the base of the wall that we saw in Section cut *2/A3.3*, Exterior Wall Section, Figure 5.4.

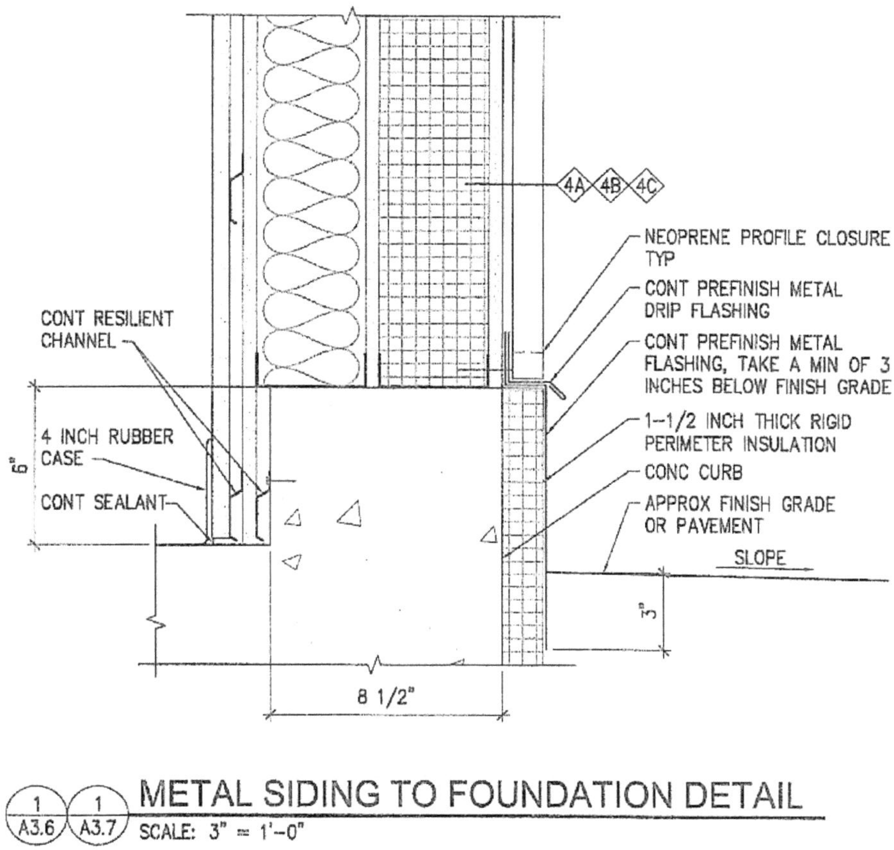

Figure 5.5 Foundation detail 1/A6.1

As you look back and forth between *2/A3.3* and this detail, you will become more familiar with how this detail is used to show more closely what the physical makeup is of this part of the building. What is important to the fabricator on this detail is that no additional steel is shown here.

The detail *8/S6.1* (see Figure 5.6) shows the framing as it exists on the 2nd floor.

We see the same information for the 2nd floor with regard to the siding, insulation, and flashing that we saw in Figure 5.5 for the foundation. We see the end view of a steel beam with another beam framing into it just under the concrete floor. This is our first look at the steel in this building and how it looks in relation to the rest of the building that the framing supports.

Because this drawing is an architectural one, the structural steel framing will not be identified here. Instead, the steel is identified in the structural set

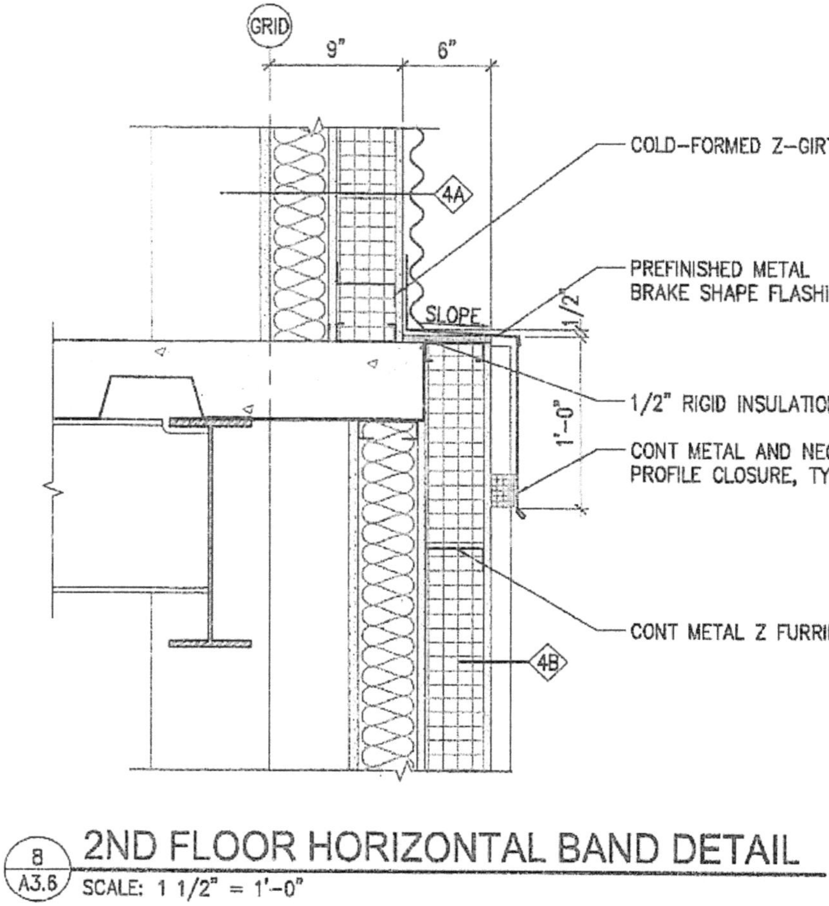

Figure 5.6 Second floor detail 8/S6.1

of drawings, set aside specifically for that purpose. The dimensioning shown for these close-up areas help the individual trades know how and where their materials are to be placed.

Having the structural steel, mechanical, and electrical drawings separate from the architectural plans allows for a more organized and specific presentation of the information, especially when the architect displays the entire make-up of the building.

Figure 5.7 shows the detail *1/A7.2*, the building's roof level. We can see the steel beam at the roof coming at us in the page, as well as the side view or elevation of the beam that frames into it. Behind the beam, we can see the metal siding, wall panels, and insulation. We can also see the roof decking

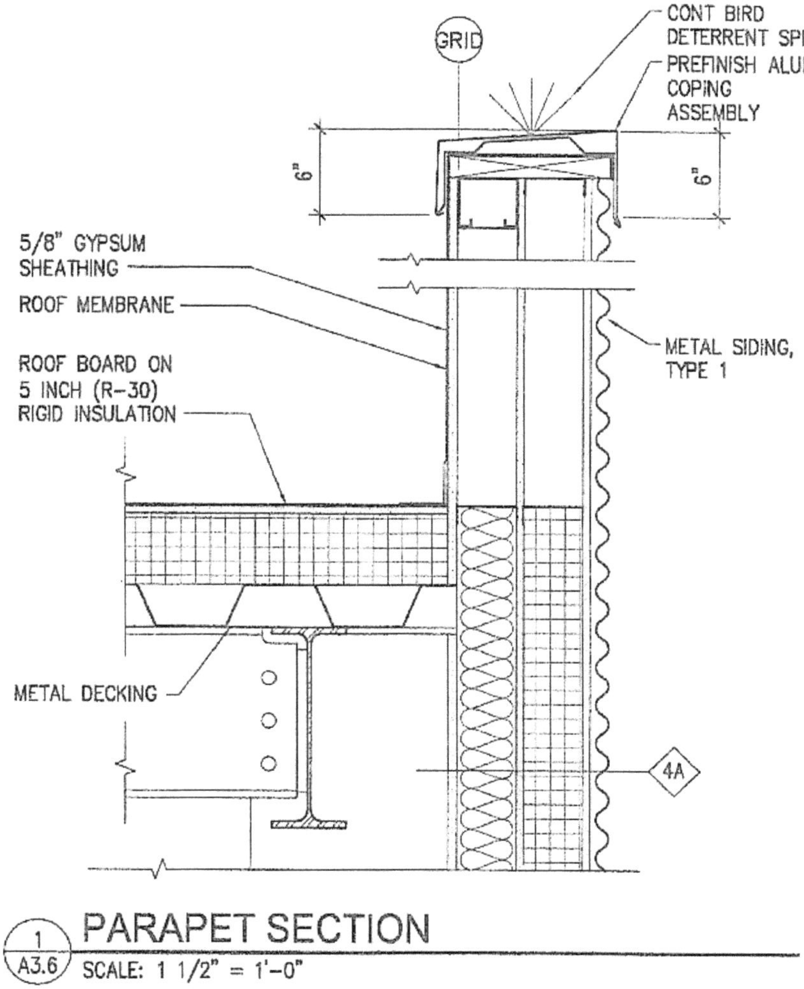

Figure 5.7 Roof level detail 1/A7.2

with the roof board on top, the 5/8" gypsum sheathing, the roof membrane, and the makeup of the parapet.

The conditions we have just reviewed are the same conditions by which we will read the structural drawings. First we will review the plan views. Next, we will go to the elevation cuts, and then the individual detail cuts so we can understand how the building is structured. This format is standard for drawing presentation. Although every building is different, how the designers show the detailed information is generally the same.

The designers do try to show all the information necessary to build the buildings that they can. Still, information which we require to build our steel may be missing. This omission is to be expected as the *application of trade specific information will not be shown* at these contract drawings; that type information requires detailing even beyond what is required of the designers.

Thus we have steel detailers and RFIs—Requests for Information. The RFI is the tool used to fill in the gaps, necessary for us to achieve the goal of completed and accurate shop detail drawings as well as the provision of the subsequent fabricated steel.

Summary

This chapter reviewed the drawing floor plan, together with the subsequent elevation view and section cuts. The discussion on detail layering should help steel estimators begin to understand how the drawings are designed. More information on reading drawings is given in Chapter 7 (Reading Structural Drawings), Chapter 8 (Structural Steel Material Listing), and Chapter 9 (Reading Architectural Drawings). The application of the basics indicated in this chapter remains the same.

6

Steel Materials

Steel that is used for structural and miscellaneous steel is carbon-based material and comes in a variety of types, shapes, and grades. The composition of steel is mostly iron that is derived from recycled metal.

Because steel is composed of recycled material, other elements or minerals are often within its content. The elements that make steel are heated to liquid form, then poured or pushed into shape, and cooled. This process is called mill rolling.

Measuring Steel

Steel materials are created at steel mills all around the world. Steel mills operate to schedules that are generally based on the salability of the product—the same supply and demand requirements as anything else. The manufacture of steel by the mills is called mill rolling. The process is much the same as pushing clay through a die plate similar to a Play-dough toy, only in larger scale. Mills will roll requested items if ordered in large enough quantities, though the timing of special rolling may not fit certain construction schedules.

Different types and sizes of steel materials are available. Steel materials are typically available in 20' to 40' lengths. Wide flange has some availability in 5' increments after 20' and up to 60', and sometimes longer. Some tube steel sizes are available in 24', 32', and 48' lengths in addition to the 20', 40', and 60' lengths that may be stocked.

Steel materials other than beams are typically manufactured in 20' lengths, but some sizes of channel, tubing, and pipe go to 40' lengths. Shops cutting their materials for a project makes better use of labor time by grouping all their cuts by member sizes.

Smaller size angles, flat bar, and round bar may be rolled only in 20' lengths. Steel warehouses have product catalogs for their customers that provide this information. These catalogs are also handy guides to other types of available materials, such as cold rolled materials, stainless steel, aluminum, and sometimes fiberglass.

When calculating material prices, the amount of drop or waste is generally a consideration—economic utilization may make a difference in the estimated budget. Once the take off on a project is completed, the steel materials are sorted and nested into groups to provide the best possible purchase for each material size.

When a project is in the fabrication process, the shop receives the nested list of materials to cut from the steel that is purchased. With nesting, several pieces for different parts are cut out of one piece of steel, length allowing. Our labor time for shop fabrication is based on this efficiency.

Estimating the material lengths to as close as the nearest inch allows for the best possible material nesting application. Making the best use of stock length materials provides the lowest possible material pricing.

Nesting materials also makes an allowance for the saw *kerf*, or thickness of the blade. When you are cutting several pieces out of one length of stock, this kerf allowance accumulates. If it not added into your calculations, you could face a shortage in the amount of stock to buy. There are material nesting programs for computers; they make allowances for the saw kerf. If you do the nesting with a spreadsheet program or by hand, find out what the kerf allowance needs to be.

Splicing of materials to achieve a part length for structural members is generally not allowed. If you have a column or beam, you need to have these parts cut from a continuous piece of steel. If splicing is required because of limited availability of a particular member type, then a splice detail will need to be engineered, engineer stamped, and sent for approval to the engineer-of-record prior to the actual use of the splice. The resulting fabrication will probably need to be shop tested; documentation will be required to verify the quality of the splice location before it can be used. This process is very lengthy and cumbersome. It should only be utilized in extreme cases.

Grading Materials

The different grades of materials are determined by the mix from which they are made. The grades are designated by the ASTM (American Society for Testing and Materials). ASTM has set standards or specifications for different grades of materials based upon the specific chemical and mechanical requirements of the products.

These ASTM requirements and specifications change the particular amounts of carbon, manganese, molybdenum, vanadium, and other various alloys that, in effect, create the recipe by which the steel is made. Increased carbon will make a material harder; less carbon, softer. Grades of material are designed for various purposes, including:

1. resistance to corrosion
2. use in a low-temperature environment
3. strength
4. flexibility

There are recipes to fit almost any need.

Depending on what is being built, the engineers make their choice of materials based on what their design requires. These choices for material types and grades will be indicated in the general notes for the drawings, as well as in the project specifications.

Shapes of Steel

The shapes and grade of steel shown in contract drawings can be found in the AISC (American Institute for Steel Construction) manual. This manual will show you the dimensions and design properties of all the different shapes available. Be aware that because the steel industry is ever-changing, some sizes that are not used very often may have been discontinued and new ones may be added to accommodate engineering advances.

W Shape

Figure 6.1 shows a wide flange shape of steel. The *W* or Wide Flange shapes range from 4" deep all the way up to 44" as a standard mill rolled size. Wide flange beams are generally used for building columns, floors, and roof framing. In understanding wide flange shapes, the first number after *W* is the nominal (or average) beam depth; the second number is how much this steel weighs per foot. For example, if the beam is a W4 x 13, then it is a wide flange beam, about 4" deep, and 13 lbs per foot.

As the beams get heavier, the beam flange and web get thicker. However, the beams' true depth dimension is not completely precise. For example: a

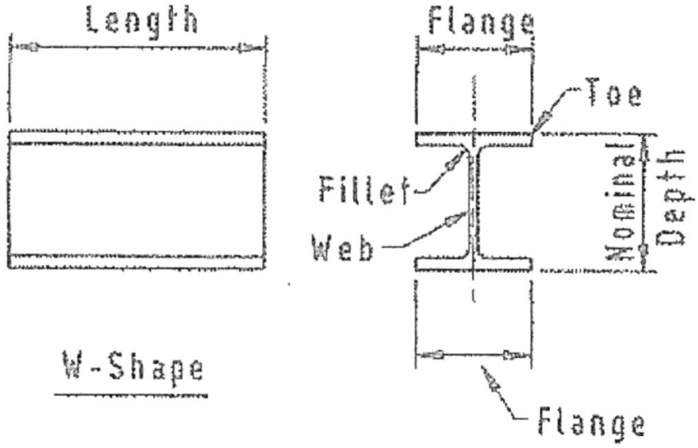

Figure 6.1 W shape of steel

W10 x 12 may actually be 9-7/8" from the top of the top flange to the underside of the bottom flange, whereas a W10 x 19 may actually be 10-1/4" deep at the same measurement.

A wide flange will generally be taller than it is wide for sizes W8 and above. W4 and W6 flanges will be close to the same in height and width. Common material grades are ASTM A992, A572, and A36, with A36 being the lowest grade available.

C and MC Shapes

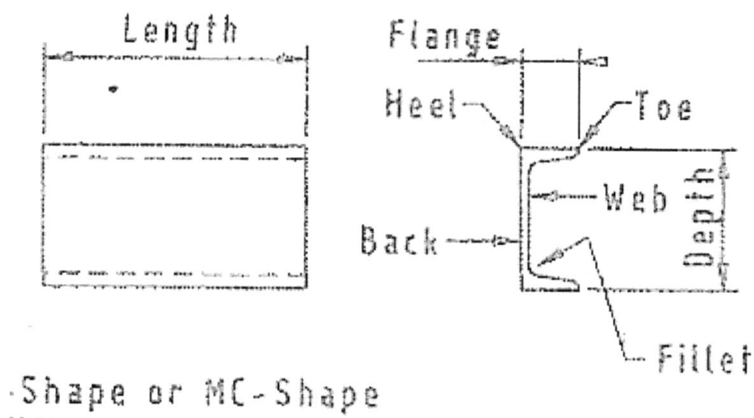

Figure 6.2 C shape of steel

Figure 6.2 shows the C shape of steel. Reading the information designations for C shapes is the same as reading the wide flange shapes. For example, a C12 x 20.7 piece of channel is 12" deep and weighs 20.7 lbs per foot. Channels will stay true to their depth—if it is a C12, it will stay 12" deep no matter if it weighs 20.7 lbs per foot or 30 lbs per foot. The same is true for MC shapes.

Miscellaneous channel, or ship channel, tends to have wider flanges than the standard C shapes of similar sizes. MC channels are used in special applications for building framing or in stairs, and sometimes ship's ladders to the roof. This material is usually provided as an ASTM A36 grade. American Standard Channels range from C3 x 4.1 up to C15 x 50; Miscellaneous Channels range from MC10 x 6.5 up to MC18 x 58.

M, HP, and S Shapes

M, HP, and S shapes for steel beams look much the same as the wide flange beam in Figure 6.1. M shapes range from M5 x 18.9 through M14 x 18, HP shapes from HP8 x 36 through HP14 x 117, and S shapes from S3 x 5.7 through S24 x 121.

The use of M, HP, and S shapes in steel building framing is rare. S beams are unique as they have tapered flanges; because of these, they are generally

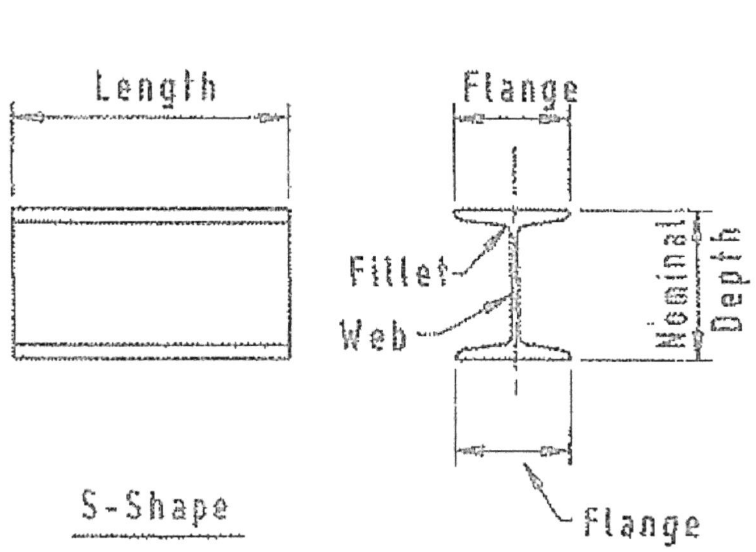

Figure 6.3 S shape of steel

used for overhead crane rails. The flanges of the W, M, and HP shapes are of an even thickness and do not taper.

For M shapes, the beam is always the same depth as it is designated. For example, an M14 is going to be 14" deep. The depth dimension of an M shape will always be the same, regardless of the weight per foot, although the thickness of the web and flange will change to allow for the weight of the beam.

Figure 6.3 shows the S shape of steel. For S shapes, the flanges taper; they are thicker at the web and thinner out to the face of the beam. Also called an I beam, they are used as crane rails and sometimes as bracing for floor, walls, and roofs.

H or HP beams are very close to being the same depth and width. If you

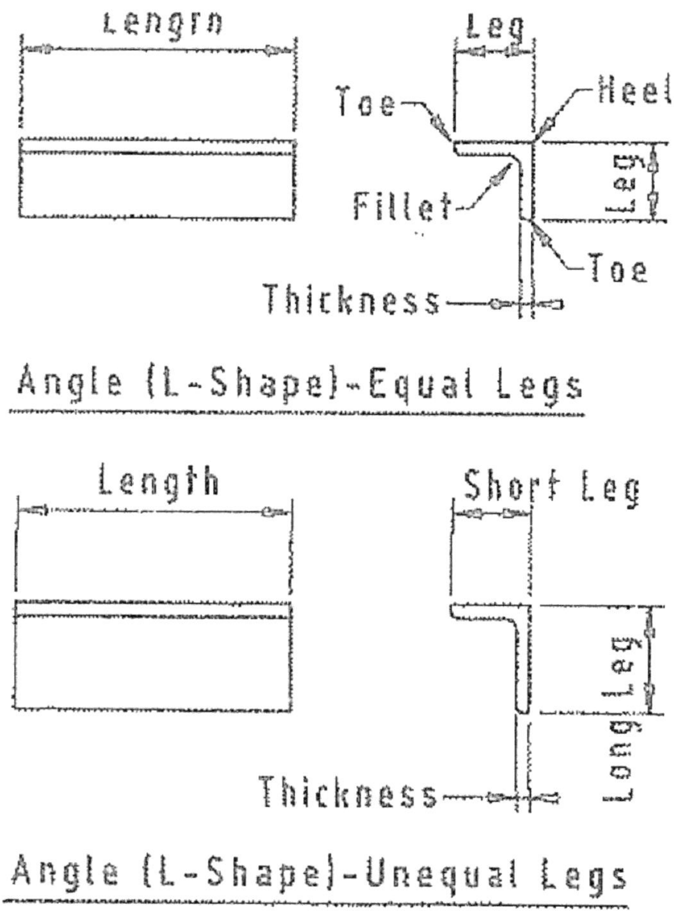

Figure 6.4 Steel angle or L shape of steel

have an HP12 x 84, it will be 12-1/4" wide by 12-1/4" deep. Although it stays close to this 12" dimension, it too may change a bit, depending on how much the beam weighs.

L Shapes

Steel angle or "L" shapes (see Figure 6.4) are seen very often in floor framing, bracing, and canopies. They are used for openings in the floor, truss framing, or bracing for beams back to the walls. They are also used as steel connection material.

Figure 6.4 shows both equal and unequal leg angles. Unequal leg angles are often used for ledger support, or for grating supports where the longer leg needs to create additional support for the framing. Like the channel, this material is generally provided as A36 and in 20'-0" lengths. Steel angles range from L 1 x 1 x 1/8 up to L 9 x 4 x 5/8. Keep in mind that not all sizes are readily available and will vary greatly in price.

WT Shapes

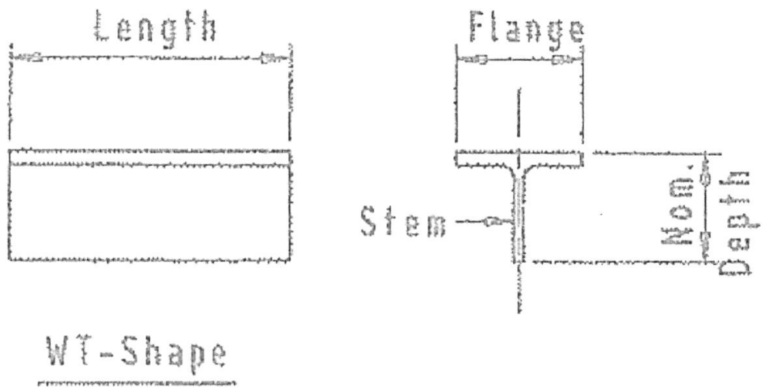

Figure 6.5A WT shape

The WT or split tee members shown in Figures 6.5A and 6.5B are really full-size beam members cut in half at the center of their web. You can buy these, split and straightened, from a steel warehouse. Their price is usually double the cost of the beam it came from because of the additional labor to fabricate it.

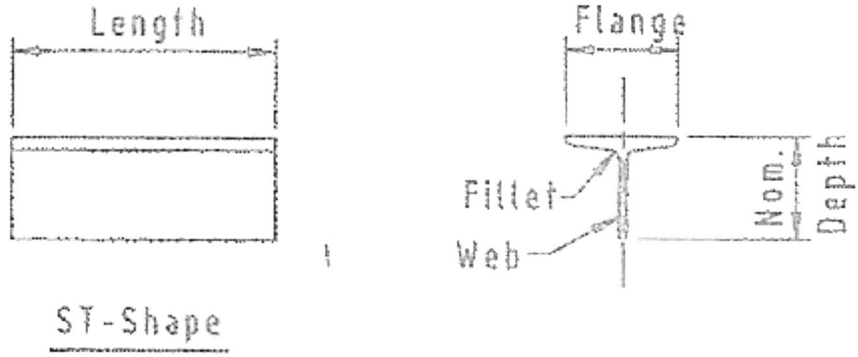

Figure 6.5B Split tee (ST) shape

When the beam is cut in half to make a WT section, it will bow and twist. Because of this distortion, the WT section has to be straightened by heat or machine prior to it being used in steel framing. You will see WT used in opening framing, canopy framing, bracing applications, and small floor supports. In some cases you might see WT sections being used in steel-to-steel connections instead of plate or angle. Grades of the WT material will be the same as the wide flange A992, A572, and A36.

Steel Tubing

Figure 6.6 shows an assortment of steel tubing. Larger sections of round, square, and rectangular tubing is typically used for columns, bracing, floor framing, stair framing, and canopy framing. Smaller sections might be used for pipe bollards, door and window framing, catwalk supports and framing, and hand railing. ASTM A500 grade is typical for tube steel material, and a standard pipe grade is A53.

Square and Round Bar Steel

Square bar and round bar comes in various sizes (see Figure 6.7). Square bar runs from 3/8" square up to 3" square and round bar can be found from 3/16" diameter up to 12" diameter. General use material grade is A36. Rebar or reinforcing bar is round bar used for embeds in concrete and has numbered diameters: #3 for 3/8", #4 for 1/2", #5 for 5/8", #6 for 3/4", #7 for 7/8", #8 for 1", and #9 for 1 1/8". (Note that each # designator increases by 1/8" with each size.) We will typically find rebar welded to plates or steel members that are set into concrete for added strength.

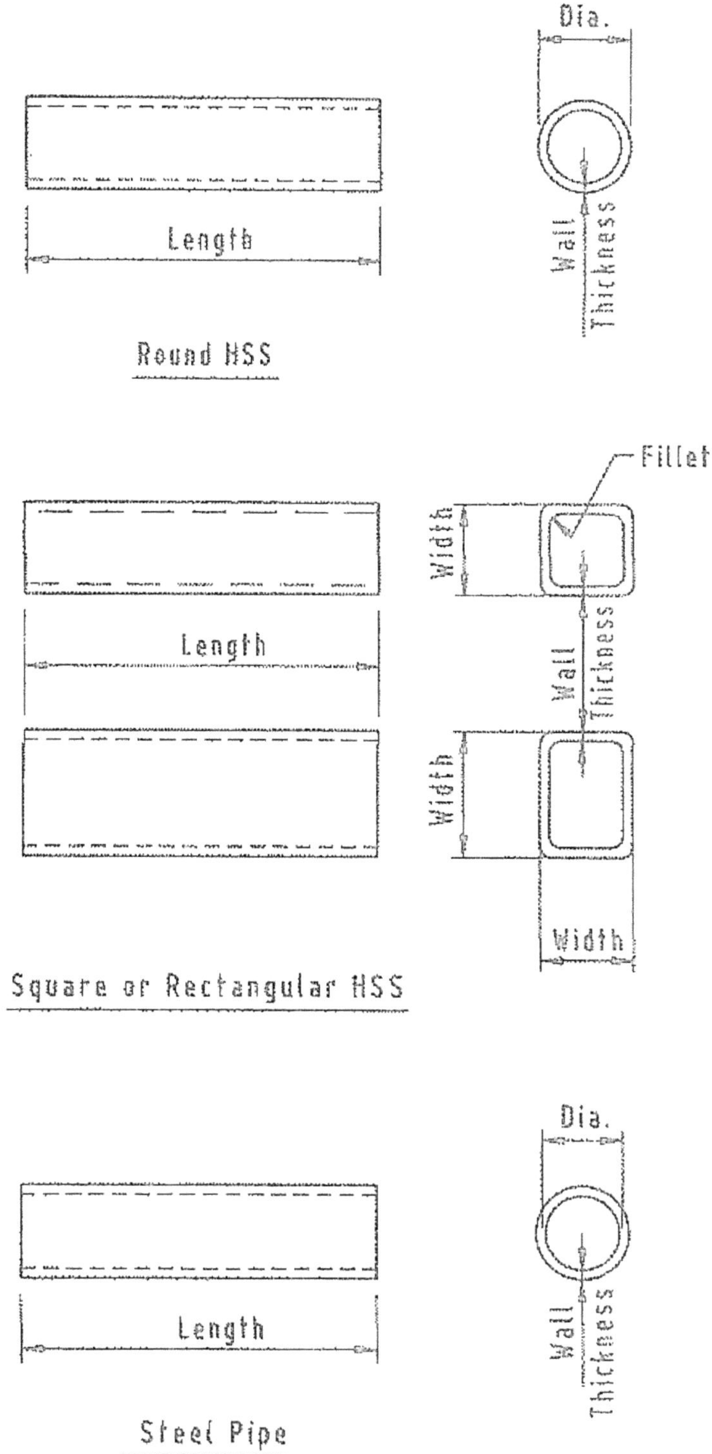

Figure 6.6 Steel tubing

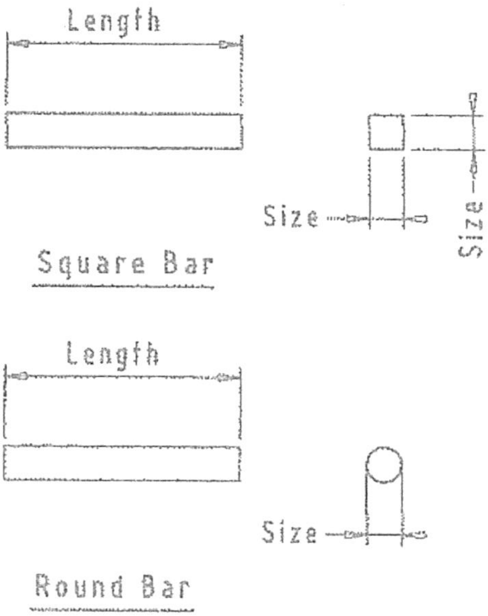

Figure 6.7 Square and round bar steel

Steel Plate and Flat Bar

Steel plate and flat bar (see Figure 6.8) is available in many sizes and thicknesses as an A36 material grade. Steel plate changes to sheet steel plate or gauge material when the material is less than 3/16" thick. Plate that is 1/8" (decimal equivalent is .125) is so close to the thickness of 11 gauge material (decimal equivalent is .1196) that typically it is referred to as 11 gauge.

As gauge materials get thinner, the numbered references get larger. Sizes you will see used in the structural drawings are 12 gauge (decimal equivalent is .1046) and 14 gauge (decimal equivalent is .0747), where these materials are used to make stair pans or deck closure plates.

Steel plates used for base and cap plates at columns, or as a significant structural support, may be of a grade different than A36. Be sure you are clear on what you are using and where it is indicated in the contract drawings.

Flat bars are used where they are called out on the plan, and where it is convenient for the fabricator to use them—as long as the material grade of A36 is designated to use. Structural flat bars will range from 3/16" thick up to 3" thick and up to 6" wide. Hot rolled strip is also available, from .065 thick up to—and here's the crossover into structural—.1875 or 3/16" thick.

The difference between hot rolled and cold rolled steel is that cold rolled steel is of a higher grade material, is much cleaner, and is used for items to be machined. Generally you will not see cold rolled materials called out for use

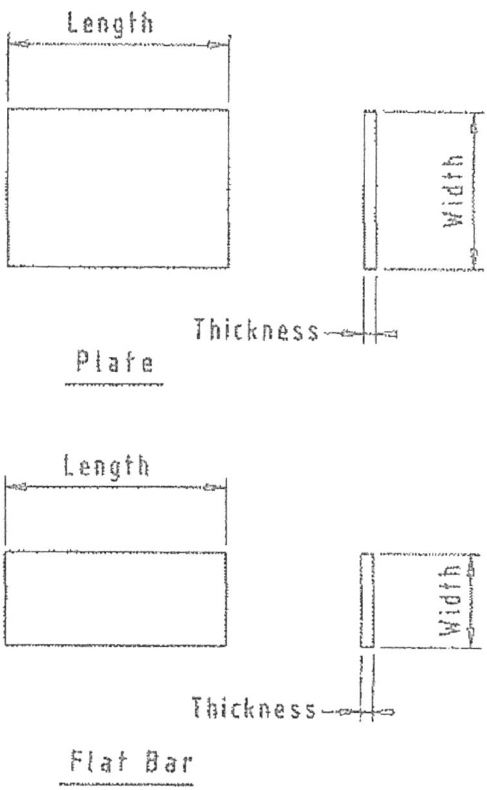

Figure 6.8 Steel plate and flat bar

in a structural or miscellaneous steel application. Cold rolled steel is much more expensive than hot rolled steel. It can be found in the square and round bar at tubing applications.

There are structural shapes available in aluminum, stainless steel, and even fiberglass, although the availability of the sizes made is limited. Nonferrous applications, such as water treatment plants, use these structural shapes most often. Aluminum and stainless steel pipe and tubing are commonly used in the fabrication of ornamental railing and architectural features.

Material availability, even for the most common sizes, varies greatly, as will the pricing. To keep the pricing accurate, steel estimators must check pricing and availability for every project, especially those that may have a long lead time.

Calculating Material Weights

Calculating the correct weight of the material is important because thematerial costs are based upon the weight of the items to be purchased. Today, computer programs do mostof the work, though we still need to know how weight extensions are performed. Materialweights for items in shape form are calculated by their weight per foot. Shape forms are wide flange, angles, channels, split tee, pipe, tubing, flat bar, round bar and rebar.

The formula used for calculating weights for shapes is the same for all material types and grades. The weights per foot for shapes can be located in a product catalog provided by a steel supplier, on the Internet, or in an AISC manual (see index for AISC).

Simply look up the weight per foot of the desired product. Using the length of the shapein feet and inches, convert increments of the inch into decimals of the foot. Charts to aidthis conversion are found in the back pages of a steel products catalog or on the Internet.For thr purpose of calculation, 10'-6 would become 10.5, or 3'-8 would become 3.6667.

To calculate a weight, multiply the quantity by the length, then by the weight per foot. If we wanted to calculate the weight of (1) W12 x 21 x 10'-6 beam we would calculate thus:1 x 10.5 x 21 = 220.5#. If we wanted to calculate the weight of (1) L 4 x 4 x 1/4 x 3'-8 (weight is 6.6# per foot), then: 1 x 3.6667 x 6.6 = 24.20#.

The weights for plate and sheet stock is based on the thickness of the item and is calculated by the square foot. Using the dimensions for the width and length of the item, convert anyincrements of the inch to decimals. Multiply the item quantity by the width, then by the length, then by the weight per square foot.For example, if we wanted to find he weight of a steel plate that is 1/4" x 1'-6 x 2'-6 (10.21# per square foot), then:1 x 10.21 x 1.5 x 2.5 = 38.28#.

The weight per square foot will be different if the 1/4" thick material is not steel, though the method for calcuting the weight would remain the same. Locate the material by type, then find the weight per square foot based on the thickness. Using aluminum 1/4" x 1'-6 x 2'-6 (3.660# per square foot) then: 1 x 3.660 x 1.5 x 2.5 = 13.72#.

Summary

This chapter has reviewed the different types and sizes of steel materials used in the fabrication of structural and miscellaneous steel. It is intended to help steel estimators become more familiar with the material callouts and descriptions, as well as how these materials are utilized.

7

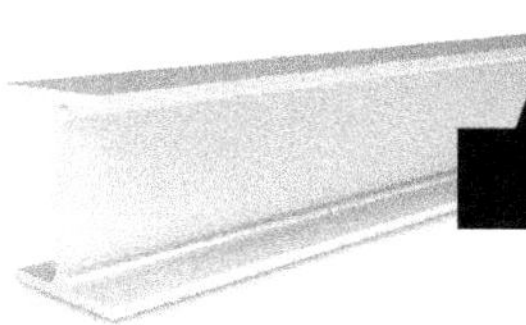

Reading Structural Drawings

The steel framing in any building is the skeleton inside the walls that hold up the building. Building support columns and beams bolt together much in the same way one would put together a toy erector set. Levels are added to make it taller; cross members are added to make the frame stronger. In addition, that framework provides support and attachment to interior and exterior finishes.

This chapter explains how to read and understand the general notes shown on the drawings and how they are applied to the plans and details shown in this structural set. We will walk through the building beginning with the foundation plan, then through to the floor and roof plans. We will review the specific elevations and section cuts in the building to help us form a working understanding of what is being shown and how that information is to be applied for the use of steel estimating.

The Basic Frame

Steel columns and beams make the basic frame for the building. This framing material is found in the first few S series drawings of the bid set called the foundation and floor framing plans. After the foundation and floor framing plans, the remaining majority of the S series drawings will show the specific details of how the framing members connect to tie it all together. There will be other steel items that connect to the basic steel frame supporting both interior and exterior walls, windows, siding, and brick.

Typical steel framing items to be found in the S drawings, in addition to the steel columns and beams, are as follows:

- Stairs in multi-story buildings to get from one floor to another.
- Steel decking that make the floors for supporting the concrete will have steel angles at the perimeter to hold in that concrete.
- Steel bracing will be between columns to make the building rigid.
- Steel joists are in between the steel beams to hold up the floors and the roof.
- Steel angles will frame floor openings to hold back the concrete.
- Steel framing for large door openings and window wall framing
- Steel supports for equipment that are supported from the floor above
- Mechanical walkways
- Catwalks
- Angles and/or channels that support brick over window and door openings or "ledger" supports
- Roof top screen walls and parapet supports

Take some time to page through all of the sheets provided in the bid set of drawings and make a mental note of the physical size of the building, how many floors it has, and what the general shape of the building is. Go through the detail sheets that follow behind the foundation, floor, and roof plan drawings. These details will help the reader better see the actual way the building is being constructed.

The steel framing columns and beams need to be delivered to the jobsite right behind the concrete pour. Some items, such as anchor bolts for steel columns and steel fabricated embeds, need to be delivered before the concrete pour is set in place. The type of building for which you are creating an estimate will designate those delivery requirements. As you become more familiar with the construction sequence, you will know instinctively what will need to be delivered first.

General Notes

The structural drawings usually begin with the general notes regarding the requirements for the type and grade of construction items used. You are mainly concerned with any information regarding your steel supply. For the most part, you will find requirements for galvanizing, type of paint and painting, steel and rebar grades, fasteners and anchor bolts, and welding requirements.

Regarding the steel, there may be notes with special inspections that will need your attention. There will be a Key for the abbreviations that engineers will use in the drawing notations.

The bid set of drawings may have notes that are both similar and conflicting to those provided in the project specifications. In the case of a conflict, the AISC manual dictates that the specifications govern over the drawings. To provide clarity for the reader, steel estimators need to list in the bid or scope letter any qualifications that were used to create the quote.

The following notes are based on samples drawn from typical projects.

GENERAL NOTES

SUBMITTALS: SHOP DRAWINGS SHALL BE SUBMITTED TO THE ENGINEER PRIOR TO ANY FABRICATION OR CONSTRUCTION FOR ALL STRUCTURAL ITEMS, INCLUDING THE FOLLOWING: CONCRETE REINFORCEMENT, EMBEDDED STEEL ITEMS, STRUCTURAL STEEL, STEEL DECK, SHEAR STUD LAYOUT, METAL GRATING, CLADDING PANELS, AND STAIRS.

IF THE SHOP DRAWINGS DIFFER FROM OR ADD TO THE DESIGN OF THE STRUCTURAL DRAWINGS, THEY SHALL BEAR THE SEAL AND SIGNATURE OF THE WASHINGTON STATE REGISTERED PROFESSIONAL ENGINEER WHO IS RESPONSIBLE FOR THE DESIGN.

DESIGN DRAWINGS AND CALCULATIONS, OR SHOP DRAWINGS FOR THE DESIGN AND FABRICATION OF ITEMS THAT ARE DESIGNED BY OTHERS, SUCH AS COLD FORM STEEL EXTERIOR FRAMING, SHALL BEAR THE SEAL AND SIGNATURE OF THE WASHINGTON STATE REGISTERED PROFESSIONAL ENGINEER WHO IS RESPONSIBLE FOR THE DESIGN, AND SHALL BE SUBMITTED TO THE ENGINEER AND THE BUILDING OFFICIAL FOR REVIEW PRIOR TO FABRICATION. SUBMITTED CALCULATIONS ARE FOR INFORMATION ONLY AND WILL NOT BE STAMPED OR RETURNED.

Figure 7.1 General notes about submitting shop drawings

This first paragraph in Figure 7.1 (regarding shop drawing submittals) tells us that steel shop fabrication drawings need to be submitted to the engineer for approval prior to fabrication.

For the typical approval process, the drawings are sent to the general contractor, who then sends them to the architect and engineer for approval review. After they review the drawings, the architect and engineer send them back to the general contractor, who then returns them to the steel fabricator.

It is important to know that steel fabrication does not begin until those drawings have been returned marked as "approved" by the architect and engineer. If fabrication does start prior to receiving approval, steel fabricators will be liable for errors and omissions that may be contained within these shop drawings. These errors and omissions will then have to be fabricated correctly at the steel fabricators' expense.

The second and third paragraphs in Figure 7.1 advise that if the building design is different than what is shown at the contract set, the drawings will have to be stamped by a registered engineer. Verification will need to be provided of the calculations that were made in the course of the re-design. This cost would be covered by the steel fabricators if they take it upon themselves to perform such a function. As a normal course of doing business, steel fabricators will not provide such services; they will look to the architect and engineer of record to provide all the information required for anything that is unclear or not designed in the contract drawings.

INSPECTION: SPECIAL INSPECTION, PER IBC CHAPTER 17 SHALL BE PERFORMED BY AN APPROVED TESTING AGENCY AS OUTLINED IN THE SPECIAL INSPECTION SCHEDULE. ALL PREPARED SOIL-BEARING SURFACES SHALL BE INSPECTED BY THE GEOTECHNICAL ENGINEER PRIOR TO PLACEMENT OF REINFORCING STEEL. SOILS COMPACTION SHALL BE SUPERVISED BY AN APPROVED TESTING AGENCY OR GEOTECHNICAL ENGINEER.

SPECIAL CONDITIONS: CONTRACTOR SHALL VERIFY ALL LEVELS, DIMENSIONS, AND EXISTING CONDITIONS IN THE FIELD BEFORE PROCEEDING. CONTRACTOR SHALL NOTIFY THE ENGINEER OF ANY DISCREPANCIES OR FIELD CHANGES PRIOR TO INSTALLATION OR FABRICATION. IN CASE OF DISCREPANCIES BETWEEN THE EXISTING CONDITIONS AND THE DRAWINGS, THE CONTRACTOR SHALL OBTAIN DIRECTION FROM THE ENGINEER BEFORE PROCEEDING. DIMENSIONS NOTED AS PLUS OR MINUS (±) OR AS "REF" INDICATE UNVERIFIED DIMENSIONS AND ARE APPROXIMATE. NOTIFY ENGINEER IMMEDIATELY OF CONFLICTS OR EXCESSIVE VARIATIONS FROM INDICATED DIMENSIONS. NOTED DIMENSIONS TAKE PRECEDENCE OVER SCALED DIMENSIONS--DO NOT SCALE DRAWINGS. DIMENSIONS OF EXISTING CONDITIONS MAY BE BASED ON RECORD DRAWINGS AND ARE TO BE FIELD-VERIFIED BY THE CONTRACTOR.

CONTRACTOR SHALL VERIFY ALL EXISTING CONDITIONS BEFORE COMMENCING ANY DEMOLITION. CONTRACTOR SHALL PROVIDE ADEQUATE SHORING AND BRACING OF ALL STRUCTURAL MEMBERS, EXISTING CONSTRUCTION AND SOIL EXCAVATIONS, AS REQUIRED, AND IN A MANNER SUITABLE TO THE WORK SEQUENCE. TEMPORARY SHORING AND BRACING SHALL NOT BE REMOVED UNTIL ALL FINAL CONNECTIONS HAVE BEEN COMPLETED IN ACCORDANCE WITH THE DRAWINGS AND MATERIALS HAVE ACHIEVED DESIGN STRENGTH. NO REINFORCING BARS IN EXISTING CONSTRUCTION SHALL BE CUT UNLESS DIRECTED TO BY THE ENGINEER OR AS SHOWN ON THE DRAWINGS.

CONTRACTOR SHALL BE RESPONSIBLE FOR ALL SAFETY PRECAUTIONS AND THE METHODS, TECHNIQUES, SEQUENCES OR PROCEDURES REQUIRED TO PERFORM THE WORK.

SOILS: EARTHWORK MATERIAL, BACKFILL AND COMPACTION SHALL BE IN ACCORDANCE WITH THE SPECIFICATIONS. BACKFILL BEHIND WALLS SHALL NOT BE PLACED BEFORE THE WALL IS PROPERLY SUPPORTED BY THE FLOOR SLAB OR TEMPORARY BRACING. ALL TOPSOIL ORGANICS AND LOOSE SURFACE SOIL SHALL BE REMOVED FROM BENEATH FILL SUPPORTING CONCRETE SLABS OR PAVING. FOUNDATIONS SHALL SIT ON UNDISTURBED NATIVE GLACIAL TILL OR COMPACTED STRUCTURAL FILL PLACED ON UNDISTURBED NATIVE FILL PER THE SPECIFICATIONS.

Figure 7.2 IBC inspections

The first paragraph in Figure 7.2 is about inspections as required by IBC (International Building Code); it refers to the requirements outlined in the special inspection schedule. These notes will alert the steel estimators to find this special inspection schedule to see what may be required with regard to the fabricated steel. Sometimes requirements for special inspection of the steel fabrication are paid for by the steel fabricator.

The steel fabricator's bid letter states that the costs concerning special inspections are excluded because these inspections are normally paid for by the owner. Special conditions of a project may require that steel fabricators change their position on inspection; therefore, the information about this provision needs to be specifically addressed.

The remaining paragraphs in Figure 7.2 are standard requirements of the general contractor or other trades; they do not concern the steel framing, unless we are also installing the structural steel.

STRUCTURAL STEEL

REFERENCE SPECIFICATIONS

STRUCTURAL STEEL	AISC SPECIFICATIONS FOR STRUCTURAL STEEL BUILDINGS, 2005
HIGH STRENGTH BOLTS	SPECIFICATION FOR JOINTS USING ASTM A 325 OR ASTM A 490 BOLTS, 2006 EDITION
WELDING	AWS D1.1 , TYPICAL AWS D1.3 FOR METAL DECK AND COLD-FORMED FRAMING AWS PREQUALIFIED JOINT DETAILS
WELDER CERTIFICATION	AMERICAN WELDING SOCIETY (AWS) WASHINGTON ASSOCIATION OF BUILDING OFFICIALS (WABO)
STEEL DECKING	STEEL DECK INSTITUTE, AISI-SPECIFICATIONS FOR THE DESIGN OF COLD-FORMED STEEL STRUCTURAL MEMBERS

STEEL MATERIALS

STRUCTURAL STEEL & WT'S	ASTM A 992, GRADE 50
CONNECTION MATERIAL, EMBEDDED ITEMS, CHANNELS, ANGLES, BASE PLATES, AND MISC. STEEL	ASTM A 36
STRUCTURAL TUBES	ASTM A 500, GRADE B
STEEL PIPE	ASTM A 53, GRADE B
STRUCTURAL BOLTS	ASTM A 325-N
ANCHOR BOLTS	ASTM F 1554, GRADE 36, UNO
THREADED RODS	ASTM A 36
WELDING ELECTRODES	70 KSI, LOW HYDROGEN, TYPICAL 60 KSI, MINIMUM, METAL DECK AND COLD- FORMED FRAMING
HEADED SHEAR STUDS	ASTM A 108

Figure 7.3 Materials used for framing

Figure 7.3 lists the standard requirements for the materials to be used in framing this building. This information is typical for most projects. It is important we check these references to make sure there isn't anything out of the ordinary. The notations in the figure are standard.

The first paragraph's first sentence in Figure 7.4 ties the steel fabrication work into the requirements of Chapter 22 of the *International Building Code* (IBC). A copy of the IBC may be obtained over the internet for a fee. A copy of a current IBC is a publication that may be good to have in your estimating library. Most likely, the design of the building has already been completed in compliance with this code. If the structural fabricators were to take it upon themselves to design any portion of the framing, then the engineer whom they hire would then follow the code requirements in addition to the fabricators' own engineering requirements.

STRUCTURAL STEEL DESIGN, FABRICATION AND ERECTION SHALL CONFORM TO THE REQUIREMENTS OF CHAPTER 22 OF THE INTERNATIONAL BUILDING CODE. ALL MEMBERS ARE TO BE ERECTED WITH NATURAL MILL CAMBER OR INDUCED CAMBER UP, UNLESS OTHERWISE NOTED ON THE PLANS. SUBSTITUTION OF MEMBER SIZES OR STEEL GRADE WILL NOT BE ALLOWED WITHOUT PRIOR APPROVAL OF THE ENGINEER. BOLTED CONNECTIONS ARE TO BE OF HIGH STRENGTH ASTM A 325 BOLTS AS SHOWN. A MINIMUM OF TWO BOLTS IS REQUIRED FOR ALL BEAM CONNECTIONS. ALTERNATIVE CONNECTIONS TO THOSE SHOWN ON THESE DRAWINGS WILL REQUIRE PRIOR APPROVAL OF THE ENGINEER.

BEAM CAMBER INDICATED ON DRAWINGS IS THE UPWARD CAMBER REQUIRED IN THE BEAM AS DELIVERED TO THE JOB SITE. CONTRACTOR TO CONSIDER CAMBER LOSS, IF ANY, DUE TO SHIPPING AND HANDLING.

THE CONTRACTOR SHALL BE RESPONSIBLE FOR ALL ERECTION AIDS AND JOINT PREPARATIONS THAT INCLUDE, BUT ARE NOT LIMITED TO, ERECTION ANGLES, LIFT HOLES AND OTHER AIDS, WELDING PROCEDURES, REQUIRED ROOT OPENINGS, ROOT FACE DIMENSIONS, GROOVE ANGLES, BACKING BARS, COPES, SURFACE ROUGHNESS VALUES, AND UNEQUAL PARTS.

Figure 7.4 Steel fabrication and the IBC

The beam camber in Figure 7.4's second paragraph is in reference to the natural arch or bow in steel beams. In some cases, to help support the floor loading, the engineer will require that additional cambering of the beam be performed as part of their design. This requirement will be noted in the floor framing plans next to the beam size by a note similar to: $C = 3/4$", or $C = 1/2$". Note that the arch in the beam created by the act of cambering will always be to the top side of the beam, as this additional arch is intended to provide additional lift, which then enables the beams to take on a heavier load.

To add additional camber to a beam, the shop either applies heat to select points on the beam or uses a hydraulic cambering machine. Some fabricators may elect to buy their beam already cambered from a steel supply warehouse.

As you are preparing your material listing, identify the cambered beams as such. Indicate the required amount of camber.

In Figure 7.5, we find the structural drawing notes. The figure provides a detail of Typical Steel Framing Symbols as well as other typical steel details found in structural drawings. This diagram helps the reader decipher the elements of the notes, and goes a long way in helping us understand the designer's nomenclature.

The information found in the third paragraph of Figure 7.4 on erection aids is not to be ignored. We don't want to be responsible for supplying additional fabrication services for the benefit of another trade, or an element that is not specifically noted as required to be provided by the steel fabricator. Because we don't know what the erector will require, any references for erection aids will need to be added to our exclusion list.

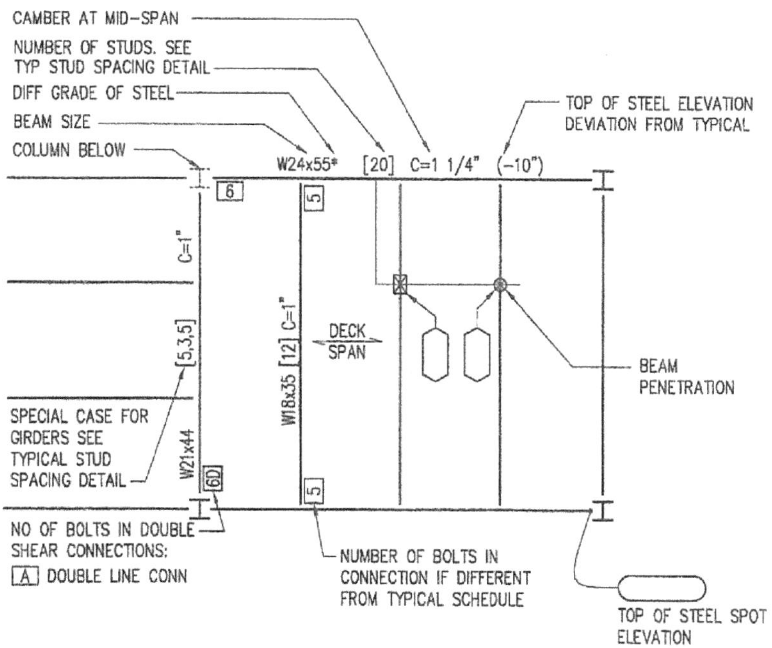

Figure 7.5 Typical Plan View

Galvanizing Steel

Figure 7.6 includes a very important note on galvanizing, which applies to all steel that is exposed to weather and is not painted. Galvanizing is most commonly done by dipping the steel parts into a vat of hot liquid zinc that is about 860°F. The part is completely submerged, then removed and hung to cool.

GALVANIZING: STRUCTURAL STEEL AND CONNECTIONS EXPOSED TO WEATHER AND NOT TO BE PAINTED SHALL BE HOT-DIPPED GALVANIZED AFTER FABRICATION IN COMPLIANCE WITH ASTM A 123. ALL FIELD WELDS ON GALVANIZED MATERIAL SHALL BE COATED WITH BRUSH APPLIED ZINC-RICH PAINT COMPLYING WITH THE SPECIFICATIONS. ALL PLATES AND OTHER STEEL ITEMS EMBEDDED IN CONCRETE SHALL BE HOT-DIPPED GALVANIZED AFTER FABRICATION.

Figure 7.6 Galvanizing

If the part is too large to galvanize in a single dip, then it will have to be double dipped to coat the part that wasn't coated with the first dunking. We need to know the size of the tank that is used by the galvanizer from whom we get the price of galvanizing. Providing the physical exterior dimensions along with the weight will help the galvanizer give an accurate price. It is not safe to assume that the price will just be double—some of the labor included in your cost will already have been performed (moving and staging) with the first dip. For this reason, it is best not to guess; better to work instead towards getting a confirmed price.

Galvanizing may also involve a hot spray, much in the same way as regular painting. This process is usually reserved for items that are extra large and difficult to ship. The term often used for this approach is "metalizing" and there are firms that specialize in this process. Some may perform the work at your shop; some firms have shops of their own. In either case, remember any shipping and handling costs that you may need to add to your estimate for transportation.

If you are in doubt about what may or may not need galvanizing with a particular project, call your customer or the architect of record noted on the contract drawings.

Welding

The welding note in Figure 7.7 discusses compliance requirements; it is standard for most projects. Weld size requirements are not always shown on contract drawings. This note advises you what is expected.

Weld size requirements directly affect your application of labor to the project. Too heavy a weld will make you apply too much labor to the project. However, not putting in enough labor for particular weld areas will lead you to low bid, but not in the way you want to be low bid. Therefore, it is impor-

WELDING: ALL WELDING SHALL BE IN CONFORMANCE WITH AISC AND AWS STANDARDS, AND SHALL BE PERFORMED BY AWS-WABO-CERTIFIED WELDERS USING 70 KSI ELECTRODES AND LOW HYDROGEN PROCESSES. ONLY WELDS THAT ARE PREQUALIFIED, AS DEFINED BY AWS, OR QUALIFIED BY TESTING SHALL BE USED. SHOP DRAWINGS SHALL SHOW ALL WELDING WITH AWS A2.4 SYMBOLS. WELDS SHOWN ON THE DRAWINGS ARE MINIMUM SIZES. INCREASE WELD SIZE TO AWS MINIMUM SIZES BASED ON PLATE THICKNESS. MINIMUM WELD SIZE SHALL BE 3/16-INCH, UNLESS NOTED OTHERWISE. THE WELDS SHOWN ARE FOR THE FINAL CONNECTIONS. FIELD WELD SYMBOLS ARE SHOWN WHERE FIELD WELDS ARE REQUIRED BY THE STRUCTURAL DESIGN. WHERE FIELD WELD IS NOT INDICATED, THE CONTRACTOR IS RESPONSIBLE FOR DETERMINING IF A WELD SHOULD BE SHOP- OR FIELD-WELDED IN ORDER TO FACILITATE THE STRUCTURAL STEEL ERECTION.

SEISMIC CRITICAL WELDS: WELDS CLASSIFIED AS "SEISMIC CRITICAL" ARE INDICATED ON THE DRAWINGS. SEE PROJECT SPECIFICATIONS FOR REQUIREMENTS.

Figure 7.7 Welding

tant to pay attention to notes like this one on the drawings.

Observe that the designers leave an option for the steel fabricator and erector to determine the best-served condition for welding. This type of situation may occur when dealing with canopies that may be shop assembled as opposed to field welded together.

If a condition of the framing like this one is found in the bid set of drawings, determine if additional shop labor time should be calculated into your price in order to create assemblies or fabricated frames. The erectors will quote the install conditions based on what works best from their point of view, assuming that the steel fabricator will provide the steel accordingly. It never hurts to have a conversation with either the general contractor or one of the erectors bidding the work to find out how the steel is to be sent to the job site. Then state those specific conditions in the included item of your bid letter.

Decking

If you are supplying decking as part of your bid package, then the notes on floor and roof decks in Figure 7.8 are important. Steel decking is used in commercial buildings the same way one might use plywood sheeting in a house.

The floor and roof deck material is attached to the top of the steel beam by welding (or button punch or puddle welds). Steel decking serves to support the concrete that will be poured for the floor. Meanwhile, insulation is installed on top of the roof deck, and then the roofing.

Figure 7.8 provides information regarding "closure plates and flashing." Closure plates and flashing are usually supplied by subcontractors that provide

STEEL DECK SHALL CONFORM TO ASTM A 653, GRADE 33, MINIMUM. MINIMUM FY = 38,000 PSI. WHERE THE DECK IS LEFT PERMANENTLY EXPOSED, GALVANIZED COATING SHALL CONFORM TO ASTM A 924, G90. IN OTHER AREAS, GALVANIZED COATING SHALL CONFORM TO ASTM A 924, G60.

MINIMUM DECK GAGES ARE SHOWN ON PLANS AND ARE BASED ON 3-SPAN, UNSHORED CONDITIONS. HEAVIER DECK GAGES MAY BE REQUIRED FOR CONDITIONS OTHER THAN THESE, DEPENDING ON MANUFACTURER'S AND CONTRACTOR'S LAYOUT. DECK SUPPLIER SHALL VERIFY DECK GAGES AND CAPACITIES BASED ON ACTUAL DECK LAYOUT AND SPAN CONDITIONS INCLUDING A 18 PSF SUPERIMPOSED DEAD LOAD ALLOWANCE FOR THE ROOF DECK. DEVIATIONS IN DECK GAGES FROM THOSE SHOWN SHALL BE SUBMITTED TO THE ENGINEER, ALONG WITH A VALID ICBO OR ICC REPORT FOR APPROVAL PRIOR TO SHOP DETAILING.

DECK WELDING SHALL BE IN ACCORDANCE WITH AWS D1.3, "SPECIFICATION FOR WELDING SHEET STEEL IN STRUCTURES." WELDERS SHALL BE QUALIFIED BY WABO LIGHT GAUGE CERTIFICATION.

CONTRACTOR SHALL PROVIDE CLOSURE PLATES, FLASHING, AND ALL MISCELLANEOUS LIGHT GAGE METAL SHAPES NECESSARY TO COMPLETE THE WORK. THE MINIMUM BEARING REQUIRED SHALL BE 2 INCHES.

COMPOSITE FLOOR/ROOF SLAB DECK: STEEL FLOOR DECK SHALL BE A COMPOSITE TYPE DECK WITH RIBS AT 12 INCHES ON CENTER OF THE SIZE AND GAGE SHOWN ON THE PLANS AND DETAILS, OR AN APPROVED EQUAL.

FINISHED COMPOSITE FLOOR SLABS SHALL HAVE A MINIMUM SUPERIMPOSED LOAD CAPACITY OF THE GREATER OF THE FOLLOWING:

A. DESIGN LIVE LOAD PLUS SUPERIMPOSED DEAD LOADS, SUCH AS ARCHITECTURAL FINISHES, TOPPING SLABS AND LOAD SUSPENDED FROM SLAB.
B. 125 PSF MINIMUM.

FLOOR DECK FASTENING SHALL BE AS FOLLOWS, UNLESS NOTED OTHERWISE, AND EXCEPT AS INDICATED IN TYPICAL COMPOSITE BEAM DETAILS.

1/2-INCH DIAMETER ARC-SPOT PUDDLE WELDS AT 12 INCHES ON CENTER AT TRANSVERSE AND PERIMETER SUPPORTS

1/2-INCH DIAMETER ARC-SPOT PUDDLE WELDS AT 18 INCHES ON CENTER AT LONGITUDINAL SUPPORTS

BUTTON PUNCH OR 1 1/2-INCH TOP OR SIDE SEAM WELD AT 18 INCHES ON CENTER AT SIDE LAP CONNECTIONS

Figure 8 Floor Decking

the roof and floor deck. However, we do need to watch out for job details where the steel fabricators supply closure angles and plates for special conditions.

When the materials used for closures is 1/8" plate or heavier, it might be something the steel fabricators would supply. Light gauge materials—less than 1/8" thick— may be provided by the same subcontractor supplying the metal deck. Notes provided with the specific details in the drawings should designate this provision. When in doubt, add these items to the list of items that are to be excluded in the bid letter.

Any special openings in the floor will need some steel supports that may or may not be shown, as well as closure plates or angles needed to hold back concrete.

On the inclusions list in the bid letter, list all the applicable drawing numbers, detail call outs, material sizes, and your linear footage as applicable. Be clear what is being supplied by the steel fabricator. Listing all of the necessary information will eliminate confusion.

NONCOMPOSITE ROOF DECK: STEEL ROOF DECK SHALL BE OF THE SIZE AND GAGE SHOWN ON THE PLANS OR AN APPROVED EQUAL. ROOF DECK FASTENING SHALL BE AS SHOWN ON THE PLANS. THE MINIMUM END LAP SHALL BE 2 INCHES CENTERED OVER SUPPORTS.

SUSPENDED CEILINGS, LIGHT FIXTURES, PIPES, DUCTS, MECHANICAL OR ELECTRICAL EQUIPMENT, OR OTHER UTILITIES SHALL NOT BE SUPPORTED BY THE NONCOMPOSITE STEEL ROOF DECK WITHOUT APPROVAL OF THE ENGINEER.

HOLES OR COMBINATIONS OF HOLES IN NONCOMPOSITE ROOF DECK, WHICH CUT TWO WEBS WHICH ARE CLOSER THAN 24 INCHES ON CENTER IN ANY DECK SPAN, MAY REQUIRE DECK REINFORCEMENT AND REQUIRE DIRECTION FROM THE ENGINEER.

COLD-FORMED STEEL

COLD-FORMED STEEL FRAMING MEMBERS SHALL BE OF THE TYPE, SHAPE, SIZE, GAGE, AND SPACING AS SHOWN ON THE DRAWINGS. MEMBER TYPES AND SIZES SHOWN ON THE DRAWINGS REFER TO MEMBERS AS DEFINED BY THE STEEL STUD MANUFACTURERS ASSOCIATION (SSMA). MEMBERS EQUIVALENT IN SHAPE, SIZE, STIFFNESS, AND STRENGTH BY OTHER MANUFACTURERS MAY BE SUBSTITUTED FOR FRAMING MEMBERS SHOWN. ALTERNATE MEMBERS SHALL BE SUBJECT TO REVIEW AND APPROVAL BY THE ARCHITECT AND STRUCTURAL ENGINEER PRIOR TO FABRICATION AND ERECTION. ALL COLD-FORMED STEEL FRAMING SHALL CONFORM TO THE AISI "SPECIFICATION FOR THE DESIGN OF COLD-FORMED STEEL FRAMING MEMBERS." ALL STUDS, TRACKS AND JOISTS SHALL BE GALVANIZED. FASTENINGS SHALL BE AS SHOWN ON THE DRAWINGS. FASTENINGS NOT SHOWN SHALL BE AS RECOMMENDED BY THE MANUFACTURER.

COLD-FORMED STEEL WELDING SHALL BE IN ACCORDANCE WITH AWS D1.3, "SPECIFICATION FOR WELDING SHEET STEEL IN STRUCTURES." WELDERS SHALL BE QUALIFIED BY WABO LIGHT GAUGE CERTIFICATION.

Figure 7.9 Roof decking

The roof deck notes in Figure 7.9 are typical of most projects. The aspect most important to us is the potential effect on the steel fabricator's scope of work with regard to supports of fixtures and openings in the roof. These items for support or openings need to be specifically included or excluded in the bid letter. If included, be clear about what you are including. List the sizes and quantities of materials that will be used, especially if no details are called out in the plans.

Cold formed framing is not a concern of steel fabricators. It deals with *C* studs for wall framing in much the same fashion that 2 x 4s are used in framing the walls of a house. All details and references to this *C* stud material are usually excluded from the steel fabricators' scope of work.

Anchors

ANCHORS

USE OF DRILLED CONCRETE ANCHORS, INCLUDING EXPANSION BOLTS, ADHESIVE ANCHORS, AND UNDERCUT ANCHORS, WHERE NOT SPECIFIED IN THE DOCUMENTS SHALL BE SUBJECT TO APPROVAL BY THE ENGINEER.

EPOXY ADHESIVE ANCHORS: ADHESIVE-TYPE ANCHORS SHALL BE INSTALLED PER THE MANUFACTURER'S RECOMMENDATION. THE EMBEDMENT DEPTH SHALL BE AS SHOWN ON THE DRAWINGS. SPECIAL INSPECTION IS REQUIRED FOR ALL ADHESIVE ANCHORS. ADHESIVES SHALL BE HILTI-HY150 OR AN APPROVED EQUAL. ICBO OR ICC REPORT SHALL BE SUBMITTED FOR ALL ADHESIVE ANCHOR PRODUCTS.

HEADED SHEAR STUDS AND DEFORMED BAR ANCHORS: ALL SHEAR STUDS SHALL BE 3/4-INCH DIAMETER HEADED STUDS. STUD LENGTHS AFTER WELD SHALL BE AS SHOWN ON THE DRAWINGS. DEFORMED BAR ANCHORS SHALL BE OF THE SIZE AND LENGTH SHOWN ON THE DRAWINGS. ALL STUDS AND DEFORMED BAR ANCHORS SHALL BE AUTOMATICALLY END WELDED IN SHOP OR FIELD WITH EQUIPMENT RECOMMENDED BY MANUFACTURER.

EXPANSION BOLTS: EXPANSION BOLTS IN CONCRETE OR SOLID GROUTED MASONRY SHALL BE WEDGE-TYPE ANCHORS MADE OF CARBON STEEL. BOLT SPACING AND EMBEDMENT LENGTH SHALL BE AS DESIGNATED ON THE DRAWINGS. BOLTS SHALL BE INSTALLED PER THE MANUFACTURER'S RECOMMENDATIONS WITHOUT SPECIAL INSPECTION. THE FOLLOWING TYPES SHALL BE USED, UNLESS NOTED OTHERWISE: HILTI KWIK BOLT II, RAMSET/RED HEAD TRUBOLT, OR A PREAPPROVED EQUAL. A CURRENT ICC REPORT SHALL BE SUBMITTED FOR ALL EXPANSION BOLTS TO BE USED.

Figure 7.10 Anchors

The section on anchors, shown in Figure 7.10, is about fasteners for steel to concrete. Epoxy anchors are adhesive anchors that supply additional attachment support into the hole in the concrete. Concrete drilling is performed by the steel erector when they install steel items like ledger angles that support steel decking or roof decking.

Epoxy anchors are more expensive than the wedge type or expansion anchors also listed in this section; they are referred to as expansion bolts. Also called wedge anchors, expansion bolts lack this additional adhesive, although their function is the same in that they are designed to support steel to a concrete wall.

The expansion bolts are used when additional adhesive strength is not needed. Be clear about where these specific bolts are being used and where they are not. Do not let your bid be too high or too low over the use or the lack of epoxy anchors. Be clear about the provision in the bid letter—epoxy anchors can be more than double the price of the wedge anchors/expansion bolts.

Headed shear studs are items that are welded to the steel which, in turn, is embedded into concrete. Headed shear studs look like a large nail, but are larger in diameter with thicker heads. They can be welded by regular welding methods when there are but a few. However, it is more desirable and sometimes required that a steel fabricator use a stud welder for welding headed anchor studs (HAS).

Related to the HAS or headed anchor studs are threaded anchor studs. Threaded anchor studs are a type of bolt where the end without threads is welded to the steel. The other end is threaded to receive a nut. This method is used for attaching wood or other steel items to a fixed part. The use of threaded anchor studs will be shown in the contract drawing details, if required.

Any information on the contract drawings about anchors is in addition to whatever might be called out in the project specifications. Remember that whenever there is a conflict, the specifications will prevail.

Inspection Schedules

It is important to review the special inspection schedule such as the one in Figure 7.11 (if there is one with the bid package). The impact of any additional costs needs to be included in the costs of the work scope. For example, special inspections required by the IBC, or by certain state, county, or city

SPECIAL INSPECTION SCHEDULE
ESTABLISHED PER 2006 IBC SECTION 109 & CHAPTER 17

ITEM	CONTINUOUS INSPECTION	PERIODIC INSPECTION	COMMENTS
SOILS			
GRADING, EXCAVATION & FILL			BY GEOTECHNICAL ENGINEER
FINAL FOUNDATION PREPARATION			BY GEOTECHNICAL ENGINEER
PREFAB. CONSTRUCTION			REF. NOTE 6
CONCRETE			
REINFORCING PLACEMENT		X	
REINFORCING WELDING		X	
REINFORCING COUPLING		X	
ANCHOR BOLTS & INSERTS		X	
PREPARATION OF TEST SPECIMENS	X		
CONCRETE PLACEMENT	X		
ADHESIVE ANCHOR PLACEMENT	X		REF. NOTE 7
EMBEDDED PLATES		X	
CURING		X	
STRUCTURAL STEEL			
FABRICATION & ERECTION		X	REF. NOTE 8
HIGH STRENGTH BOLTING	X		REF. NOTE 9
SINGLE PASS FILLET WELDS ≤ 5/16"		X	REF. NOTE 10
FILLET WELDS > 5/16"	X		REF. NOTE 10
PARTIAL/COMPLETE PENETRATION WELD	X		REF. NOTE 10
OTHER WELDING			
WELDING OF ANCHORS AND STUDS		X	
WELDING-STAIRS/RAILING SYSTEMS		X	
METAL DECK WELDING		X	
LIGHT GAUGE METAL FRAMING WELDING		X	

Figure 7.11 Inspection schedules

INSPECTION SCHEDULE NOTES:

1. THE ITEMS CHECKED WITH AN "X" SHALL BE INSPECTED IN ACCORDANCE WITH IBC CHAPTER 17 BY A CERTIFIED SPECIAL INSPECTOR FROM AN ESTABLISHED TESTING AGENCY. FOR MATERIAL SAMPLING AND TESTING REQUIREMENTS, REFER TO PROJECT SPECIFICATIONS, THE STRUCTURAL NOTES AND THE NOTES BELOW. THE TESTING AGENCY SHALL SEND COPIES OF ALL STRUCTURAL TESTING AND INSPECTION REPORTS DIRECTLY TO THE ARCHITECT, ENGINEER, CONTRACTOR AND BUILDING OFFICIAL. ANY MATERIALS WHICH FAIL TO MEET THE PROJECT SPECIFICATIONS SHALL IMMEDIATELY BE BROUGHT TO THE ATTENTION OF THE ENGINEER. SPECIAL INSPECTION TESTING REQUIREMENTS APPLY EQUALLY TO ALL BIDDER DESIGNED COMPONENTS.

2. PROVIDE SPECIAL INSPECTION PER SECTION 1704 OF THE INTERNATIONAL BUILDING CODE AS AMENDED AND ADOPTED BY THE CITY OF TACOMA. THE SPECIAL INSPECTION AGENCY AND THE SPECIAL INSPECTORS SHALL BE CERTIFIED THROUGH THE WASHINGTON ASSOCIATION OF BUILDING OFFICIALS (WABO) CERTIFICATION PROGRAM FOR THE INSPECTIONS TO BE PERFORMED. THE SPECIAL INSPECTION AGENCY WILL BE HIRED BY AND AT THE EXPENSE OF THE OWNER AND WILL BE RESPONSIBLE TO THE BUILDING OFFICIAL FOR THE AREAS OF SPECIAL INSPECTION APPLICABLE TO THE PROJECT.

3. SPECIAL INSPECTION IS NOT REQUIRED FOR WORK PERFORMED BY AN APPROVED FABRICATOR PER IBC SECTION 1704.2.2.

4. CONTINUOUS SPECIAL INSPECTION MEANS THAT THE SPECIAL INSPECTOR IS ON THE SITE AT ALL TIMES OBSERVING THE WORK REQUIRING SPECIAL INSPECTION (IBC 1702). PERIODIC SPECIAL INSPECTION MEANS THAT THE SPECIAL INSPECTOR IS ON SITE AT TIME INTERVALS NECESSARY TO CONFIRM THAT ALL WORK REQUIRING SPECIAL INSPECTION IS IN COMPLIANCE.

5. INSPECTION REQUIREMENTS FOR SYSTEMS DESIGNED BY OTHERS SHALL BE DEFINED BY THE REGISTERED DESIGN PROFESSIONAL RESPONSIBLE FOR THEIR DESIGN.

INSPECTION SCHEDULE NOTES: (CONT.)

6. INSPECTION FOR PREFABRICATED CONSTRUCTION SHALL BE THE SAME AS IF THE MATERIAL USED IN THE CONSTRUCTION TOOK PLACE ON SITE. CONTINUOUS INSPECTION WILL NOT BE REQUIRED DURING PREFABRICATION IF THE APPROVED AGENCY CERTIFIES THE CONSTRUCTION AND FURNISHES EVIDENCE OF COMPLIANCE.

7. INSPECTION OF DRILLED CONCRETE ANCHORS, INCLUDING EXPANSION AND ADHESIVE GROUTED ANCHORS, SHALL INCLUDE VISUAL VERIFICATION OF DRILLED HOLE DEPTH, SPACING, EDGE DISTANCES AND HOLE CLEANING. FOR GROUTED ANCHORS, GROUT INSTALLATION SHALL BE OBSERVED AND GROUT PRODUCT SPECIFICATION AND PREPARATION SHALL BE VERIFIED.

8. INSPECTION OF STRUCTURAL STEEL SHALL BE IN ACCORDANCE WITH IBC SECTION 1704.3. THE STEEL FRAME SHALL BE INSPECTED FOR COMPLIANCE WITH APPROVED CONSTRUCTION DOCUMENTS INCLUDING BRACING, STIFFENING, MEMBER LOCATIONS AND PROPER APPLICATION OF JOINT DETAILS AT EACH CONNECTION.

9. INSPECTION OF BOLT INSTALLATION FOR PRETENSIONING IS PERMITTED TO BE PERFORMED ON A PERIODIC BASIS WHEN USING THE TURN-OF-NUT METHOD WITH MATCHMARKING TECHNIQUES, THE DIRECT TENSION INDICATOR METHOD, OR THE ALTERNATE DESIGN FASTENER (TWIST-OFF BOLT) METHOD. JOINTS DESIGNATED AS SNUG TIGHT NEED ONLY PERIODIC INSPECTION.

10. SPECIAL INSPECTION WILL BE PROVIDED FOR ALL STRUCTURAL WELDS, INCLUDING THOSE DONE IN FABRICATION SHOPS WHICH ARE NECESSARY FOR THE STRUCTURAL STABILITY OF THE STRUCTURE OR FOR THE SUPPORT OF ARCHITECTURAL ELEMENTS, MECHANICAL EQUIPMENT, OR OTHER BUILDING FEATURES. SPECIAL INSPECTION OF WELDING SHALL INCLUDE VISUAL INSPECTION OF ONE-HUNDRED (100) PERCENT OF ALL WELDS AND NON-DESTRUCTIVE TESTING OF A MINIMUM OF TEN (10) PERCENT OF ALL WELDS. IF MORE THAN FIVE (5) PERCENT OF WELDS NON-DESTRUCTUVELY TESTED FAIL, THEN ALL WELDS SHALL BE NON-DESTRUCTIVELY TESTED.

Figure 7.12 Inspection schedule notes

codes, are normally paid for by the owner.

The two columns in Figure 7.11 show the required occurrences of the inspections. Some items require continuous inspection, others periodic. Periodic inspections may involve one site visit by the inspector as part of an informal review of your shop and process, especially if the inspector is already familiar with your work. If the inspector gets an idea that a process performed by your shop may be incorrect, there will be more than one visit for a particular function.

Note 10 in Figure 7.12 requires 100% visual and non-destructive testing of 10% of the welds. This note is particularly important for the inspection of partial and complete penetration welds. This requirement of non-destructive testing (NDT) will still be paid for by the owner, until discontinuities are discovered. A discontinuity is a construction term used to indicate any condition found in the welding that should not be there. Repair and retesting of welds that have been rejected will be paid for by the steel fabricator.

Any costs to re-inspect welds after shop repair is the concern of the steel fabricator's project manager. It is not something for steel estimators to be concerned about at the time of quote. There are companies that specialize in inspections and non-destructive testing; they can provide costs for their services should the need arise to include this pricing with your fabrication quote.

Some of these inspection services will base the price they provide you on your fabrication schedule, which is based on the total shop labor hours on the job. The people pricing the inspection need to know how many weeks the project will be in the shop; they will probably the number of beams and columns you have. Inspectors may even need a copy of the contract drawings to be able to quote the inspection.

Structural Drawing Symbols

The reference symbols shown in Figure 7.13 are used throughout the contract drawings. Almost every set of contract drawings has a list of abbreviations to be used; most are the same and are used from job to job. Always check these lists for new or special information. The symbols are the key to reading the structural drawings that the structural engineer has created for use in quoting this project. Although these notations are very small, they may have a huge impact on the way you view the steel supply and fabrication labor you are about to list and price.

Some information given in the general notes in the structural plans may

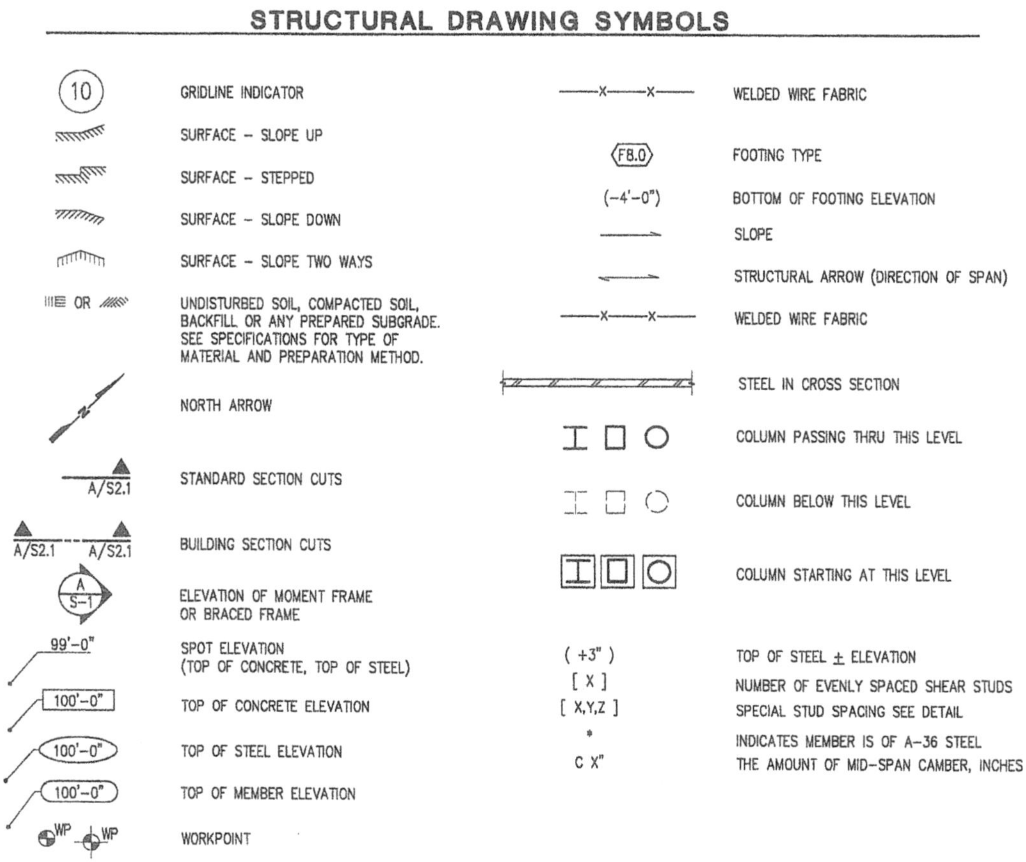

Figure 7.13 Structural drawing symbols

be duplicated in the specifications. As you write your bid letter, note the requirements by which you will perform the fabrication. Per AISC Code, the specifications govern over the plans. If you are clear in your qualifications in your bid letter about the requirements you are following for your fabrication, then you will eliminate any confusion about how you structured your pricing.

Foundation Plans

The drawings following the general information sheets should be the foundation plans. They should match the foundation plans in the architectural drawings.

In the foundation plans, you will find notes that designate the steel columns. There will also be detail references for the type of base plate and anchor bolt requirements needed for each column. These detail references will be on drawings that follow the floor and roof plan drawings, so you may have to page through several sheets to reach these details.

In addition to the steel columns, items to look for in the foundation plans include embedded items that need to be installed prior to the concrete pour. These items include steel beams needed to support the floor, the elevator pit frame angles, and grating. Embeds that may be needed in the foundation for any reason will be called out in detail bubbles.

Sometimes steel embedded items are shown only in the architectural foundation plan. Placement depends on the design team who put these drawings together. Note that if steel fabricated concrete embeds are not shown in the structural drawings, they will need to be picked up in the architectural drawings.

There may be foundation framing notes. Be sure to read them for any detail information that is not be called out or sectioned in the foundation plan itself. Some architects and engineers use both formats on the same sheet to show information you need. Read through it all. Highlight if you need to, but make the concentrated effort not to miss anything.

The foundation plans will be followed by the subsequent floor plans and then the roof, in the same order as the architectural drawings. In these floor plans, you will find the framing beams that hold up the floors. Main steel members frame from column to column. The columns are located generally at grid line locations, and look like an H (for wide flange shapes) or a box (for tube steel shapes) as you look down on the plan view of the floor.

Sometimes you will see the column at locations other than on grid lines. The beams that are framing into all of the columns are what hold up the floor and roof. Steel framing members that hold up the floor will be shown as a W if a wide flange beam, *C* if a piece of channel, or TS if a tube steel member.

Be sure to read through all the notes on the drawing. As a rule, architects and engineers mean everything that they put on the drawings to be used. Be sure you understand what they are saying in the way that they say it. The only stupid question is the one not asked. Even the architect and engineer can unintentionally leave information unclear on the drawings. You are creating an estimate based on the information they show. Therefore, it is imperative that you get what you need from these drawings. Work hard to get clarification on that which is not clear, not understood.

Drawings are made to a certain scale that allows the reader to dimension the size of the building or items within the building framework. The scale of the drawings is typically in feet and inches, though some are done in metric.

An Architect's scale is a triangular-shaped ruler that has different dimensioning on each of its three sides. The scale is about 12 inches long; it has increments from 3/32 of an inch up to 3 inches to the foot. There is also the Engineer's scale, the Mechanical scale, and the Metric scale—all offer different applications for use with the contract drawings. It is helpful to the steel estimator to have one of each type ready for use as the need arises.

First Floor Foundation

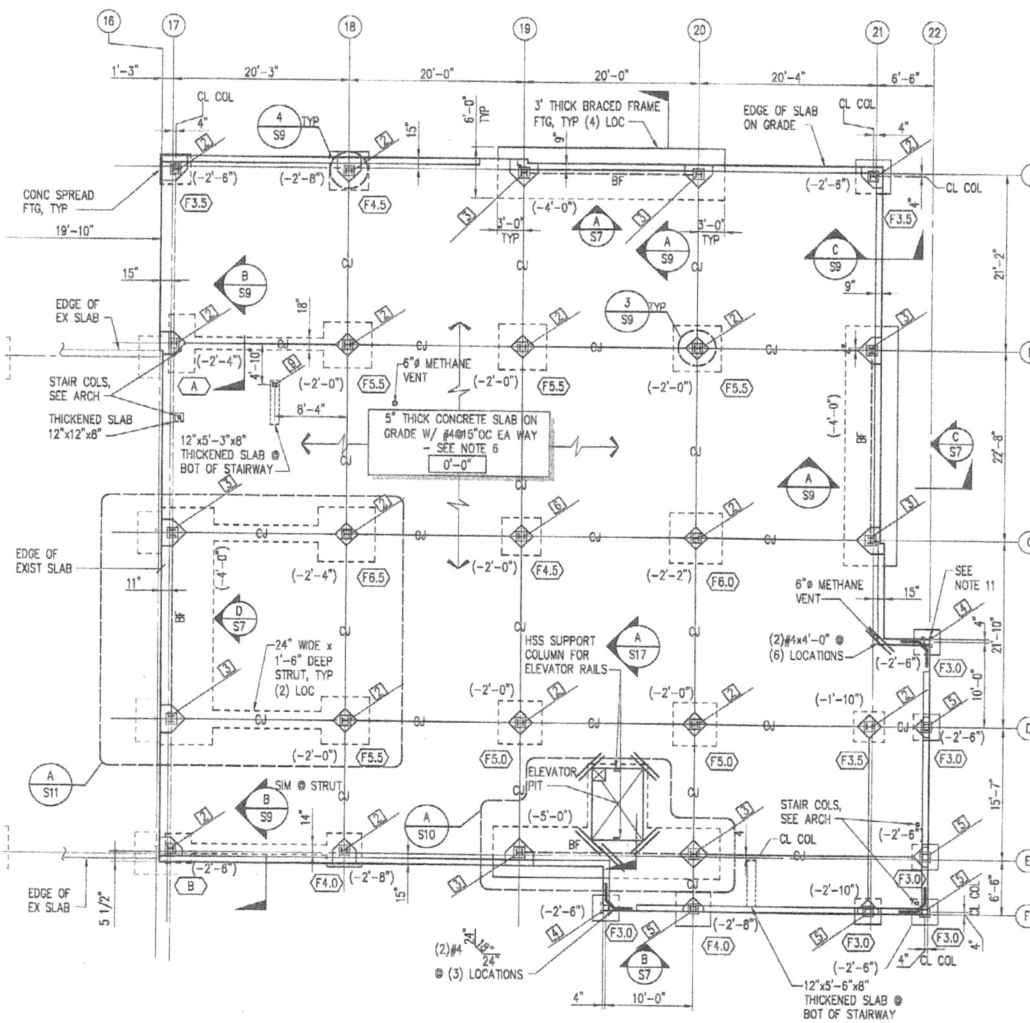

Figure 14A First floor foundation plan

Figure 14B Scale

Figure 14C North indicator

Figure 29A is a first floor foundation plan. The scale information for this first floor foundation plan, indicated in Figure 29B, shows that this drawing is prepared on a scale of 1/8" = 1'-0. Therefore, 1/8" in length on the drawing is equal to 1'-0 in real life. Many other scales may be used in contract drawings: 1/4" = 1'-0, or 1/2" = 1'-0, or 1-1/2" = 1'-0, and so on. Be sure you know what scale you are working with if you have to measure anything. Use the dimensions written on the drawings whenever possible. The North indicator arrow on the drawing, as shown in Figure 29C, is important to know for elevation references that you will be using in the set of drawings.

Drawings are often copied and used in different formats and sizes. As a result, it would be unsafe to take measurements by scaling the drawings. It is best to use the dimensions shown on the plan whenever possible. If the drawings are accessed by computer programs online, there may be a scale tool that is accessible for use in dimensioning.

When we work with reduced or enlarged drawings and want to be sure that any dimensioning is correct, we have to know on what size of sheet these drawings were drawn to know if our scale is accurate. Another way to scale a drawing is to pick a dimension and find a scale that works with that dimension for a conversion, using the dimensions at the plans for a match.

For example, if you find a 10' dimension on the drawings, mark that length using tick marks on the edge of a separate piece of paper. You can then assume that half that length would be 5', and a fourth of that as 2'-6", and so on. Double the 10' space for 20', or triple the space for 30'. These marks on the paper could then be used as a temporary ruler that could help you ascertain a required length for something in the drawings that is not shown.

Because we are estimating, we can get away with giving our best guess—as long as we don't go so crazy with our assumptions that our numbers are too high or too low.

Key Plans

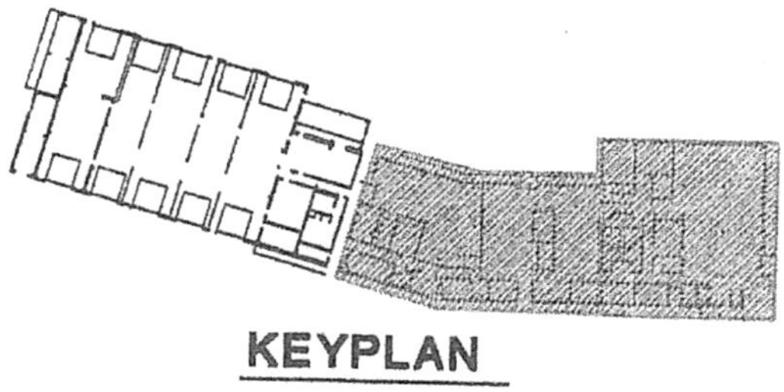

Figure 7.15 Key plan

Drawings often include *key plans* like the one in Figure 7.15. Key plans help us see which section of the building is being shown to us in the plan and its relationship with other areas. Some projects may have several buildings and many floors; having a key is helpful. Most likely there will not be a key if the designers can show the building or buildings on one sheet.

The foundation plan in Figure 7.14A has specific material type designators regarding the columns. The columns in this plan view—for plan views, you are always looking down on the building or part—are designated by what looks like an *H* or an *I*, depending on how it is shown at this plan. Wide flange columns or *W* shape are used here. If you go back to the Structural Drawing Symbols in Figure 7.13, you will see the different designations that are used to indicate the columns, also shown in Figure 7.16. Note here that *I* is for wide flange, the box is square tubing, and the circle is steel pipe.

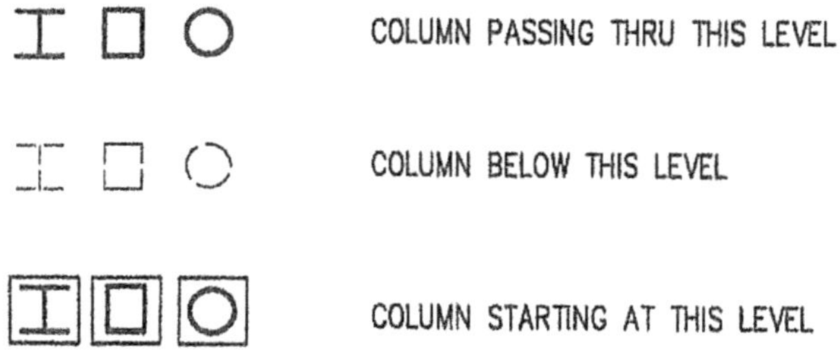

Figure 7.16 Column symbols

The pictures in Figure 7.16 look down on the columns from above, as columns are shown to us as standing up on the page. When we look at the floor plans, the beams are horizontal, connecting at the ends of the beams to the columns.

Getting back to the Structural Drawing Symbols in Figure 7.13, we see the symbols shown in Figure 7.17 being used in the foundation plan:

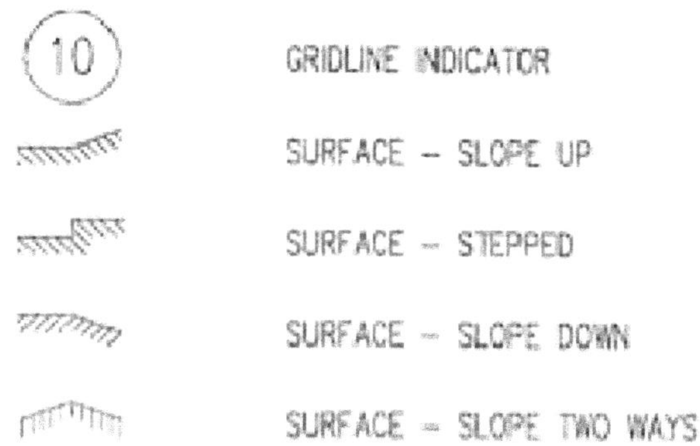

Figure 7.17 Structural drawing symbols and foundation plans

The grid line indicator in Figure 7.17 identifies column locations when preparing your material listing. Referencing grid line locations while describing a condition makes it easier for others to understand the location of what is being discussed. Grid line indicators are also helpful when creating written documentation referencing them, such as an RFI or Request for Information.

Grid lines are generally given for centerline locations of the columns; the dimensions that are given between grid lines are used in creating the detail drawings for the steel framing. The steel beam framing is generally set at the centerline of grid. Columns and beams that are offset from grid will be dimensioned specifically from that center of grid location. The entire framework of the steel is dimensioned from the grid line locations.

For an example of the grid lines, take a closer look at the foundation plan in Figure 7.18.

The grid lines in Figure 7.18 are located at the top of the page and the right side. We see grids 16 through 22 on the top, grids A through F on the right. The columns are identified by the number in the box located on the diagonal tag line drawn from the column centerline up to the right or down to the left. The footings are indicated by the letter followed by the number that is in the rectangular box under the column. The depth of the footing is the minus dimension (–2.5, for example) that is next to the footing identifier.

The *surface level indicators* in Figure 7.19 that follow the grid line indicator are important for determining the top of steel elevations. These indicators will note a change in elevation as they are required from the foundation through to the roof. This information will be used in our material take off that is included in our quote, and also with the steel shop detail drawings following our successful bid. Although these indicators are not used in the foundation plan,

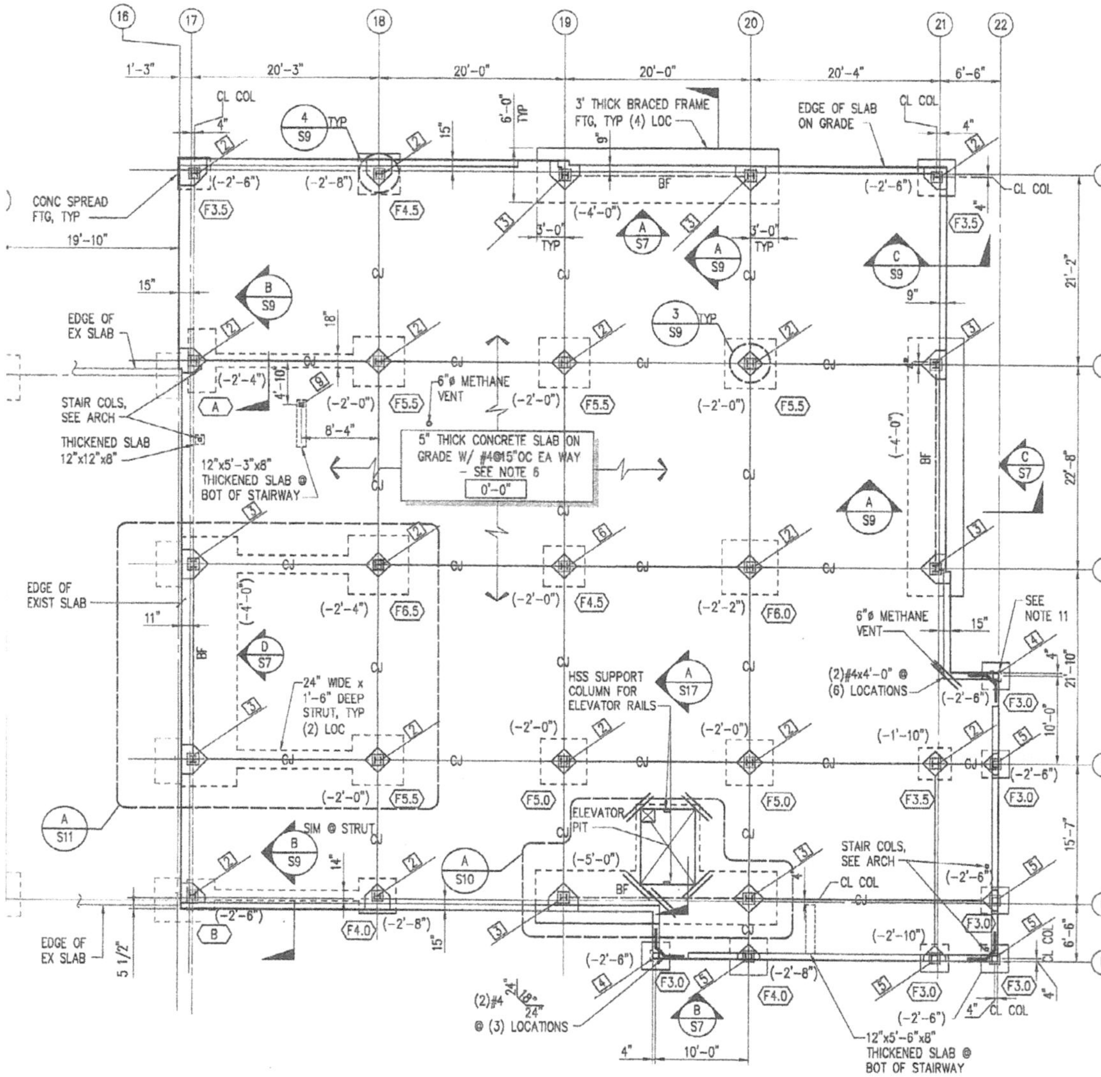

Figure 7.18 Foundation plan

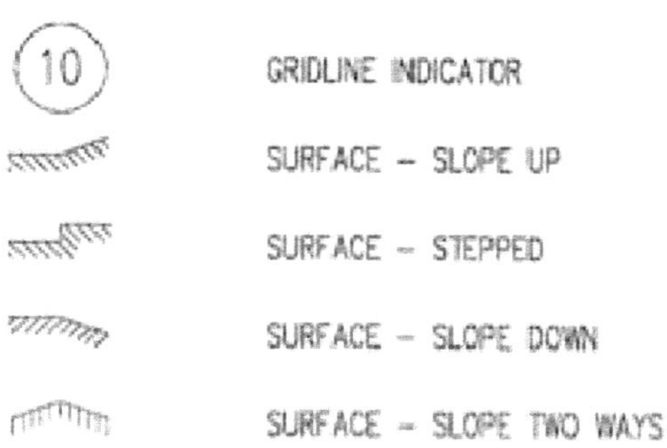

Figure 7.19
Surface level indicators

which has been used as an example here, it is important to watch for them as we perform our material take off listing.

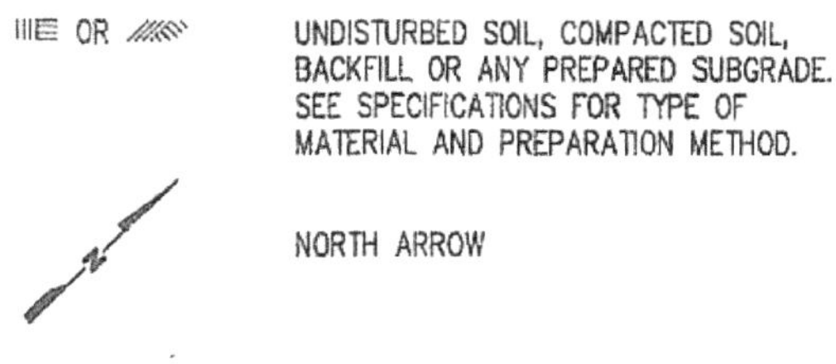

Figure 7.20 Soils indicator

The *soils indicator* in Figure 7.20 is for other trades, but the North Arrow is an important reference when we look at building elevations. It is also important to the steel detailer; the direction of the column is noted on the shop detail drawings so that the field erector will set the column in the right direction. The steel that is detailed may not fit in with the rest of the steel framing if the correct column direction isn't noted on the column prior to shipment to the field.

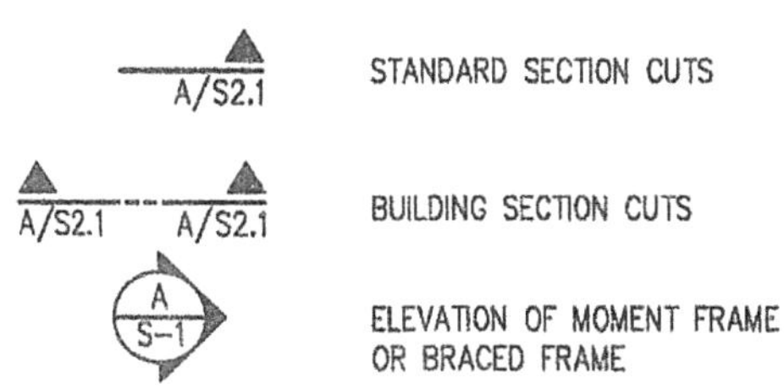

Figure 7.21 Section cuts

Figure 7.21 shows examples of the type of section cuts used on this set of plans. The *Standard Section Cut* (also called a detail cut) is designed to show you an enlarged view of a certain spot on the drawing plan. The *A* before the slash is the specific detail that you need to find, the *S2.1* after the slash is the drawing on which you will find the detail. Detail numbers are reused from sheet to sheet, starting over from *1* or *A* on each sheet— be careful to go to the right sheet to find the information you want.

Building Section Cuts or elevation section cuts generally will show the entire area through which the section is cut, starting from the first note of A/S2.1 and carrying all the way through to *A/S2.1* on the other end. The actual detailed view may or may not extend through an entire section of the building. Therefore, particular attention must be paid to the start and stop of the section, for you to be able to see what the designer is trying to show you.

This elevation section will be shown either in the same scale or size as the drawing from which the detail was cut, or may be shown to you a little bit larger. When these building section cuts are reviewed, additional detail or section cuts may be found leading to more drawings that will also need to be reviewed.

Designers perform this detail layering to help us see in more detail specific conditions within the building design. In the same way we change to higher levels of magnification in a microscope to get a closer view of the subject, designers lead us through a series of section views and detail cuts that get us closer and closer to all the parts and pieces that make up this building. We can then identify more clearly the steel items being used that may need to be included in the material take off listing. We can also see how their use affects our labor application.

The section for the elevation of moment frame or *braced frame* is used not only in the foundation plan, but also the subsequent floor and roof plans. It is intended to maintain your awareness of the moment frame locations. These moment frames may be built using deeper footings, special type materials, heavier welds, and additional bracing not used in other areas of the building.

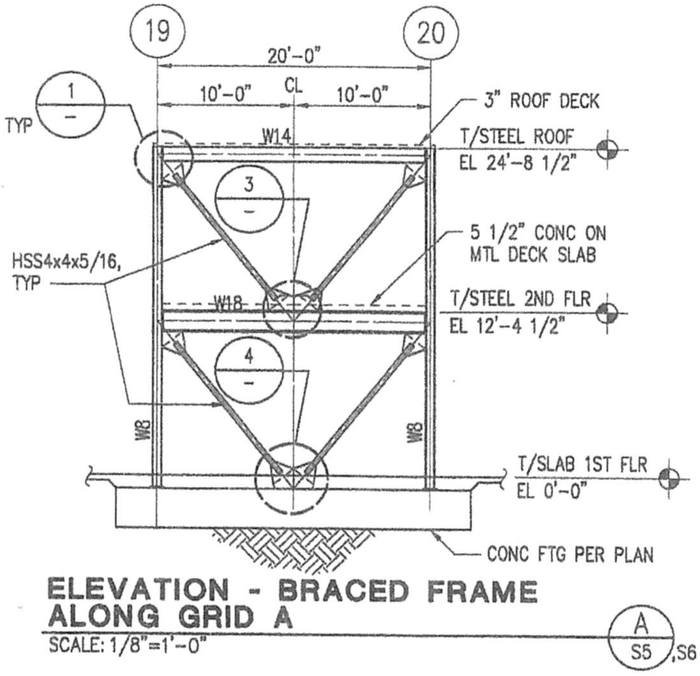

Figure 7.22 Braced Frame

Most commonly, the braced frames will be shown as a complete unit, with the details concerning the particular conditions called out from there. Refer back to Figure 7.14, Foundation/First Floor plan, for the braced frame. It is called out at grid line A between grids 19 and 20. Now you are able to locate this detail using drawing references you have learned. This is braced frame A/S7 in Figure 7.22.

Braced frames like the one in Figure 7.22 create additional strength for the building structure; they also add to the construction cost of the building. As we are performing the material take off listing, we will find additional labor in the steel detailing because of the complexity of creating shop detail drawings. The columns will go deeper into the ground because the footings are set deeper than the other column footings. The beams that frame into those columns have special connections that require additional labor.

The bracing that is connected to the columns and beams requires heavy plates, known as gusset plates, which are added to the columns and beams in the shop by special welding. The field will have additional welding in the bracing and the gusset plates. This welding adds to their costs of erection. Much like the column schedule, the braced frame elevations show each type of frame. The correct quantity of each frame type will need to be verified by going back to the foundation/first floor framing plan (Figure 7.18) to see exactly how many of each type there are.

Notice also the detail layering that began with the foundation/first floor framing plan, then continued to the braced frame elevation. From the braced frame elevation detail callouts, views are shown closer up.

As you begin to understand this process, the drawings will become easier to read. You will readily know where to go to locate what you want in the set of contract drawings—for example, whether it is in the structural or the architectural plans.

The *elevation indicators* shown in Figure 7.23 have specific notes as to their application. Be careful to be accurate in their use. These will be found in plan view drawings in particular. The work point survey marks on the drawings are used as the start and ending points with figuring angles or diagonals in bracing or certain building radius applications. These are necessary in steel fabrication for the sake of creating accurate shop detail drawings and also in figuring the math for correct dimensioning for a particular part or piece. Steel estimators use them in calculating correct lengths for material take off listings.

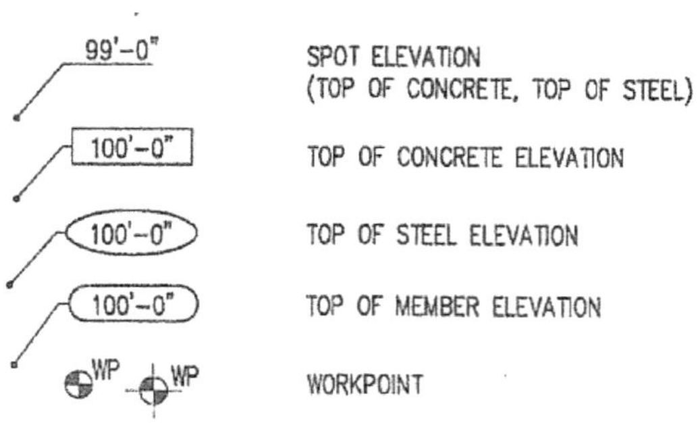

Figure 7.23
Elevation and work point indicators

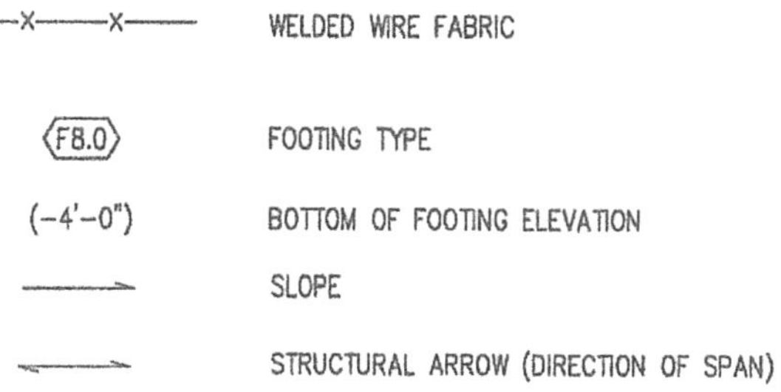

Figure 7.24
Welded wire fabric, Concrete, Span and Slope Indicators

In Figure 7.24, the *welded wire fabric* is a material used for reinforcing concrete. It is not supplied by the steel fabricators in this capacity. However, there are cases where welded wire fabric may be used for certain handrail styles and trellis work. If welded wire fabric is used for these conditions, then the steel fabricators would include it in their supply.

Footing type indicators and bottom of footing elevations in Figure 7.24 are important to steel estimators when calculating the length of the columns. We see both of these notes being used in the foundation plan next to each column location.

The slope indicator shown in Figure 7.24 gives you the direction of the slope. It also lets you know that the materials in your take off listing may need to be a little longer than the dimension in the plans may show you. For steel in slope areas, additional labor time will be needed to fabricate parts because the ends may not be square cut. Allow more time for steel detailers to run the calculations, plus more time to draw this part because it is no longer square.

The *span indicator* in Figure 7.24 is usually used in reference to the direction of the decking being laid in a given area, or the direction of the bearing bars for grating. It may even be used in conjunction with other indictors, like the span of a sloped area.

(+3")	TOP OF STEEL ± ELEVATION
[X]	NUMBER OF EVENLY SPACED SHEAR STUDS
[X,Y,Z]	SPECIAL STUD SPACING SEE DETAIL
*	INDICATES MEMBER IS OF A-36 STEEL
C X"	THE AMOUNT OF MID-SPAN CAMBER, INCHES

Figure 7.25 Elevation, stud spacing, material grade, and camber indicators

With Figure 7.25, the *top of steel* elevation dimension inside the parentheses gives you the special condition of an elevation that is different than what is called out for the rest of the building area. This information is important to note because it may change the length of a column or it may add to the detailer's work. It may also add to shop labor for the fabrication of the part due to the special conditions that make the difference in elevations.

Shear studs in Figure 7.25 (also known as headed anchor studs or HAS) are welded to the tops of the beams through the decking by the steel erector. Steel fabricators will normally exclude shear studs in their supply of the steel, and steel detailers will not include them as part of their work because they are not welded to the beam in the shop. However, steel erectors normally perform the take off for the quantity and calculate the labor to install them with their scope of work.

Steel fabricators may include shear studs for steel items embedded in concrete or threaded anchors studs for wood attachment.

The * note of the A-36 steel indicator in Figure 7.25 is important for steel material pricing. The A-36 material is one of the cheaper grades of steel material used.

In Figure 7.25, the note *c x"* indicates the beam should be cambered or slightly bent. The camber indicator is shown here, with information on beam camber requirements in the general notes. Camber supports a certain amount of upward lift for the weight of the building and is common to see in buildings with multiple floors. To camber a beam adds labor time to the beam's fabrication or added cost if the beams are purchased already cambered. Steel detailers will have extra work because they must put these notes on the shop detail drawings.

When the material take off listing is completed, it is important not only to note that the beam be cambered, but also to note the amount of camber required. The amount is never very much—generally 1/2" to 1 1/2"—but it does make a big difference to the engineering of the building.

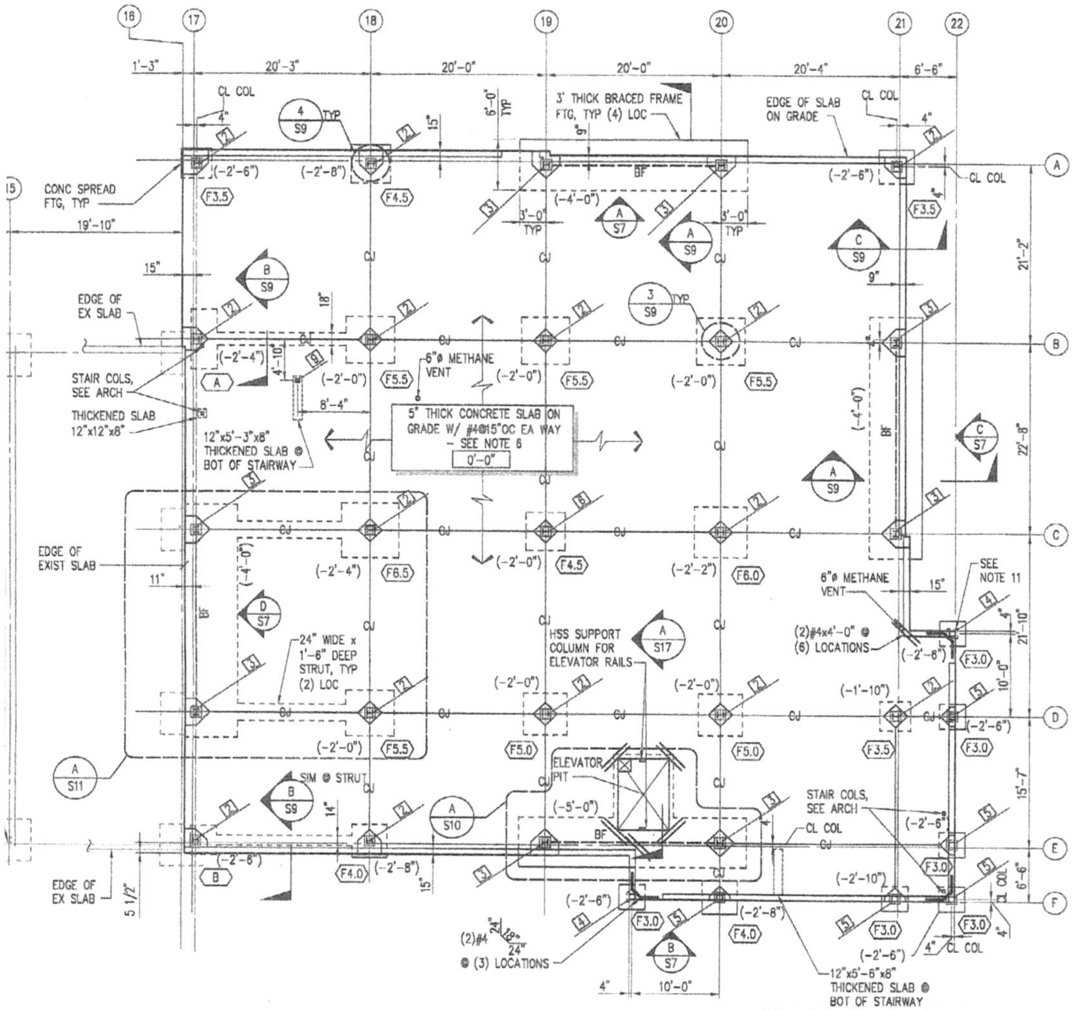

Figure 7.26 A closer look at foundation plan notes

The foundation plan, like the one in Figure 7.26, gives us the footprint of the building and a general feel for the size and shape of what is being built. In this foundation plan, the column has a line drawn at an angle with a number inside a square set on the line. To determine what the number inside the square is supposed to indicate, we need to look at the rest of the drawing for notes that reference to it—this indicator was not shown to us in the structural drawing symbols explained earlier. Figure 7.27 shows the notes taken from the same page as the foundation plan in Figure 7.26.

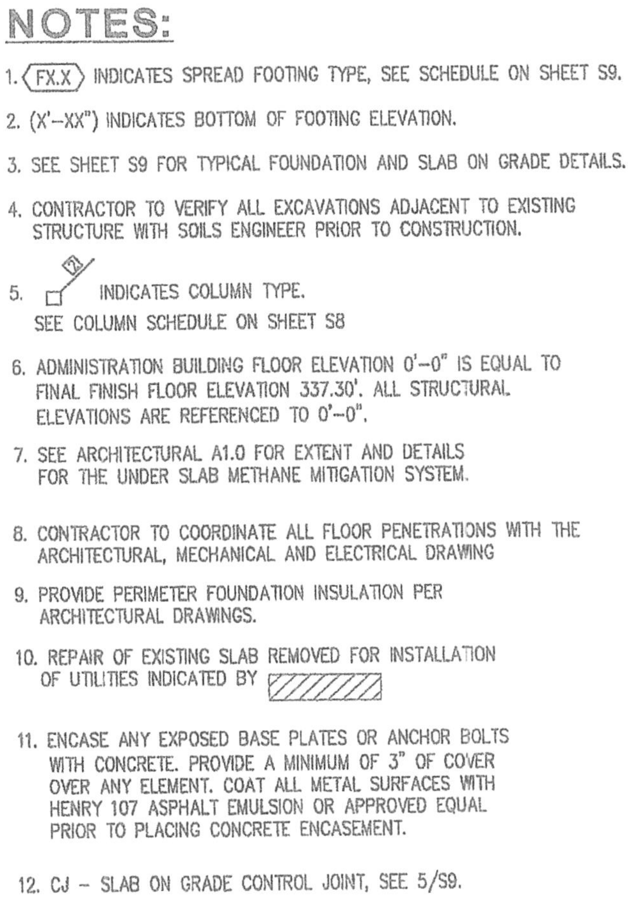
NOTES:

1. ⟨FX.X⟩ INDICATES SPREAD FOOTING TYPE, SEE SCHEDULE ON SHEET S9.
2. (X'–XX") INDICATES BOTTOM OF FOOTING ELEVATION.
3. SEE SHEET S9 FOR TYPICAL FOUNDATION AND SLAB ON GRADE DETAILS.
4. CONTRACTOR TO VERIFY ALL EXCAVATIONS ADJACENT TO EXISTING STRUCTURE WITH SOILS ENGINEER PRIOR TO CONSTRUCTION.
5. INDICATES COLUMN TYPE. SEE COLUMN SCHEDULE ON SHEET S8
6. ADMINISTRATION BUILDING FLOOR ELEVATION 0'–0" IS EQUAL TO FINAL FINISH FLOOR ELEVATION 337.30'. ALL STRUCTURAL ELEVATIONS ARE REFERENCED TO 0'–0".
7. SEE ARCHITECTURAL A1.0 FOR EXTENT AND DETAILS FOR THE UNDER SLAB METHANE MITIGATION SYSTEM.
8. CONTRACTOR TO COORDINATE ALL FLOOR PENETRATIONS WITH THE ARCHITECTURAL, MECHANICAL AND ELECTRICAL DRAWING
9. PROVIDE PERIMETER FOUNDATION INSULATION PER ARCHITECTURAL DRAWINGS.
10. REPAIR OF EXISTING SLAB REMOVED FOR INSTALLATION OF UTILITIES INDICATED BY
11. ENCASE ANY EXPOSED BASE PLATES OR ANCHOR BOLTS WITH CONCRETE. PROVIDE A MINIMUM OF 3" OF COVER OVER ANY ELEMENT. COAT ALL METAL SURFACES WITH HENRY 107 ASPHALT EMULSION OR APPROVED EQUAL PRIOR TO PLACING CONCRETE ENCASEMENT.
12. CJ – SLAB ON GRADE CONTROL JOINT, SEE 5/S9.

Figure 7.27
Specific notes shown on foundation plans

Review each note in Figure 7.27 and look to the description for what each means.

- **Note 1** tells us about the spread footings, the special section of concrete on which the column will sit to support the load of the building. The load is transferred to these columns through the steel beams. We will see all the different types of footings on sheet S9 (see Chapter 8).

- **Note 2** gives us (X'-XX") as the indicator for feet and inches to the bottom of the footing below the zero mark used for this building. Now we have to know how thick the footings are in order to determine the lengths of the columns. We already know by looking at this plan that many columns don't stop at the top of the finished floor.

- **Note 3** leads us to sheet S9 (Figure 8.11) for more foundation information. We will review this sheet for other possible embedded steel items that the fabricator needs to supply.

- **Note 4** talks about the contractor doing the excavating. It warns about verifying soil conditions next to the existing part of the building. However, soil conditions in this case will have no effect with regard to our steel takeoff.

- **Note 5** points out that the number inside the box is the reference for the column type. Then it says to see the steel column schedule on sheet S8 (see Figure 7.28).

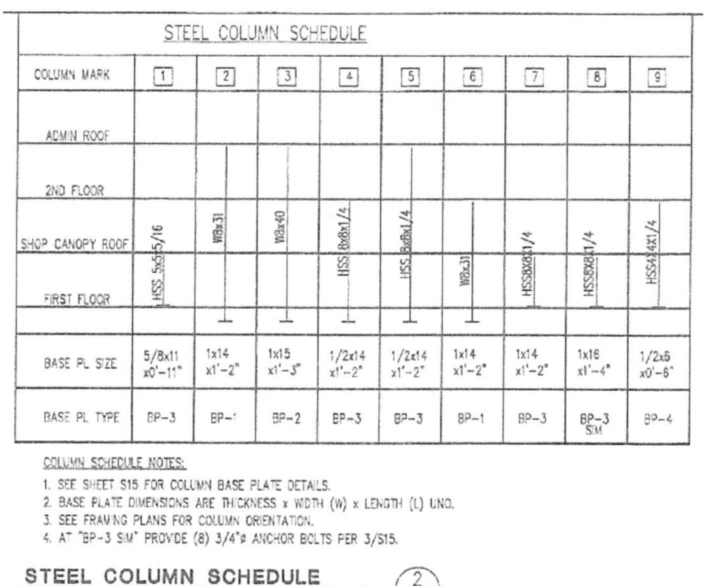

Figure 7.28 Steel column schedule

The column schedule shows the main columns being used at the foundation plan, but may not necessarily show *all* of the columns being used. Therefore, we will need to make sure by cross referencing and checking back. What the schedule does not show is the quantity of each type of column; we will need to count them. The column schedule in Figure 7.29 summarizes the size of the beams being used for the columns.

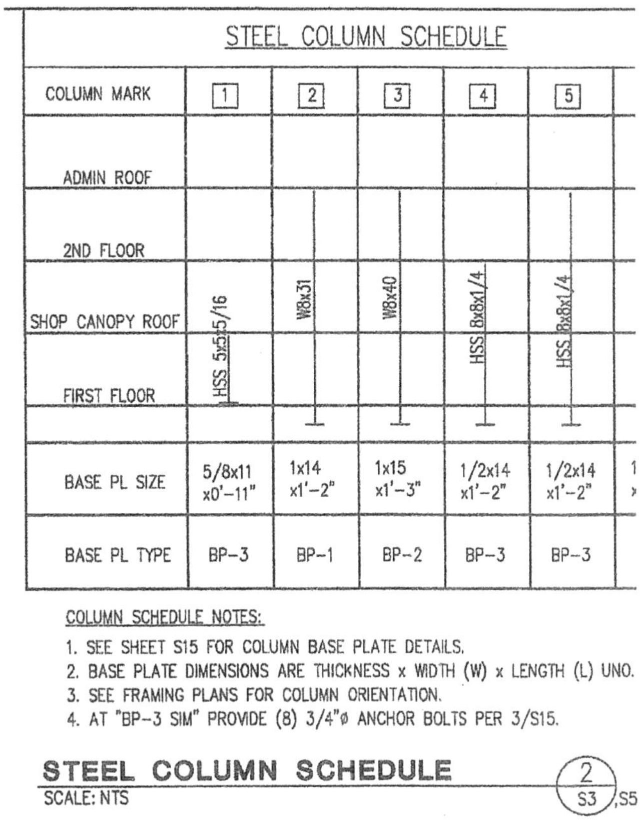

Figure 7.29 Size of column beams

Two different types of columns are shown in Figures 7.28 and 7.29: wide flange columns (the *I* at the chart from the general notes) and tube steel columns (indicated by the box in the general notes). The HSS (hollow structural steel or tube steel)—designated as TS 5 x 5 x 5/16—is best explained as tube steel that has an outside physical dimension of 5" x 5". The 5/16 is the thickness of the wall of the tube. The same is true of the HSS 8 x 8 x 1/4, where the outside dimensions of the tube is 8" x 8" and 1/4" is the wall thickness. Designers often refer to this shape of steel as HSS or Hollow Structural Steel. Other industry professionals, such as steel salespeople, steel fabricators, and steel erectors refer to this shape as TS or Tube Steel. It is important to know that the different references mean the same thing.

In Figure 7.16's structural drawing symbols, we saw how the end of this tubing appears looking down on it. The steel warehouses sell this tubing priced per foot. You can look up how much it weighs per foot in the AISC manual, or in a steel supplier's products manual.

For W8 x 31 and W8 x 40, 8 is the depth of the member in inches; therefore, this section is 8" deep. The second number, 31 or 40, is how much this steel weighs per foot. The steel warehouse price wide flange has a cost per pound or 'hundred weight' (meaning per hundred pounds). Much of what we do with regard to pricing is based on the weight of the product we are selling, so the weight of the steel is important. We will work more on this in Chapter 8, *The Structural Steel Material Listing*.

- **Note 6** in Figure 7.27 gives us the finish floor elevation. This information is important to us in calculating our column lengths and the other dimensions relative to this floor. For example, the roof top of steel elevation is based on the height above finished floor. The designers usually do not show elevations that are specific to steel locations. We will base whatever dimensional calculations we need with performing our take off from the elevation of the finished floor.

For benefit of the information, the second set of elevations given in this plan may be based on sea level (with this set of drawings, we are referring to 337.30). Sometimes all of the elevations for a project may be based on a mean sea level reference elevation. For this reason, do not ignore such elevations.

Designers also show additional building elevations based on the finished floor elevation of 0'-0", which enables us to stop looking further for a basic point of reference. Such is not the case for all jobs. Therefore, we note the alternatives that need to be available for other projects. All the information shown on all the drawings may be seemingly unrelated, yet the relationship becomes apparent later. As we move through the drawings, more will be revealed.

- **Note 7** in Figure 7.27 tells us there may be important foundation information in the architectural drawings.

- **Note 8** in Figure 7.27 lets us know that additional floor penetration information is available in the architectural, mechanical, and electrical drawings. We may find that the need for additional floor penetrations is based on assumptions by other trades rather than what is shown on the drawings. If so, we should exclude all references to supply additional floor penetrations in our bid letter.

- **Note 9** refers to insulation, which is not in our section, but is important to other trades.

- **Note 10** refers to the installation of utilities.

- **Note 11** tells us that the columns are below the finished slab, which also tells us that we need to pay attention to the scheduling of that pour. The reason is our columns will need to be in place prior to that date. This schedule may impact the need for overtime in both creating the shop drawings and fabricating the steel columns; any overtime will affect our pricing.

- **Note 12** covers the allowance of control joints (for expansion of the concrete). This allowance is important to know even though it does not directly affect our steel.

In the plan view for the foundation/first floor plan in Figure 7.26, we learned how to identify the columns and where to find the column sizes. We also established the finished floor elevation to help us calculate the length of the columns.

Because the footings are set below finished floor, we know they have an effect in the column length. We will need to go to the chart on S9 (Chapter 8, Figure 8.11) to find out how far the bottom of the footing is below the finished floor. This measurement will help us determine the length of the column.

We know the dimensions of the building grid line to grid line going north and south as well as east and west. We have learned about all of the structural notes in the general information sheets as well as the special notes on the foundation plan. All these notes are important to us as we work to create our material take off listing.

We further understand the detail layering the designers use to clarify the project's details.

First Floor Framing Notes

We can now move up to the next floor to see what kind of steel framing is being used for this building. The notes in Figures 7.30 through 7.34 will help us. We start with the notes in Figure 7.30.

NOTES:

1.

LEVEL	TOP OF SLAB ELEVATION	TOP OF STEEL ELEVATION
2	+12'-10"	+12'-4 1/2" *
R	–	VARIES

* UNLESS NOTED OTHERWISE

Figure 7.30
Elevations at top of slab and steel

- **Note 1** in Figure 7.30 indicates the elevation at the top of slab (concrete) and to the top of steel. The elevations are different because the steel decking is on top of the steel framing. In turn, the concrete on top of the steel decking makes the elevation 12'–10". Therefore, the top of concrete level is 5 1/2" above the steel—and then we have to watch for the * which indicates that the 12'–4 1/2" elevation is true, unless otherwise noted.

2. SEE SHEETS S14 THRU S16 FOR TYPICAL STEEL DETAILS.

3. BEAMS ARE TO BE EVENLY SPACED BETWEEN GRIDS OR ADJACENT FRAMING WHERE NO DIMENSIONS ARE SHOWN.

Figure 7.31
Steel details and locating beams

- **Note 2** in Figure 7.31 tells us which drawings have the steel details that apply to this floor plan.

- **Note 3** in Figure 7.31 tells us how to locate the beams when there are no dimensions at the plan view.

4. SECOND FLOOR & ROOF SLAB TO BE 2 1/2" NORMAL WEIGHT CONCRETE OVER 3" 18 GA TYPE W COMPOSITE STEEL DECK. REINFORCE WITH WWF-6-6-W2.0xW2.0.

5. STEEL ROOF DECK TO BE 3" 16 GA TYPE N-24 ROOF DECK.

6. (XX'-XX") INDICATES TOP OF STEEL ELEVATION AT ROOF.

Figure 7.32 Floor and roof deck

- **Note 4** in Figure 7.32 tells us the depth of the floor deck, with the concrete thickness on top of it. Now we know what makes up the 5 1/2” difference from Note 1.

- **Note 5** gives us the depth and type of roof deck: 3” deep x 16 gauge material type N-24 roof deck. There probably will be insulation on top of that, but this is not a concern of ours now.

- **Note 6** indicates the top of steel elevation at the roof.

7. BEAM DESIGNED TO SUPPORT A 12 PSF FULL LENGTH MOVABLE PARTITION.

8. SLAB AND ROOF DECK EDGES SHOWN ON THIS SHEET ARE INTENDED TO ACCOMMODATE EXTERIOR WALL SECTIONS AND OPENINGS PER ARCHITECTURAL AND MECHANICAL DRAWINGS. CONTRACTOR SHALL COORDINATE WITH ARCHITECTURAL AND MECHANICAL DRAWINGS PRIOR TO CONSTRUCTION.

Figure 7.33 Additional floor framing notes

- **Note 7** in Figure 7.33 is an engineering design note. It should not bother us unless we get involved with design requirements of the beam that is supporting the movable partition.

- **Note 8** informs us that we may need to look for additional details that would provide for more steel framing for some of the supports. Because of this note, we will be looking for them. However, if none are found, be sure to add this note to the list of exclusions.

9. TOS EL 12'-8 1/2" AT HSS BEAMS AT STORE FRONT. ALIGN OUTSIDE EDGE WITH GRID LINE.
10. THE EDGE OF CONCRETE AND/OR DECKING AT OPENINGS SHALL BE LOCATED AT THE EDGE OF THE BEAM FLANGE UNLESS SHOWN OTHERWISE.
11. X INDICATES HSS4x4 PARAPET SUPPORT.

Figure 7.34 More floor framing plan notes

- **Note 9** in Figure 7.34 tells us that the top of steel (TOS) elevation will be 12'-8 1/2" at HSS (hollow structural steel, or preferably called tube steel) in the store front (windows). The note also says to align the outside edge of the tubing with a grid line (this means that the tubing is not centered at gridline). We will watch for note 9 to be called out in the building plans. The steel being located off the grid line will affect dimensioning for our material listing, and will also be of concern to the steel detailers when they draw this steel.

- **Note 10** tells where the closure plate will be in the floor openings relative to the steel beams. If no details are called out with different information, we will use this information as a default.

- **Note 11** shows that X indicates HSS 4 x 4 tube steel support (TS 4 x 4). The fabricator will need to supply this steel. Details will show how this TS 4 x 4 will be attached and what the wall thickness is for this tubing.

Reviewing the 2nd Floor Plan

After having read the applicable notes on the drawings, we now review the floor plans. The floor plan in Figure 7.35 has the same grid line locations and dimensions as the foundation plan previously reviewed. It provides a lot of information regarding how this steel is put together to complete the framing of the building.

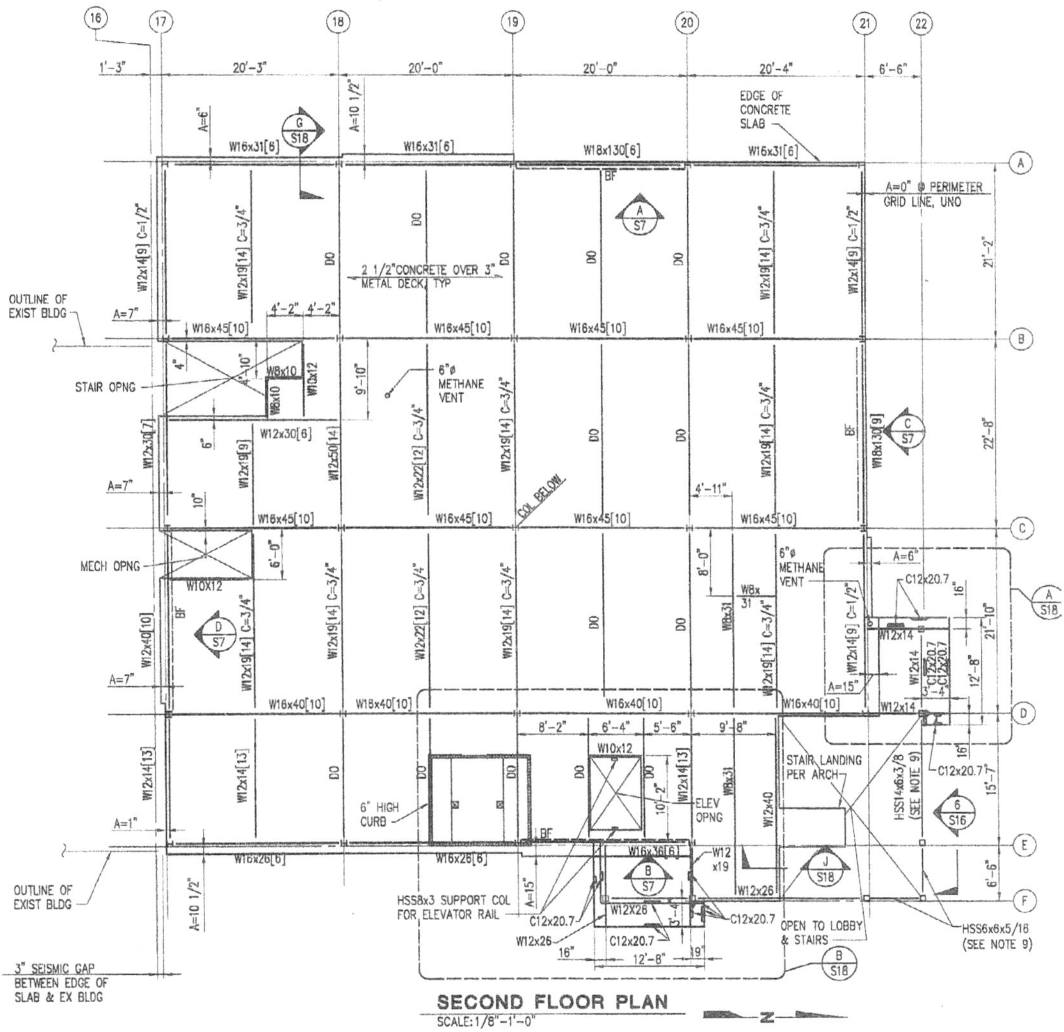

Figure 7.35 Floor plan

The plan view of the second floor in Figure 7.35 will help us utilize information given us in previous sheets. Grid lines and their dimensions should be the same as the floor below. We will find additional dimensions for specific areas, like opening framing for elevators or mechanical work, or additional beam locations for specific framing conditions.

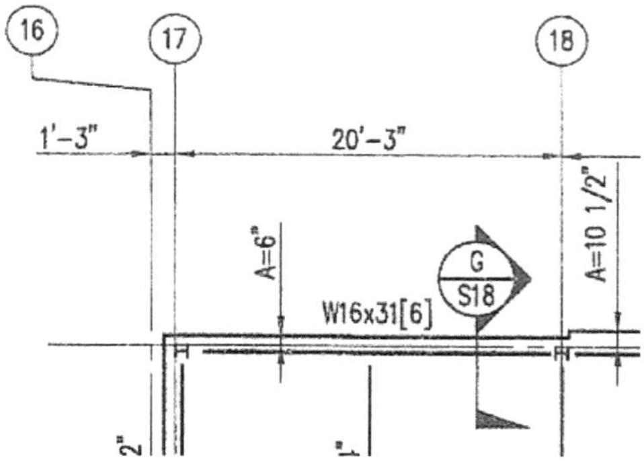

Figure 7.36
Wide flange beam in floor plan

Look at the first beam—grid A between grids 17 and 18 in Figure 7.36. This first beam, W16 x 31, is a wide flange beam about 16" deep. (Refer to either the AISC manual or the steel products handbook supplied by the steel warehouse for the actual depth of beam dimension.) It weighs 31 pounds per foot.

The (6) lets us know how many shear studs must be applied to this beam. Shear studs will be listed in the exclusions if they are not to be supplied as part of the bid package.

The dimension to grid centerlines, which in this case are *offset* to the center of the columns, is 20'–3". A = 6" represents the distance to the outside edge of the concrete floor extending from the centerline of grid A. Note that at grid 18, this dimension changes from 6" to 10 1/2".

Section G/S18 (shown in Figure 7.36) is cut through this W16 x 31 beam. Therefore, we need to look at detail G on drawing S18 (Figure 7.37) to see what the designer is trying to show us. We are looking for any additional steel fabricated material that must be added to the material take off, but is not and cannot be shown at the plan view.

Detail G/S18 is shown as G/S6 in Figure 7.37. It "bubble" references back to the earlier drawing, which is handy information because not all details provide this information. When they do refer back, it is greatly appreciated. Figure 7.37 gives the close up of the beam looking through the body of it—

the I in this case is the W16 x 31—but it may pertain to any size beam that might be used at the perimeter of the building. We can also see the edge of the concrete.

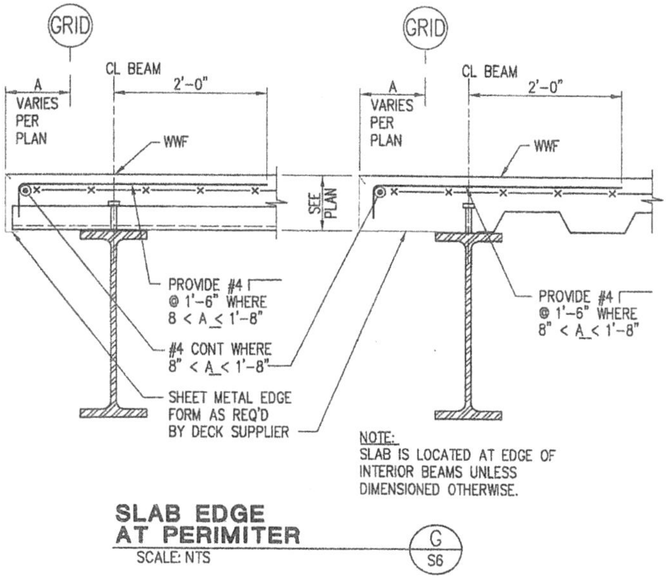

Figure 7.37 Detail of beam from Figure 7.36

In Figure 7.32, the note below the beam (and above the detail cut identification) indicates the detail is used at the edge of the interior beams, unless noted otherwise. We use this detail for all conditions of the perimeter, unless another cut is made to show us something different. The detail shows the following, as provided by the deck supplier.

- Edge form to hold in the concrete
- Size of rebar, #4 (for rebar, the # sign indicates the number of 1/8" thickness of material to be used; in this case, it's 4 x 1/8" or 1/2" rebar material)
- Occurrence (indicated as @ 1'–6" centers)
- Special conditions of that occurrence (indicated as < or > the distance from A, where the A dimension represents the distance between the grid line and the closure for the concrete)

This variance of the dimension means the edge form will need to be wider or narrower at different locations. Review the floor plans to determine this A dimension application correctly. The plate being used as edge form will also have different pricing for the forming because of the varying widths.

Because this detail is typical, the designers did not locate the grid line to the beam's center. This condition may also apply where the centerline of the beam is not at centerline with the grid. Most of this information doesn't apply to our supply of the steel. What is important is what this detail doesn't show—there is no closure angle or plate that steel fabricators need to supply with their steel. If steel fabricators needed to supply closure angle or plate, that information would be shown at a detail cut similar to this one; it may be shown to be shop welded to the beams.

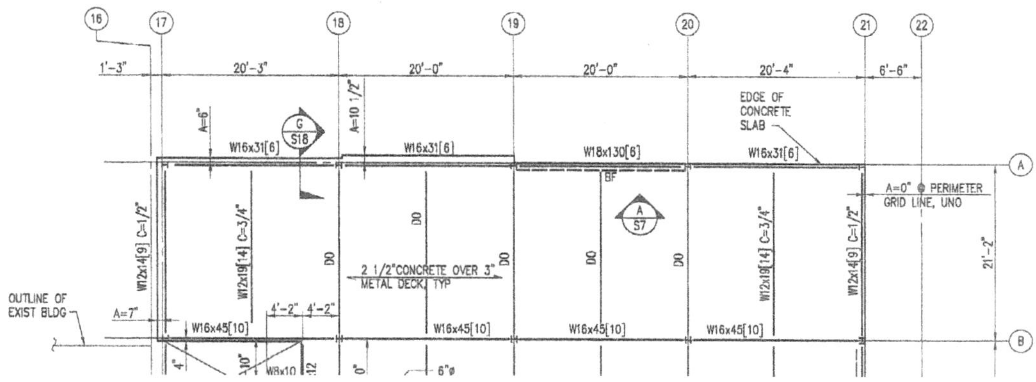

Figure 7.38 Detail of 2nd floor building framing

This portion of the 2nd floor plan in Figure 7.38 shows the building framing between grids A and B from grid 16 to 22. Between grids 19 and 20, we see the *A/S7* callout for the braced frame that occurs there. We see four beams at grid A: three of them are W16 x 31 and one W18 x 130. At grid line B, we have four W16 x 45 beams that span grids 17 through 22. We see the note for the concrete, which is 3" deep over the 2 1/2" metal deck that spans perpendicular to the way the steel beam are shown in this plan.

Notice also that the deck edge is of different dimensions (6" between grids 17 and 18, 10 1/2" between grids 18 and 19) off center line of grid A. These locations are where the dimension A variance comes in.

Now let's examine Figure 7.38 further, looking at the beams that span between grids A and B, and the notes applicable to those beams. (Grids A and

B run north and south on the page, not to be confused by the "true" north, as indicated by the building compass. We are using this orientation to see what beam we are referencing.)

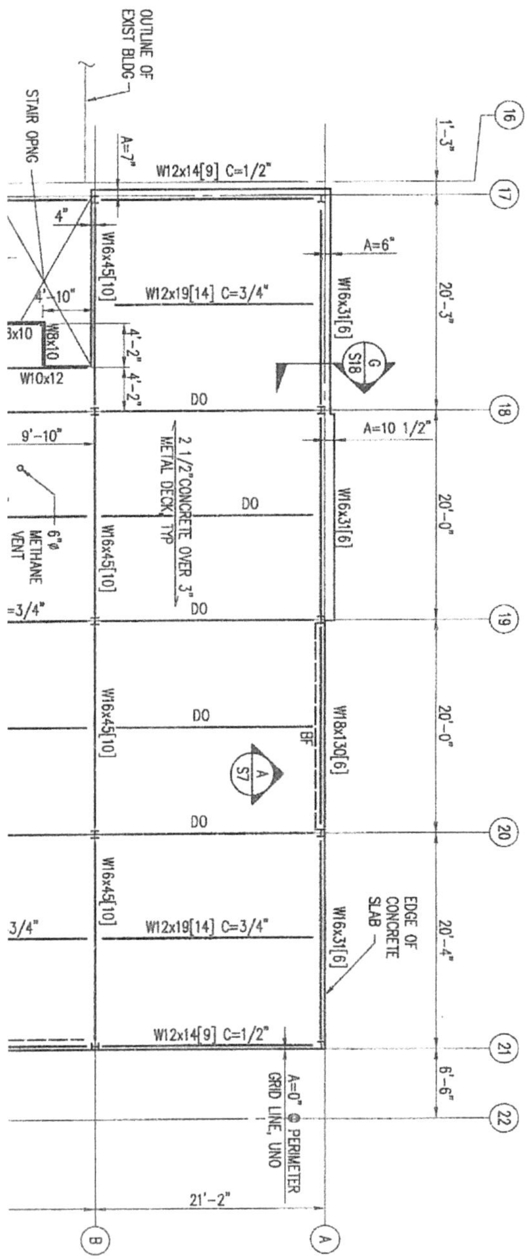

Figure 7.39 A beam spanning grids A and B

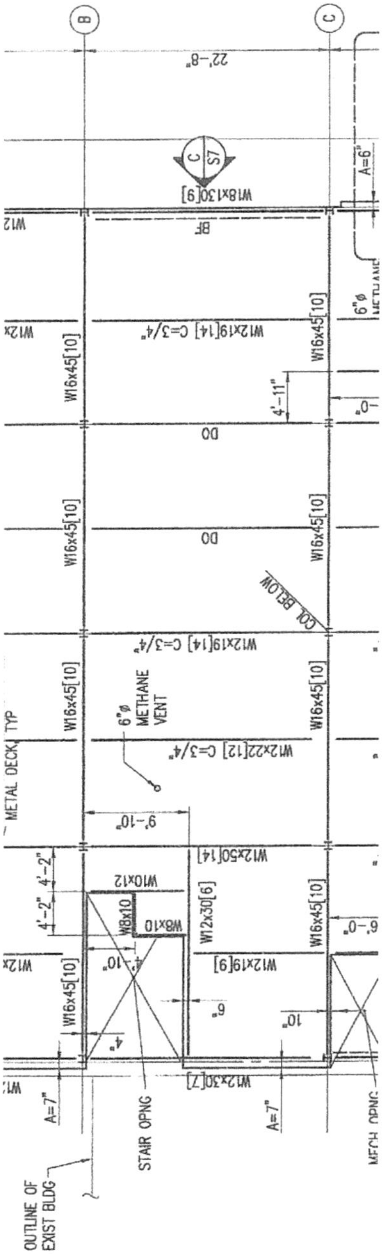

Figure 7.40 2nd floor framing and stairs

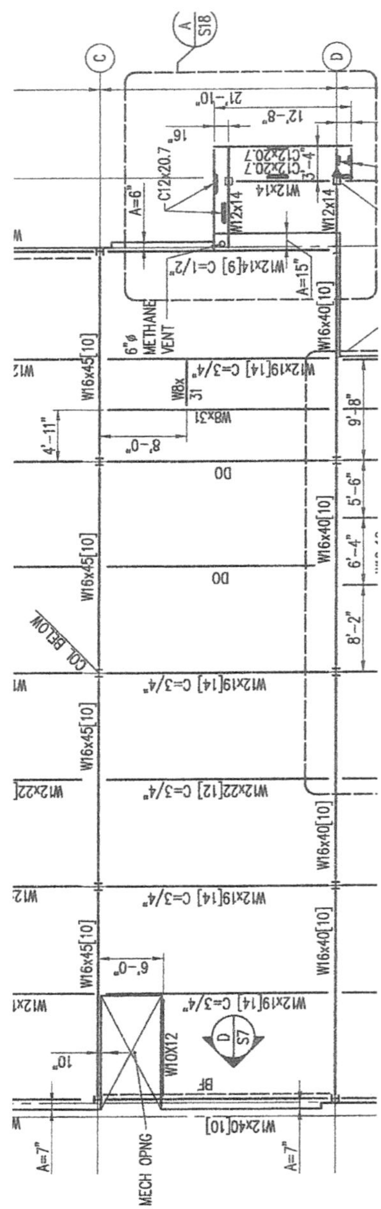

Figure 7.41 Mechanical opening frame and braced frame

In Figure 7.39, we see that the beam at grid 17 (spanning between A and B) is W12 x 14; it has (9) shear studs, but also must be cambered 1/2". The dimension between grids is 21'–2" to the centerline of columns. Below that beam is a W12 x 19 beam with (9) shear studs and a 3/4" camber. The following five beams below are all designated by *do*, which means that the information for these beams is the same as what is required for the beam above. The beam that follows the five beams marked *do* in this series is the same as the first one. The row ends with a beam at grid 21 between grids A and B; this beam is the same as the first one in the row. The outside edge of the concrete is at grid 17. It is 7" beyond the centerline of that grid and is at 0" at grid 21.

Looking at Figure 7.40, in this 2nd floor plan between grids B and C from 17 to 22, we see more of the same type of framing and information, but we also see a stair opening with short W8 x 10 beams that frame the opening. There are dimensions showing the length of these beams; they will help determine the length of the beams that frame into this opening. The dimension between grids B and C is 22'-8". Subtract the 9'-10" dimension for the opening to calculate the actual use length of the beams that frame into the opening north of grid C. For the short W8 x 10 beam that also runs north and south of the stair opening, we can use the 9'-10", then subtract 4'-10" to determine that estimated beam length.

On the right side of Figure 7.40, between grids B and C, we see another callout for a brace frame, designated by *C/S7* and the note *BF* instead of a beam size call out. This same type of indicator (BF) was used at the braced frame located at grid A as well. Note *COL BELOW* at the center of the picture at grid C. Apart from all the other columns, this column will stop with its top of steel elevation at this 2nd level.

We could go back and locate this column at grid C-19 in the foundation plan (Figure 7-26) to find the column type, then cross check it with the steel column schedule in Figures 7.28 and 7.29. The column is type 6. The steel column schedule shows the column stops at the 2nd floor, but we are not sure of the quantity. We will verify the quantity when we perform our material take off listing.

Look at Figure 7.41 between grids C and D and from grids 17 to 22. We find a mechanical opening to the left, another braced frame designated by *BF* and section D/S7, more cambered beams with shear studs, and some typical beams indicated by DO. We also find a new item: section A/S18 for a close up of some framing that is beyond the initial footprint of the building. For this building, the footprint is the box made by grids 17 to 21 and A through E. Yet we see the framing to the right of grid 21 and below E. This framing is more clearly defined by the additional section callouts that are provided by the designers. Figure 7.42 shows section A/S18.

Although the dimensions shown in Figure 7.42 are the same as those shown in the 2nd floor plan, notice that the scale is a bit larger (1/4"—1'-0 here whereas 2nd floor framing is 1/8"—1'-0). This partial plan shows more clearly the framing condition that is a bit tight in the second floor plan. We see additional channel framing here not previously seen. There is also a new sec-

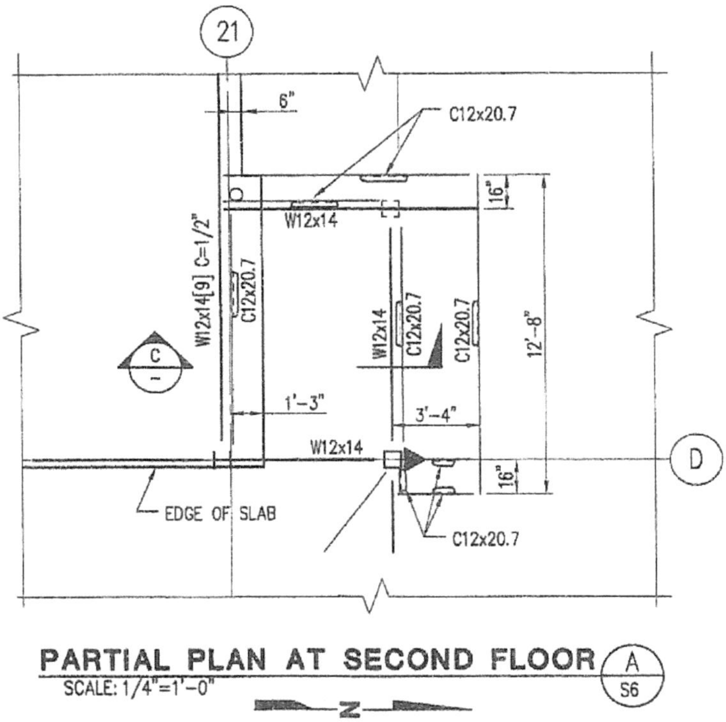

Figure 7.42 Detail of section A/S18

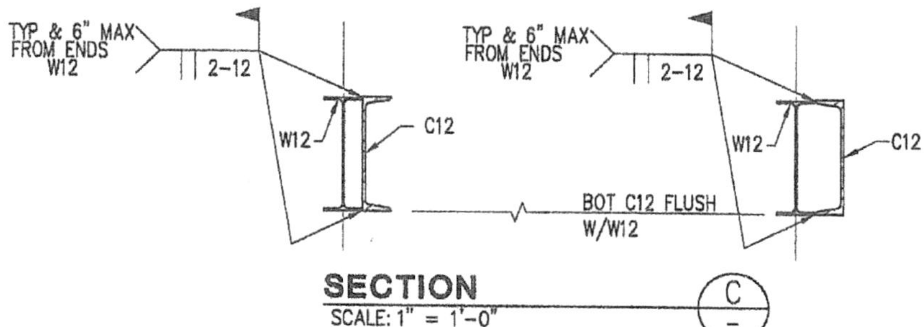

Figure 7.43 Section cut through canopy framing

tion C in this partial 2nd floor plan that is not shown in the 2nd floor plan at S6 (Figure 7.43):

Looking at section C/S6 in Figure 7.43, we see the cut through this framing. Starting with the W12 x 19 at grid 21 to the left of this picture (which is actually a building framing beam), the C12 x 20.7 channel is field welded right up against the beam flanges. This happens again with the W12 x 14 and C12 x 20.7 shown out by the tail of the cut in the partial 2nd floor framing plan.

This detail does not include the C12 x 20.7 that is out at the face of this framing. Detail C shows us how this framing is field welded in place. What is important to note here is that this framing may have to be assembled in the shop and sent out as a completed frame. Therefore, its shipping size and weight will need to be determined. This measurement will more accurately cover the more expensive costs of handling and shipping the framing as a completed unit to the field.

When it comes to our fabricated steel supply, conditions that *are not* shown in a detail may mean more to us than what is shown. For example, shop welding canopy frames to ease field installation are a feature that the designers are not likely to show. Steel fabricators and steel erectors know that it is cheaper and easier for all concerned to shop fabricate the canopy frames if they are of a shippable size. Knowing what is and what is not important is something that is learned both by asking questions and by practical experience.

Figure 7.44 provides an example of another item that is shown but not explained.

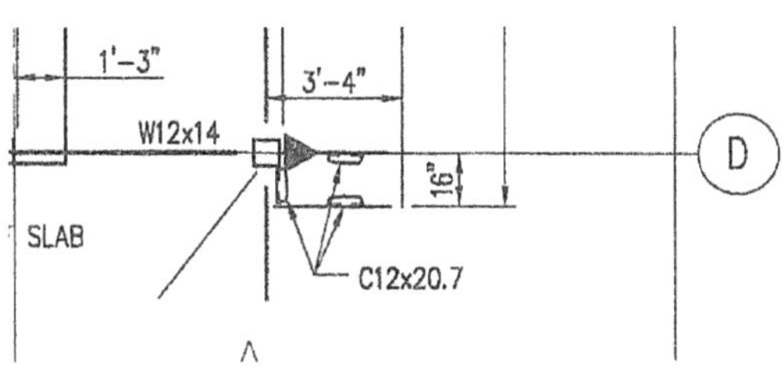

Figure 7.44
Moment connection indicators

The little black triangle located at the right side of the column at grid D in Figure 7.44 is not identified by any notes on the drawings. This triangle is a common indicator for what is called a "moment" connection. A moment is an industry trade word meaning to reference a seismic (earthquake) resistance condition. The connection may require additional welding, additional connection plates, or a special end condition of the beam that frames into this side of the column.

Because this single requirement for a moment connection is small within the vast scope of the project, it probably has little overall effect from a cost standpoint. However, it will matter to steel detailers when it's time for them to draw the connections and send out an RFI. An addendum may be issued during bid time to clarify this connection. You will need to watch for the issue of new documents while you are working on your quote. Prior to bid time, you may even call the engineer of record directly for clarification. To be safe, you could include this moment connection requirement within your bid letter's exclusions.

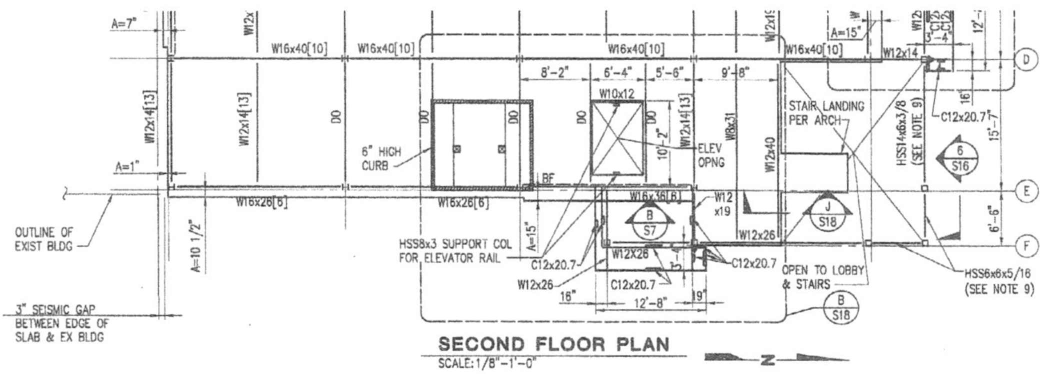

Figure 7.45 Elevator framing

Looking at the last bay for the 2nd floor plan in Figure 7.45, we see more beam framing than seen previously, with the same indicators for shear studs and the same indicator (DO) for similar beams. We see another floor opening for the elevator near the center of the bay with an area that has a 6" high curb. Two HSS 8 x 3 vertical tube steel supports are shown at the elevator opening. The thickness of the TS 8 x 3 is not shown here, so we will look for that information to be shown at details yet to be shown.

The vertical tubes in Figure 7.45 are for the elevator rail that starts at the Foundation/First Floor plan. These tubes will continue above the second floor here, and then will ultimately terminate at the roof. They will support a beam that spans the opening for the elevator; that beam is designed for supporting the elevator itself.

We also see a stair landing, indicating that more information for it will be shown in the architectural drawings. An area that includes this elevator opening and the 6" high curb is found in the call out for B/S18, shown in Figure 7.46, giving us a close up of the framing.

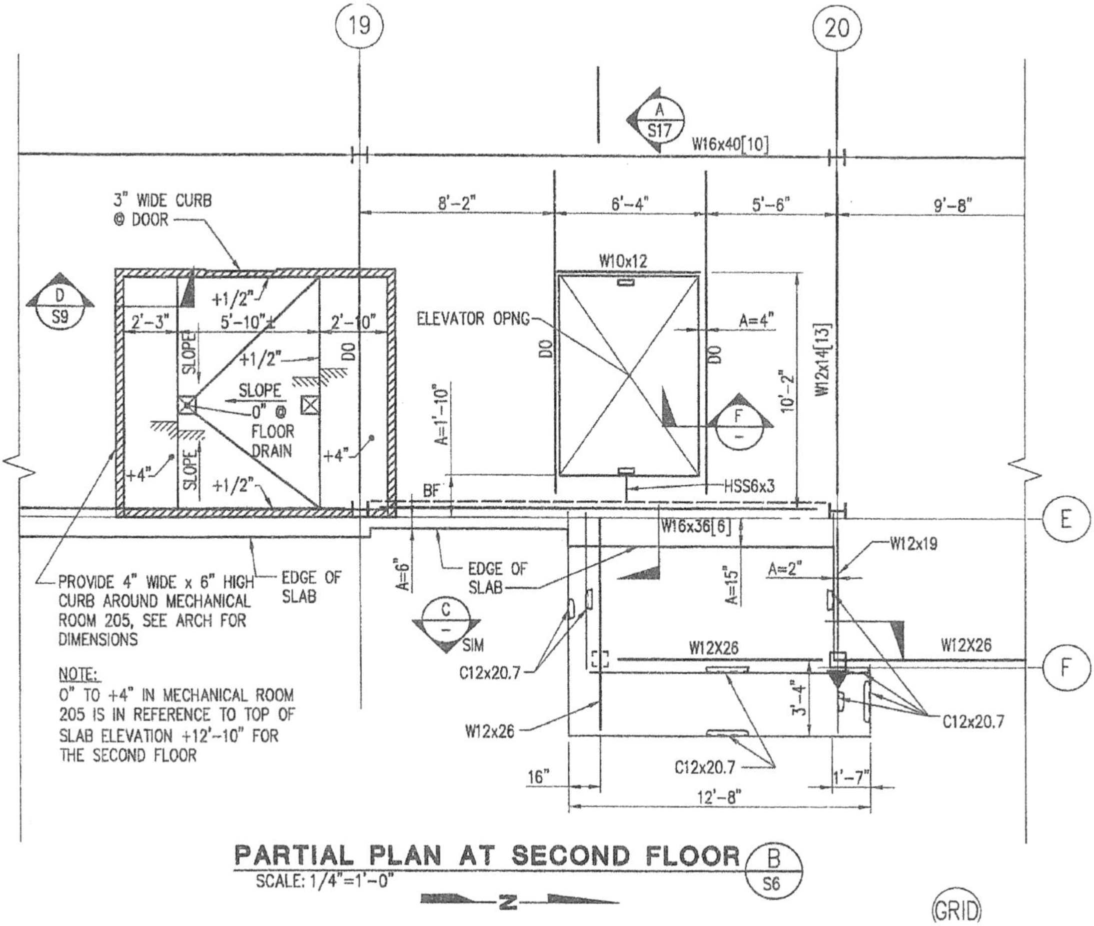

Figure 7.46 Mechanical curb and elevator details

Figure 7.46 is a close up of the framing for this area. The mechanical room 204 area has a 4" wide curb except at the door area, where the curb narrows to 3". We can see the difference in floor elevation indicators, the direction of the slope. However, we don't find any steel beam that supports this change in the floor. We see TS 6 x 3 support beams (though they neglect to call out the wall thickness for this steel) to the elevator opening. We see section C called out again (Figure 7.47), running horizontally through the entire framing area in between grids E and F.

We need to refer back to the original 2nd floor framing at S6, or Figures 7.35 and 7.45, to get the dimensions we need for calculating the beam lengths in between grids E and F.

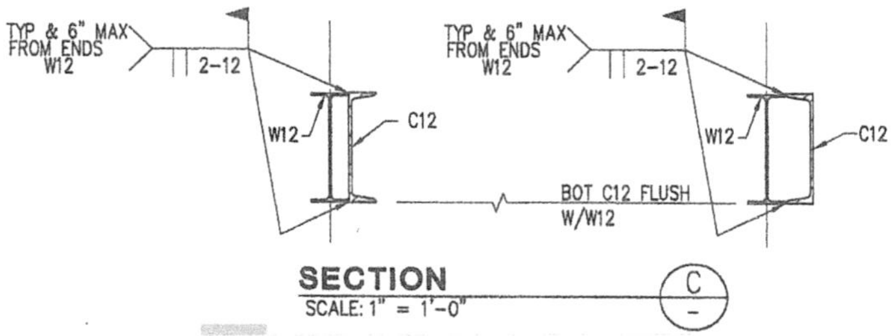

Figure 7.47 Section C: Canopy framing

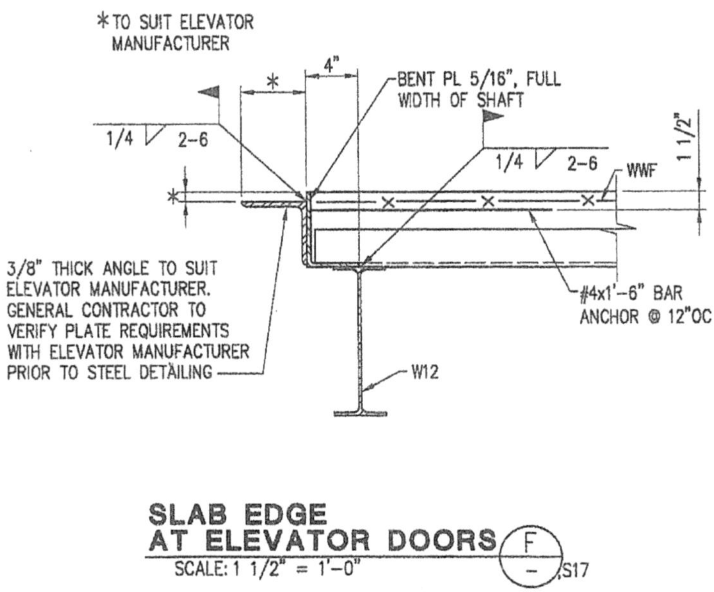

Figure 7.48 Elevator sill detail

Section F in Figure 7.48 is cut from the partial 2nd floor plan in Figure 7.48 to the left of grid 20 and above grid E. This 5/16" thick bent plate, extends the length of the elevator opening (provided only at this side of the opening where it is called out by detail F). It will be provided by the steel fabricator, even though it field welds to the floor framing beam and is not shop attached. The 3/8" thick angle that is "to suit elevator manufacturer" may be provided by the elevator manufacturer. It is important that these responsibilities be listed in the bid letter's exclusion. Then the customer is not confused about who is supplying this material.

The Roof Plan and Roof Plan Details

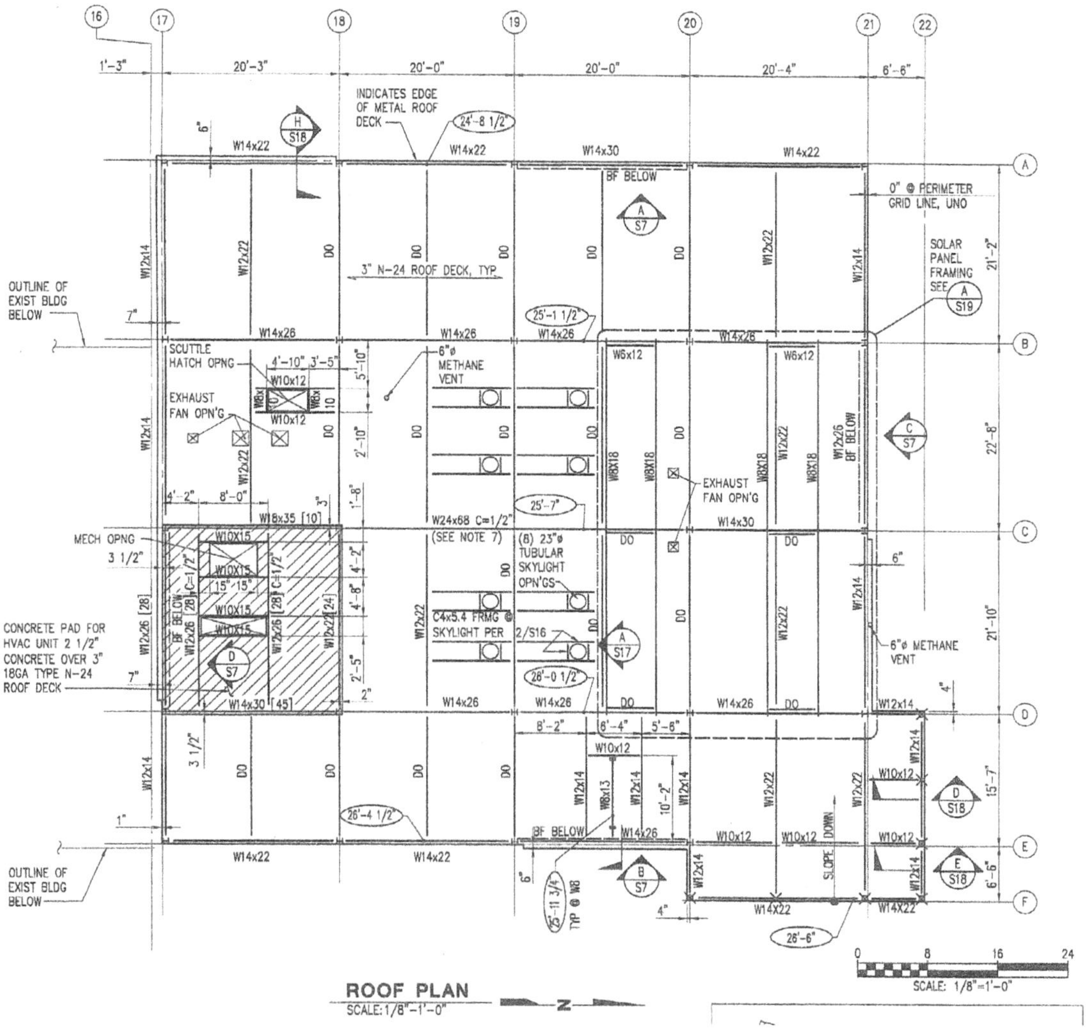

Figure 7.49 The roof plan

In Figure 7.49, we can see the roof plan. We verify first that the grid line locations are the same as we saw for the foundation and the 2nd floor framing. The scale is the same at 1/8" = 1'-0. We see that the top of steel elevations vary (see the elevations in the oval). It is normal for the roof elevations to vary for water runoff.

Roof framing may also have steel framing and supports for items that would not be found at the foundation and floor levels, such as HVAC supports (heating, ventilation, air conditioning units) and the steel framed roof screens to conceal them from view. Sometimes there will be steel walkways or "cat-walks" from the roof hatch openings over to the HVAC units, plus ladders to get to the different elevations of the roof, and miscellaneous items like window washer tie downs, or supports for antennas or signs.

Figure 7.50 looks at the steel in the first bay between grids A and B, 16 to 22.

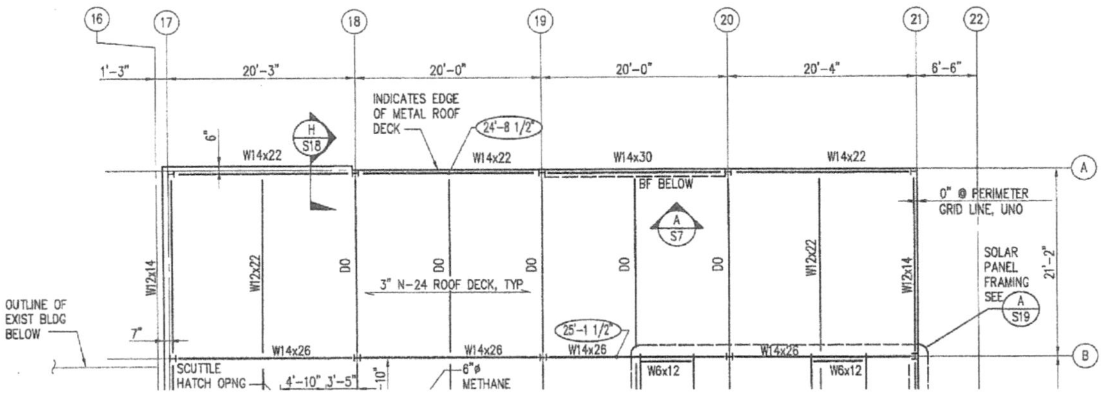

Figure 7.50 First bay close up at roof plan

Wide flange beams running east and west on the page along grid A are the W14 x 22 and W14 x 30. Along grid B, we see W14 x 26. (Actual north is to the right, but when we talk about steel placement on the drawings, the top of the sheet is north.)

These beams are running column to column. For their approximate length, you can use the centerline of column dimension. However, their actual length will be a bit shorter than that because half the width of the columns will need to be subtracted as well as the clearance allowance of about 1/2" each end. (Information on beam clearance will be addressed in Chapter 8, *Structural Steel Material Listing.)*

Also notice in Figure 7.50 that we see the detail A/S7 called out again for the braced frame. The beams running north and south are W12 x 14 and W12 x 22. The beams that span column to column will have a different "set back" or deduction for the length of the beam—these beams stretch to the inside of the web of the column instead of stopping at the flange. We will find these details in the framing sheets later. But in Figure 7.51, we see detail H/S18 along grid A at the roof edge.

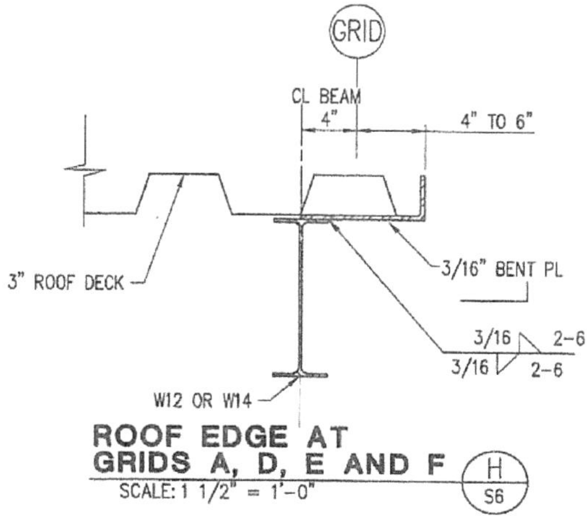

Figure 7.51 Roof edge closure plates

In this detail in Figure 7.51, we find that the Roof edge bent plate or "closure" plate is 3/16" thick x 8" or 10" from centerline of the beam to the edge of the bend. We also see we will need to add for the depth of the vertical bend of about 3" because the deck is 3" deep. This bent plate is shop welded to the beam with a 3/16" fillet weld 2" long every 6", both sides, staggered. In addition, this detail tells us that this same detail is to be used at grids A, D, E, and F. Grids A, D, E, and F are all east-west grids; now we know they intend something different for the north-south grids. Remember, we have to read these details for what they don't say as well as what they do. Making these distinctions becomes easier to understand as we move along.

Going back to the roof plan at grid A in Figure 7.52, we see the 6" from the centerline of grid to the edge of deck. Therefore, we know that between grids 17 and 18, the outside edge of the bent plate will be 6" off grid. However, also notice that the beam is located 4" off grid. From this, we now know that the plate between grids 17 and 18 is 10" wide plus the 3" vertical, making the purchase width of this plate 13". We also know that the bent plate for the remaining grids will be 4" from centerline or 8" wide plus the 3" vertical, making the purchase width of the plate 11".

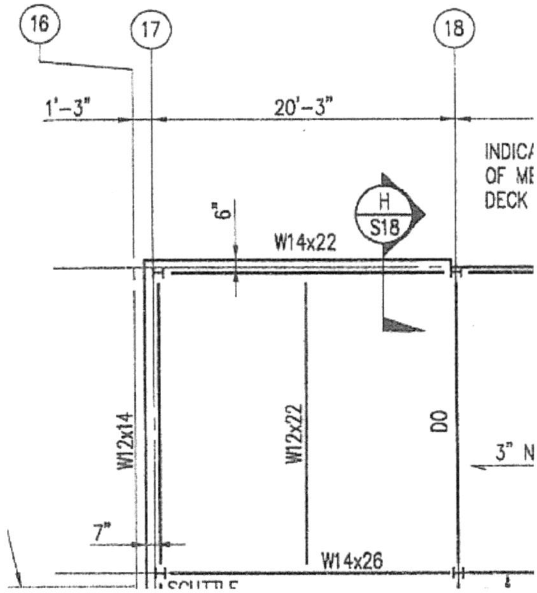

Figure 7.52 Closure plate offset from centerline location

In Figure 7.52, when we look at the north/south condition at grid 17 between A and B, we see 7" from centerline of grid to the edge of the plate. This measure will be important whenever we find out what the edge plate looks like. We will continue to look at this plan to find it.

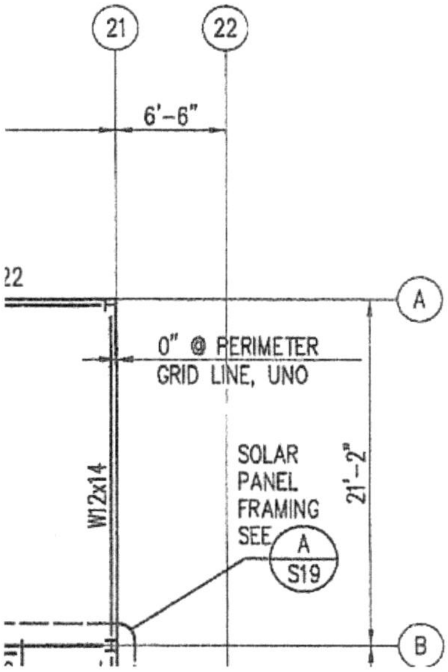

Figure 7.53 Closure plate equal to centerline location

Looking at grid line 22 between A and B in Figure 7.53, we see the note 0 at the perimeter grid line, unless noted otherwise. This means that the bent plate will stop at the grid's centerline, or 4" from the centerline of the beam. This bent closure plate then will be 4" plus 3", for a total of 7" wide at the rest of the locations where we will be using it. Next, we look in Figure 7.54 at the next bay of the roof between grids B and C, from 16 to 22.

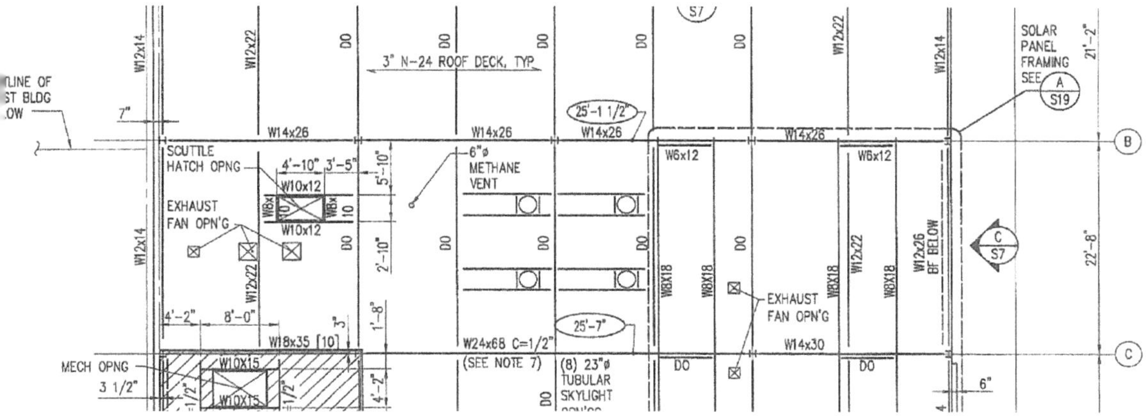

Figure 7.54 Roof scuttle and roof hatch openings

Located to the left side of Figure 7.54 below grid B, we see the framing for the "Scuttle" hatch opening. A roof scuttle or roof hatch is just a door that opens to the roof. Exhaust fan openings are also shown; these openings may need bent plate for the concrete hold back. There is some framing that is sized elsewhere between grids B and C, between grids 18.5 and 19.5.

The beams running east and west are W14 x 26, W18 x 35, W24 x 68, and W14 x 30. See the *(10)* next to the beam call out—this note is for those evenly spaced shear studs that we saw at the general notes, also known as HAS or headed anchor studs that are field installed. Next to the W24 x 68 is the note: c = 1/2". This note means that the W24 x 68 beam will need to be cambered 1/2". Do you see note 7 in parentheses under the W24 x 68 beam designation? Finding this note 7 is intended to lead us back to the notes that are on this drawing. We need now to refer back to the Additional floor framing notes. See Figure 7.33, Note 7 that stated "beam designed to support a 12 PSF full length movable partition." The reference to this Note 7 also explains why the W24 x 68 at this location is to be cambered. The camber of the beam adds the extra lift it needs to support the partition below.

Beams running north and south are W12 x 14, W12 x 22, W8 x 18, and W12 x 26. At grid 22 between grids B and C we see the note *BF Below* (braced frame) and detail call out C/S7. We will look at these braced frames again later; this one is probably much the same as the detail we saw at A/S7 in Figure 7.6.

In looking at the roof plan in Figure 7.55—between grids C and D, 16 to 22—we see that they are pouring a concrete pad for the HVAC unit areas. This pad explains why they need the HAS to weld through the decking in the field. It secures the deck to the beams and strengthens the concrete support.

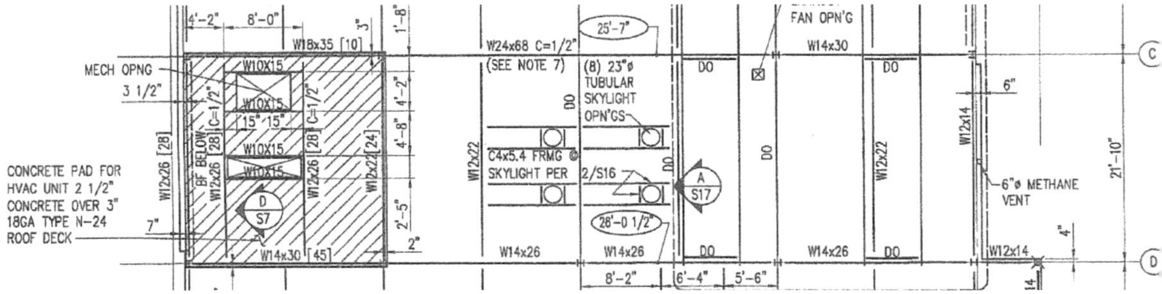

Figure 7.55 Conditions for HVAC concrete pad

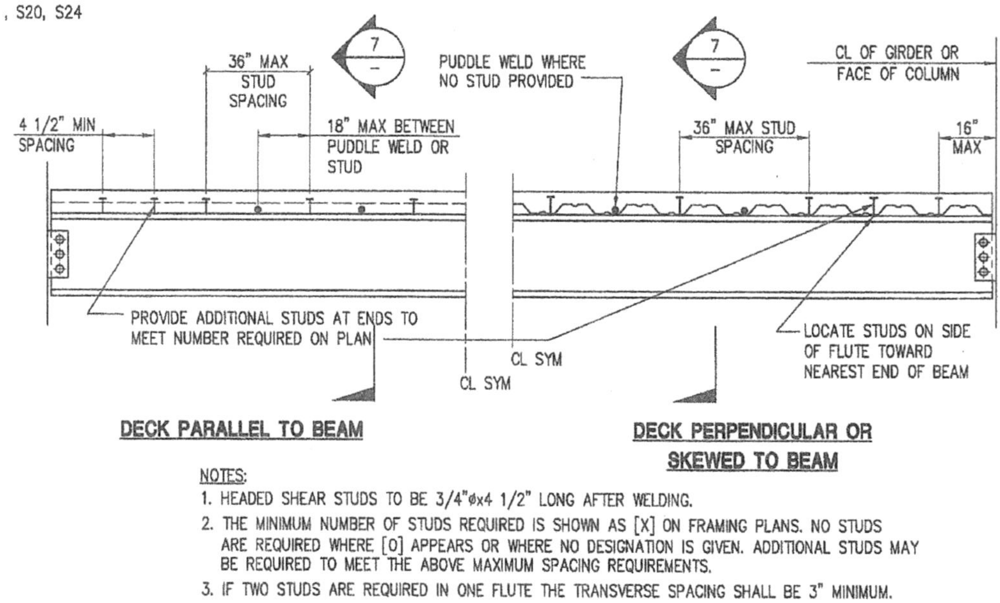

Figure 7.56 Shear studs at top of beams

Figure 7.56 shows what the beams look like with the deck on top and the shear studs welded in. The notes give us the diameter and required finished length of the studs, and how they show the minimum number required. All this work is done in the field. The erector will typically supply the studs with their price to install.

You can see in Figure 7.55 that most of the other beams for the concrete pad are cambered, and have the call out for the shear studs. Section D/S7 that is called out just above grid D between grids 17 and 18 is for another braced frame—see *BF Below* right at grid 17.

The framing that we didn't know about in Figure 7.54, between grids 18.5 and 19.5 between grids B and C, now reveals itself between grid C and D. The C4 x 5.4 framing used is typical for this "tubular skylight" framing, which will be shown to us at detail 2/S16. In going to detail 2/S16, we find the steel framing for new and existing roof deck openings, but at detail 3/S16 we find the opening framing we are looking for. Sometimes the designers will get detail call outs mixed up—there are so many—so we may expect to see an addenda to clarify this detail call out.

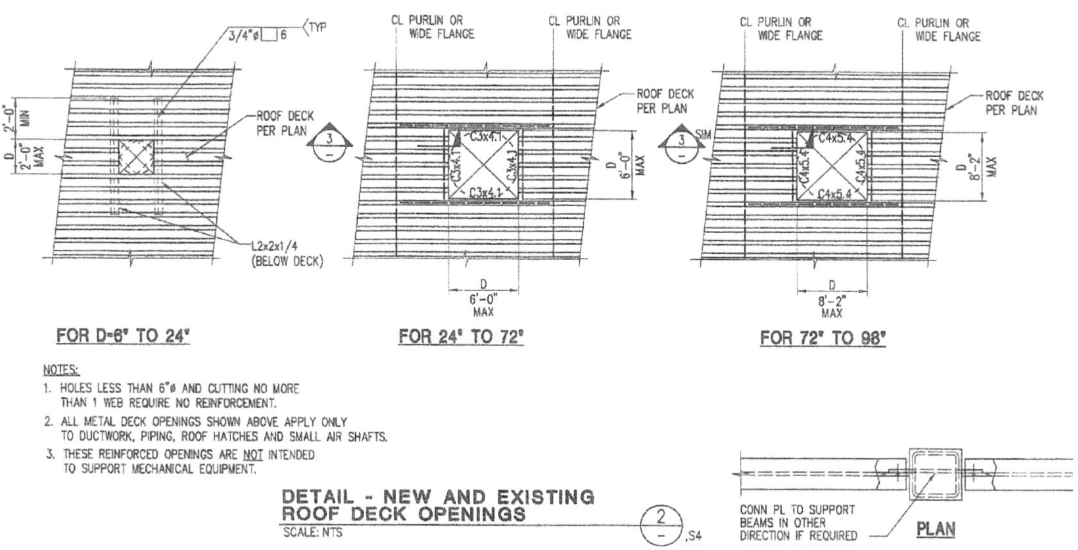

Figure 7.57 Roof deck openings

The details in Figure 7.57 are used for the roof openings where there are no materials shown in the floor and roof plans. We know what to use based on the size of the opening.

Because we know that the callout is wrong for detail 2, we look to other details that might be used for the condition we need. We find detail 3/S16 in Figure 7.58.

In Figure 7.58, we can tell this detail is really the one the engineer wants because the material sizes used are the same as those called out in the plan. This detail is an elevation view (a view from the side, as opposed to a plan view, which is looking down on the area). We can see the opening, the wide flange beam, and the C3 (C4 at similar) channels coming at us on the page.

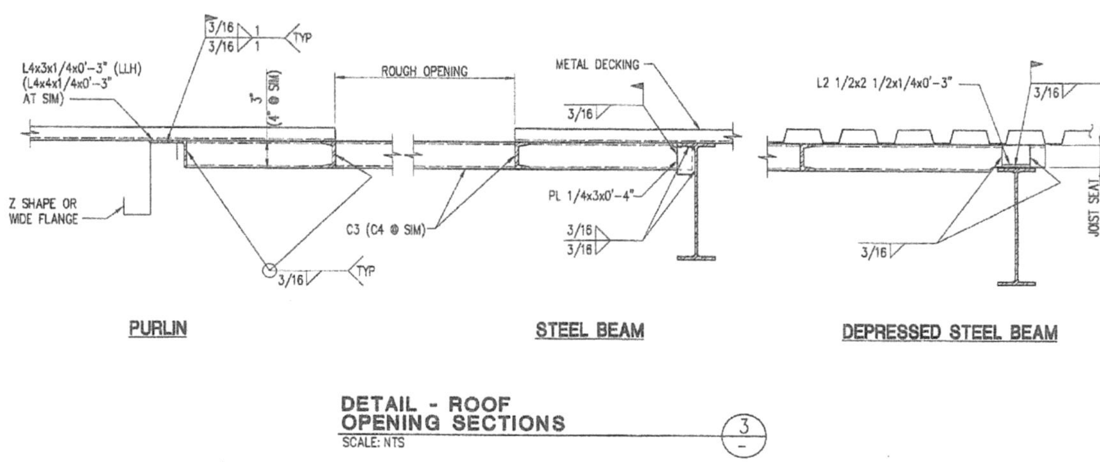

Figure 7.58 Roof opening sections

In Figure 7.58, we can see the metal deck for the roof (or floor if applicable), the connection materials as the plate 1/4", and the L 4 x 3 x 1/4 (4 x 4 x 1/4 at similar). This typical detail shows us the connections required not only at the steel beam, but also at the "purlins" and at the depressed steel beam. A purlin is a light gauge Z shape for floor or roof support that is not provided by the fabricator. However, our steel may need to connect to them. A depressed steel beam is one that is below floor/roof elevation.

In this case, the connections at the depressed steel beam have L 2 1/2 x 2 1/2 x 1/4 x 3" long clips to connect the C4 channel to the beams. Because this detail is a typical one, we use adjust the lengths of the parts to an "as needed" basis. We expect fabricators will provide this material at all locations where the condition requires it. We are responsible for verifying the use of the openings with the mechanical requirements. This responsibility will need to be qualified in the inclusion part of our bid letter, identifying where we are using the frames, what size the framing members are that we used, and how many times. The detail NTS or Not to Scale applies, as is the case for most details that are considered typical.

In the roof plan in Figure 7.58, we see the same beams at grid C running east and west we saw in the first section of the building that we reviewed. At grid D between grids 16 and 22, we see W14 x 30 (45), W14 x 26, and W12 x 14 (which is in the entry area). Beams running north and south between these grids include W12 x 26 (28), W12 x 22 (24), and W12 x 14. We also see some intermediate beam framing that are labeled do. (These are beams that go from beam to beam, but are not located on the grid lines). In reviewing the area between the grids B and C above this location, between grids 19.5 and 20.5, we can see these are W6 x 12.

Looking more closely at this area with beams marked *do*, we see a dashed line that carries to the area between grids B and C between 19.5 and 20.5. This line tells us to look at detail A/S19, *Solar Panel Framing*. We go to this plan view at S19 and we find more sections called out. This layering of details will help us understand better how the framing goes together.

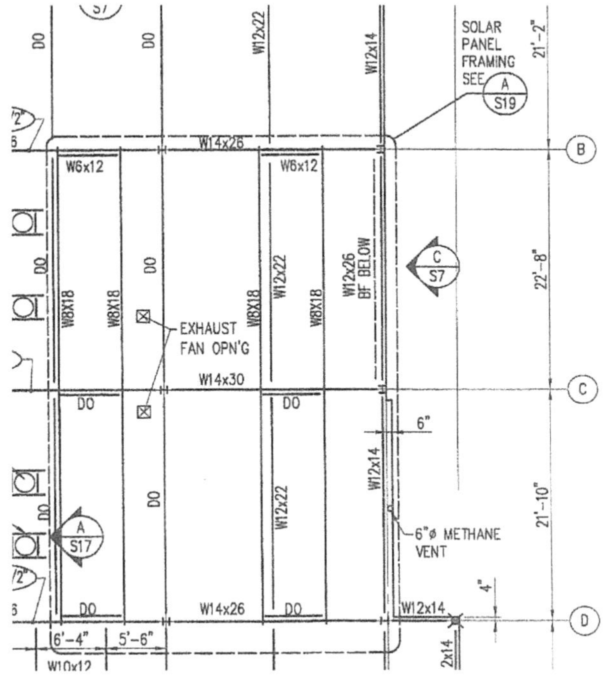

Figure 7.59 Area section cut: Special framing conditions

Figure 7.59 also reminds us to be careful not to duplicate information in our take off. The example used here includes the W8 x 18 as well as the W6 x

12 beam framing—these steel beams are shown in Figure 7.55's plan view, and are also shown in the close up of this area in Figures 7.59 and 7.60.

In plan view A/S19, we are most interested in the detail cuts B and C. These details will show us the elevation view of these areas.

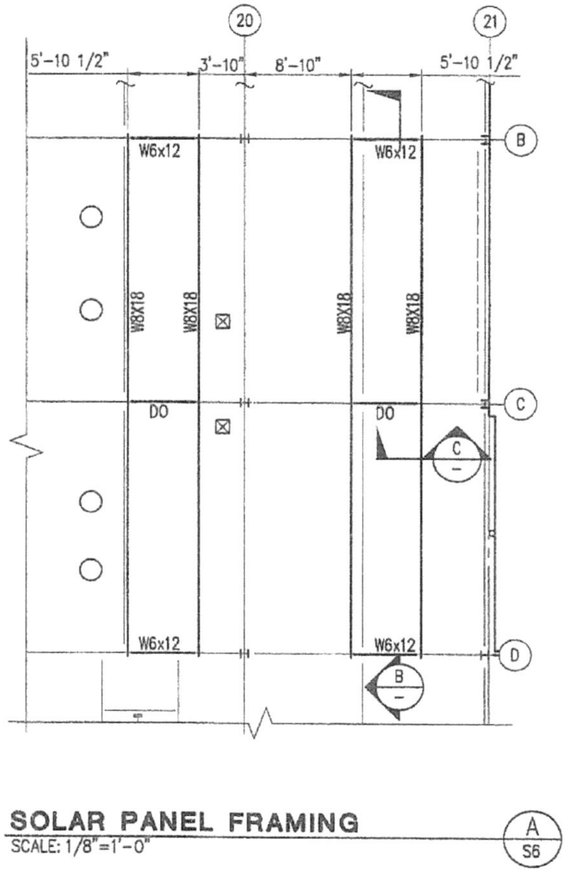

Figure 7.60
Solar panel framing

In Figure 7.60, the detail bubble says A/S6. The S6 in the bottom half of this callout lets us know where this detail was cut. The framing is drawn to scale at 1/8" = 1'-0".

In Figure 7.61, Section B is the elevation view that is cut at grid 20.5, looking to the left, and is a bit larger (see the scale at 1/4" = 1'-0). Section A/S19 is a plan view. We take that plan view as if we were holding it in our hands like a box; we then turn that box and look to its left hand side.

We can see the ends of the steel beams W14 coming at us. These are the same as what we saw at the roof plan. We see the sideways view of the W8 x 18 and the ends of the W6 x 12. We can see that the beams are on the grid line locations where we see them on the plan.

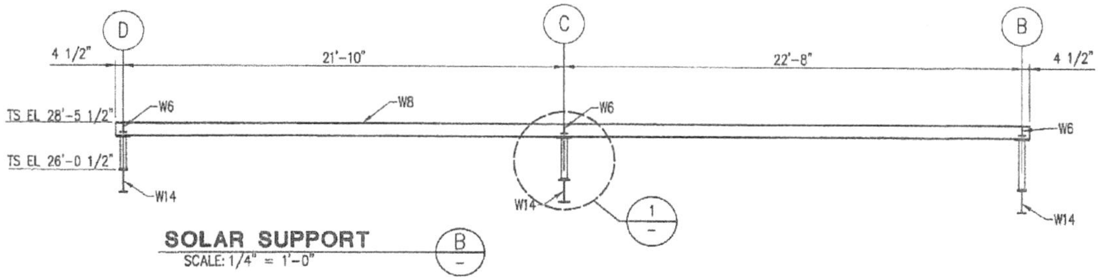

Figure 7.61 Elevation view of solar panel framing steel

What this section in Figure 7.61 is showing us specifically is that the W8 x 18 beams and the W6 x 12 beams are above the W14 beam. They use detail 1/S19 to show us how the W8 x 18 and W6 x 12 beams are being supported on top of the W14 beams. We also know by looking at the plan that section B will need to be used at four locations—one for every span of the W8 x 18 at this framing condition.

The key to using this detail is at elevation B. From this display of the grid lines, we know that detail 1/S19 is needed at all grid line locations for grids B, C, and D. Therefore, this detail 1/S19 is needed at 12 locations. In your bid letter, include the quantity of locations when conditions are confusing like this for customer verification.

What is *not* clearly shown is that, because of the roof slopes, the lengths of these supports will be different at each grid line location. This is not specifically indicated by elevation dimensions, although it is slightly noticeable in the picture from section B.

Looking back at section B (which was taken from the Solar Panel Framing section A), we find that the top of steel elevation shown by grid line D is 28'-5 1/2", and the top of the W14 is shown as 26'-0 1/2". Go back to the roof plan and you will find that the top of steel at elevations for grids B and C varies because of the slope of the roof. Therefore, the lengths of the supports will change accordingly.

Looking at detail 1 in Figure 7.62, we see that the W8 x 18 beam is supported above the W14 by the 3" diameter standard pipe. This piece of pipe is like a mini-column, called a post, with a cap and base plate welded to it (1/4" fillet continuous—the little circle at the end of the weld line that joins with the arrow lines means "weld all around"). We also can see that the post is bolted to both the W14 and the W8 x 18, centered with the beams both above and below. There is the W6 x 12 that bolts into the W8 x 18. For the length or height of this post, we see they have given us elevations to work with at the Solar Support detail B, over by grid D.

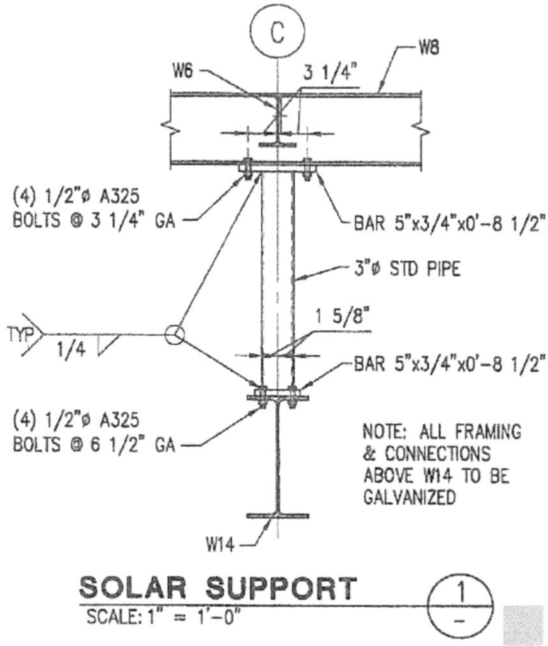

Figure 7.62
Solar support posts

The note to the right of the beam elevation lets us know that all the framing and connections will be galvanized, including the bolts that hold all this together.

Now we have worked through three subsequent details from the roof plan to finally see how this framing is to work. Collectively, these details provide a good example of detail layering.

We don't want to forget to look at detail C, which is cut at grid 20.75 at grid line C in Figure 7.63. We take that plan view like a box from the looking down position, and turn it up to look at the end closest to us.

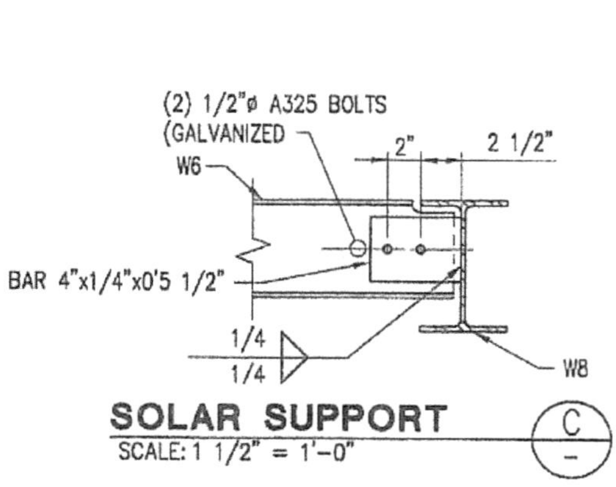

Figure 7.63
Beam to beam connection

Section C/S19 in Figure 7.63 shows the connection of the W6 x 12 into the W8 x 18. We can see that the bolts are galvanized. Typically when materials are galvanized, the fasteners will need to be as well. Still, it is good that the designers clarified this information for us in the detail. The locations of the bolts are dimensioned from the centerline of the W8 x 18. The flat bar or plate is 1/4" thick and is 4" x 5 1/2". It is then shop welded to the W8 x 18 beam with a 1/4" fillet weld on both sides. See the 1/4 with the triangle at both top and bottom of the line (that's the weld indicator); the arrow points to where the plate joins the beam. This condition will be used at all the locations where the W6 x 12 meets with the W8 x 18.

Now go back to the rest of the information that is shown at the concrete pad for HVAC between grids C and D between grid 17 and 18 in Figure 7.48. You can see the dimensions off the grid line at the perimeter as 3 1/2", 2", and 7". Remember: this dimensioning is to the outside face of the perimeter bent plate from the centerline of the gridline. You will use the detail G/S18 or Figures 7.37 and 7.65 for the slab edge at interior beams (this detail does not require any material from the steel fabricator). You will also use H/S18 or Figures 7.51 and 7.66 for the perimeter of the building.

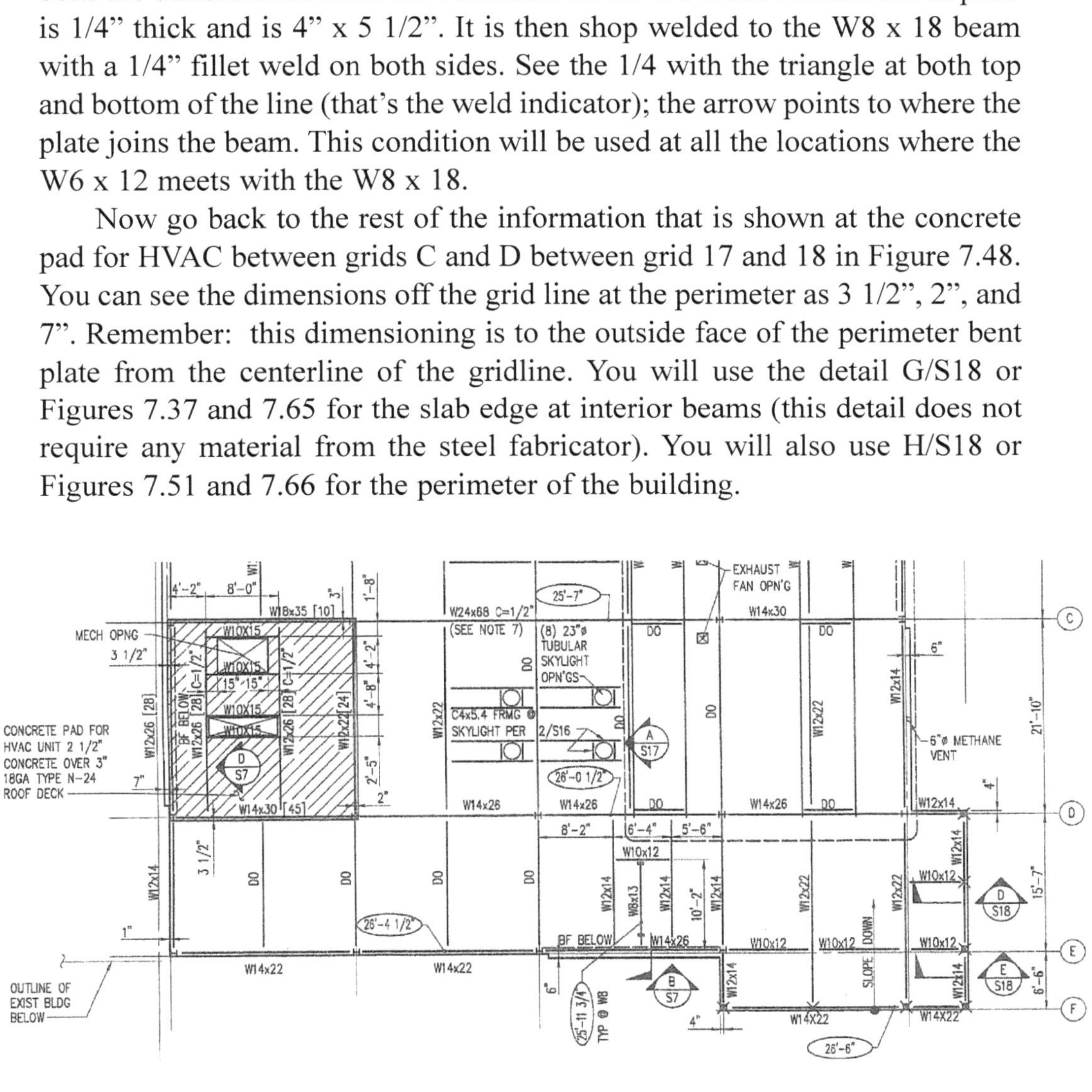

Figure 7.64 Roof edge closure plate

There are no section cuts showing roof edge perimeter closure plates; they are not called out in this plan (see Figure 7.64). We know they are needed by the information provided in the typical detail, Figure 7.65. When in doubt, a call to the engineer is in order to get the necessary information about the use of particular details shown at the detail sheets, but not called out specifically in the plan views of the building.

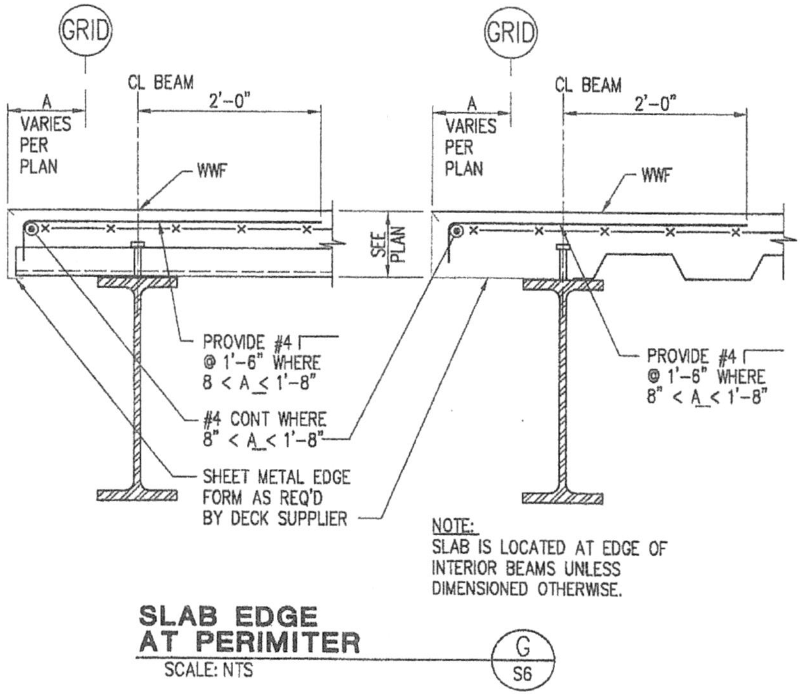

Figure 7.65 Slab edge at perimeter

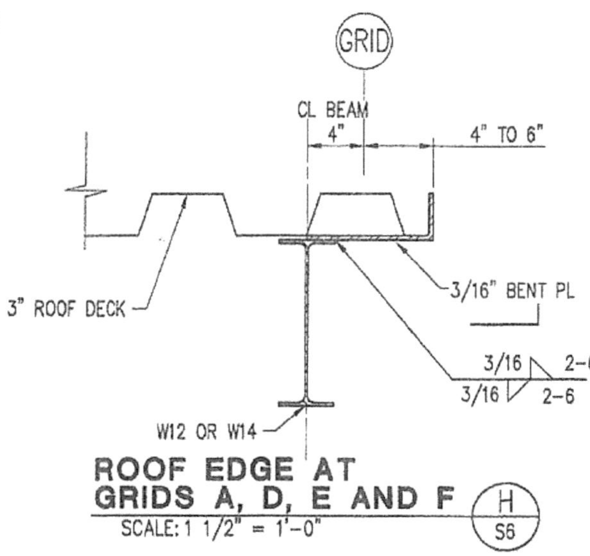

Figure 7.66 Closure plate for roof edge by fabricator

Most projects that have decking on top of the roof framing will require closure plates. However, not all projects will require closure plates that need to be supplied by the steel fabricator. If the material is 12 gauge or 1/8" thick or heavier, then it may be something that the steel fabricator supplies. Light gauge closures that are identified by the deck supplier will typically be of a thickness less than 12 gauge. It never hurts to ask questions to be sure.

Looking at the last bay of the roof in Figure 7.67, we see the beams are W14 x 22, W14 x 26 (braced frame detail B/S7), and W10 x 12. They run east to west at grid E. Beams running north to south between grids D and E are the W12 x 14 and the W12 x 22.

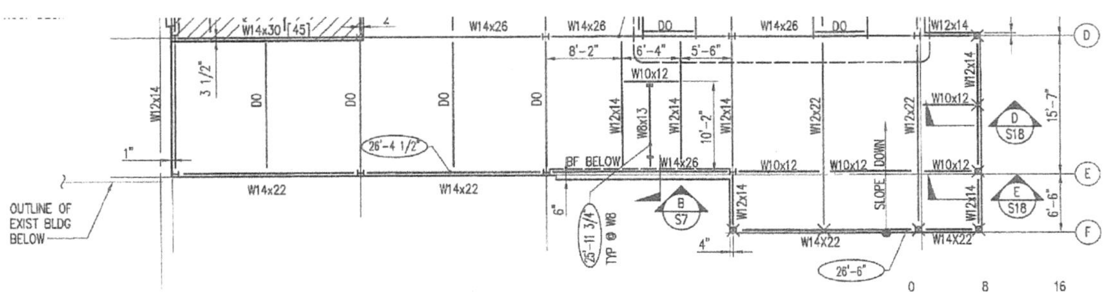

Figure 7.67 Close up of roof framing

In Figure 7.68 we look at the details D and E at grid 22 between grids D and F.

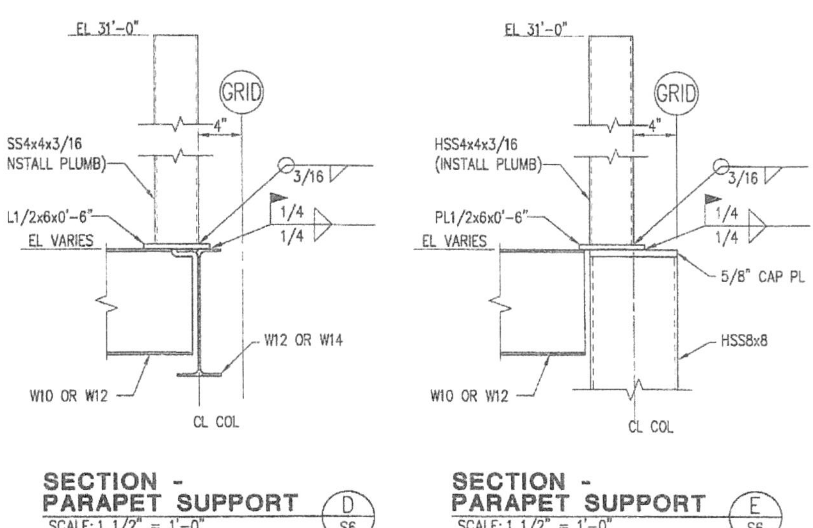

Figure 7.68 Parapet support framing

The details in Figure 7.68 show a parapet support—TS 4 x 4 x 3/16 posts. There appears to be seven of them in the plan, as indicated by the *X* designating their location. (See note 11, which we had reviewed earlier when we looked in the plan views for the 2nd floor and roof.) A parapet is the short wall that goes above the roof about 18". It hides all the roof top mechanical units when you look up from the ground and makes the building more attractive.

Roof scuppers—holes in the parapet—provide a way for the gutters to connect to the downspouts, which move the roof water to the ground. Not all parapets require steel framing. This one has been designed to do so and we will list this material in our take off. This material is not called out as galvanized, although it may be exposed to the weather. If we don't find in the Specifications Section 055000 Miscellaneous Metals that this item is to be galvanized, then we should exclude its galvanizing in the bid letter so that our customer is clear on the provision.

Braced Frames and Their Fabrication Details

Now that we have gone through the floor and roof plans, we can look at the details in more depth. Our purpose is to become more familiar with how this building looks from the steel perspective. This process will clarify the framework, giving us a better feel for the designers' perspective, and helping us prepare our take off and material listing.

Looking at the foundation plan in Figure 7.69, we can locate a braced frame section A/S7, Figures 7.20 and 7.69, located at grid A between grids 19 and 20:

The frame will be found at sheet S7 as an elevation view shown in Figure 7.70.

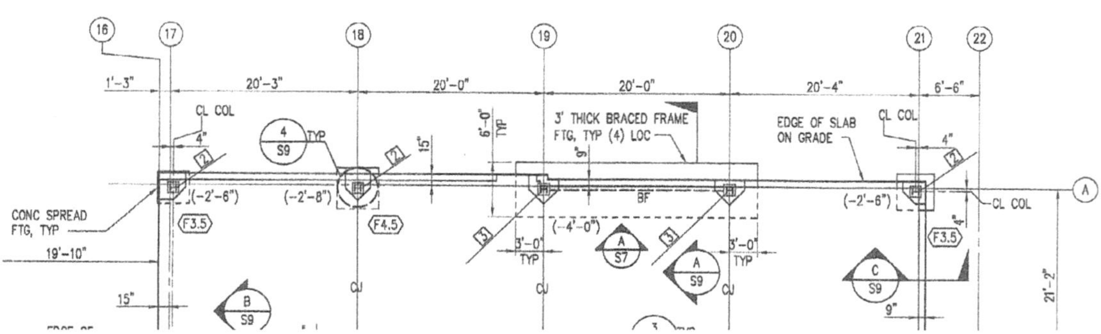

Figure 7.69 Braced framing call outs in the foundation plan

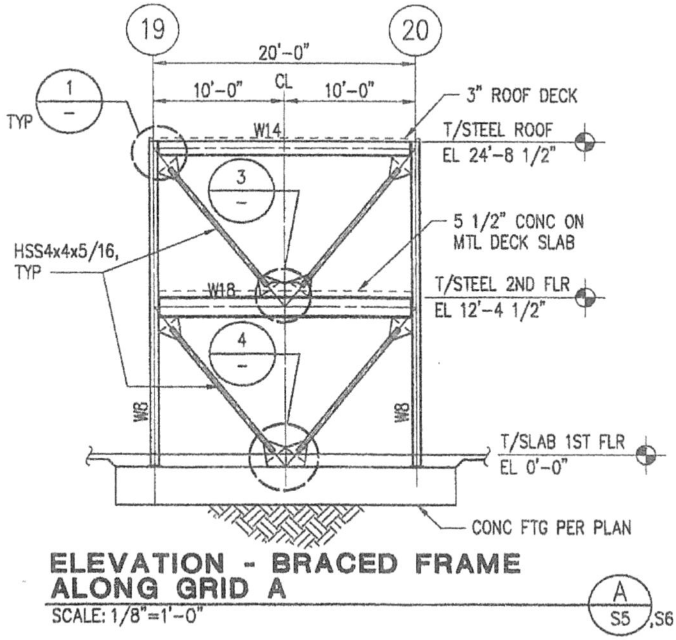

Figure 7.70 Elevation of braced frame from Figure 7.69

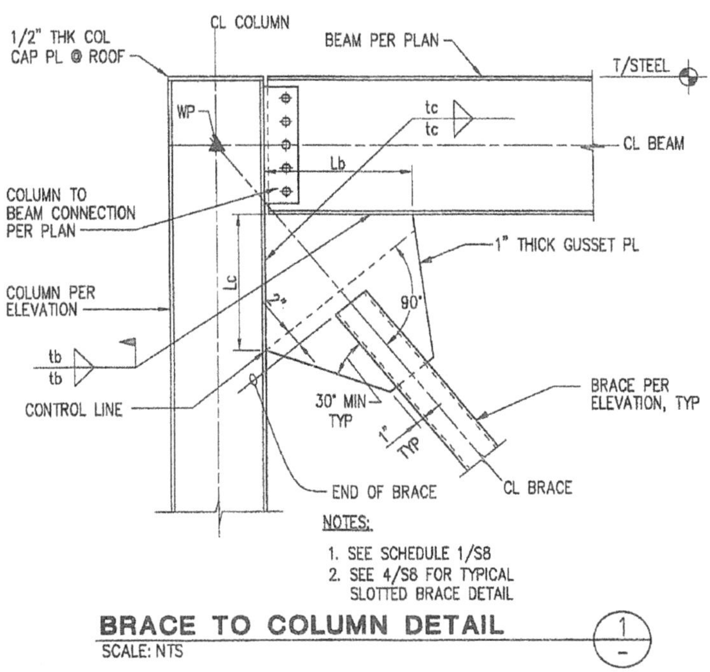

Figure 7.71 Bracing gusset into beam and column

In the elevation shown in Figure 7.70, we can see the concrete slab at the 1st floor, the steel at the 2nd floor, and the roof. The diagonal pieces that run from the foundation to the underside of the 2nd floor beams, and again from the top of the 2nd floor steel to the underside of the roof beam, are called diagonal bracing. Here they are sized at TS 4 x 4 x 5/16. What holds that bracing in place? The details that are called out as details 1, 3, and 4 show us in close up what the connections look like.

At detail 1/S7 in Figure 7.71, we can see the column with the cap plate, the beam that frames into it, and the connection that bolts the beam to the column. We see the survey mark that indicates that the top of steel elevation in the roof's plan view applies to the top of steel at this braced frame as well. The *WP* next to the black triangle at the centerline of the column indicates the work point location. This location is what steel detailers will need for setting the angle of the bracing; it is used to calculate the correct length of this bracing.

Remember, all the information that we use as estimators will have a more in-depth application when the time comes to draw the steel. If information like work point is missing from the contract documents, then clarification will have to be asked for later, slowing the detailer down. Missing work point locations are only of concern for steel fabricators if they are the successful low bid.

The gusset plate that holds the diagonal bracing in place is shop welded to the column. You can see the note tc next to the fillet weld symbol, on the line off the arrow pointing to the weld location. At the other side of this gusset, the weld note *tb* points to the weld location on the underside of the roof beam. These indicators *tc* and *tb* are notes for the specific weld size to be used.

We see the Notes below the detail in Figure 7.72. Note 1 gives the grade of material to be used for the gussets. Note 2 indicates specific weld information that may require additional costs associated with inspection. Note 3 warns of the potential application of compressible joint filler (which is rarely used). Note 4 directs us to the typical slotted brace detail—this detail is where we will see how the bracing will be welded and the tube steel brace is to be slotted. Note 5 provides more information regarding the dimensioning of the gusset plates, letting us know that the lc dimension is to the highest point of the gussets shown in Figures 7.73 and 7.74.

We see that the gusset plate is 1" thick, but what size is this plate for width and length? The dimensions are indicated with *lc* and *lb*. These dimensions are found at the schedule on 1/S8, shown here as Figure 7.72.

As we review the details in Figure 7.72, we find out what our reference indicators *tb* and *tc* mean. These welds are to be 3/8" fillets. Length *lc* is to be 16-1/2" and length *lb* is to be 17". We can see by this chart that the gusset plate sizes will change with their location, because the dimensions for lb and lc are different at the foundation, as well as above and below the 2nd floor. It appears that all the weld sizes are the same at all applications. The notes in this detail give us the specifics on material grade (A573 Grade 50—a high strength steel), inspection requirements (continuous visual for all bracing gusset welds—this will need to be included or excluded in our bid letter), the requirement for compressible material (does not affect the steel fabrication or supply), the slotted brace detail reference (we will need to look at that), and the note for the "shared edge" of the bracing (this will most likely involve material that we will need to supply).

BRACED FRAME ELEVATION	DETAIL	LOCATION	Lb	tb	Lc	tc
A/S8	1/S8	ROOF	17"	3/8	16.5"	3/8
	3/S8	ABOVE 2ND FLOOR	22"	3/8	12.5"	3/8
	1/S8	BELOW 2ND FLOOR	18"	3/8	14"	3/8
	4/S8	FOUNDATION	17"	3/8	18"	3/8
B/S8	1/S8	ROOF	25.5"	3/8	17.5"	3/8
	2/S8	ABOVE 2ND FLOOR	27"	3/8	17"	3/8
	2/S8	BELOW 2ND FLOOR	17.5"	3/8	15.5"	3/8
	4/S8	FOUNDATION	17"	3/8	18"	3/8
C/S8	1/S8	ROOF	17"	3/8	16.5"	3/8
	3/S8	ABOVE 2ND FLOOR	23.5"	3/8	12.5"	3/8
	1/S8	BELOW 2ND FLOOR	19.5"	3/8	14"	3/8
	4/S8	FOUNDATION	16.5"	3/8	17"	3/8
D/S8	1/S8	ROOF	26.5"	3/8	18.5"	3/8
	2/S8	ABOVE 2ND FLOOR	26.5"	3/8	18.5"	3/8
	2/S8	BELOW 2ND FLOOR	17"	3/8	16.5"	3/8
	4/S8	FOUNDATION	16.5"	3/8	17.5"	3/8

NOTES:
1. ALL GUSSET PLATES AND BASE PLATES SHALL BE ASTM A572 GRADE 50.
2. ALL BRACED FRAME WELDING IS SEISMIC CRITICAL. CONTINUOUS VISUAL INSPECTION IS REQUIRED FOR ALL BRACE TO GUSSET PLATE WELDS AND FOR ALL GUSSET PLATE TO BEAM COLUMN WELDS.
3. PROVIDE 1" COMPRESSIBLE JOINT FILLER AROUND BOTH SIDES OF GUSSET PLATE WHERE NOTED ON DETAIL.
4. SEE 4/S8 FOR TYPICAL SLOTTED BRACE DETAIL.
5. FOR DETAILS 3/S7 AND 4/S7, Lc REFERS TO THE HEIGHT OF THE SHARED EDGE.

BRACED FRAME CONNECTION SCHEDULE
SCALE: NTS

1 / S7, S8, S10, S11

Figure 7.72 Braced frame connection schedule

Detail 4/S7 in Figure 7.73 is taken from the foundation level where the diagonal braces meet. In this detail 4/S7, we see that the gussets for the braces come together at the center, with a 1/2" plate on each side as a stiffener.

Figure 7.73 tells us that the gusset plate is a single plate, cut as a continuous piece, with 1/2" plates welded to both sides at the centerline location of this plate. We can see the work point at the underside of the 1/2" plates and at the bottom of the underside of the base plate. We can see that the base plate is 1 1/4" thick by 13" wide "into the page." (We are looking at a side view; therefore, the length dimension of this plate would be the same as the overall length of the gusset. The dimension we don't know—the width of this plate—is given to us here as 13".)

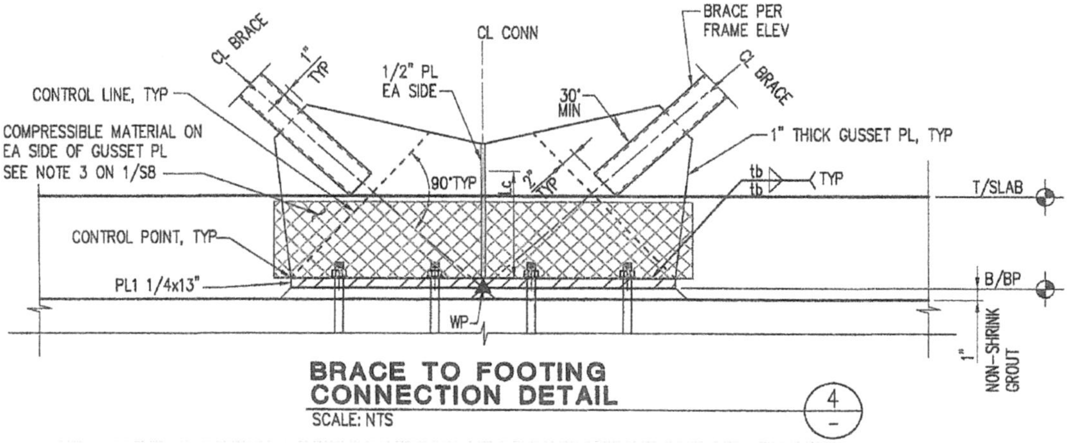

Figure 7.73 Brace to footing connection detail

We know that our gusset plate is 1" thick. The weld indicator means we are to use a 3/8" fillet weld, both sides, typical (which means where welds are not shown, the same weld will be required at this detail).

The detail 3/S7 in Figure 7.74 is taken from the 2nd floor level, where the diagonal bracing meets with the top of the floor beam. We can see all the same gusset plate requirements and weld information that we saw at detail 4/S7 in Figure 7.73. What's different here is that we have stiffener plates that weld at specific points to the beam, as the stiffener plates slide in between the top and bottom flanges of the beam and up to the beam web. These 1/2" thick stiffener plates will be welded on the side, top, and bottom with 5/16" fillet welds typical. If we turn this beam to the side, the stiffeners will look like the sample in Figure 7.75 as they are set into the beam.

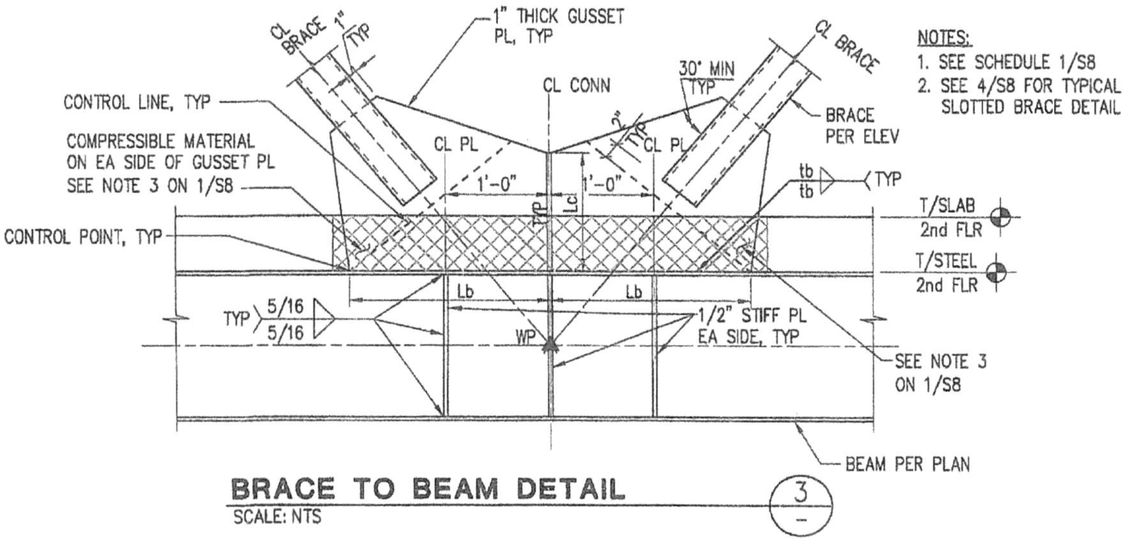

Figure 7.74 Brace to beam detail

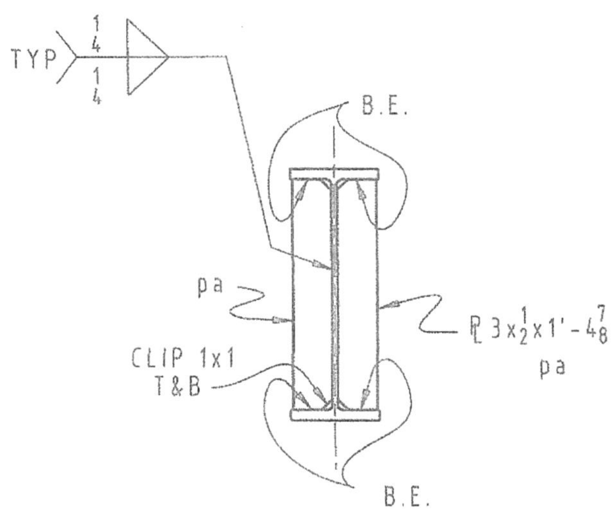

Figure 7.75 Beam stiffener plates and welding

In Figure 7.75, you can see the end of the beam coming at you. The stiffeners are at either side of the beam. Note that the stiffeners are clipped at the corners to allow the beam web to flange transition, also called the *k* distance. The welding of this stiffener is not continuous, but starts and stops as you change direction from the web to the flange and web again.

As we look at the individual details for bracing connections, other items we find on S7 are Notes 1 through 4 in Figure 7.76. We have seen Notes 1 and 2 before, but Note 3 gives us information on the "typical foundation bolts" or anchor bolts that we did not see before—the size of the anchor bolts that were shown at 3/S7. See Figure 7.77 for the detail F/S11.

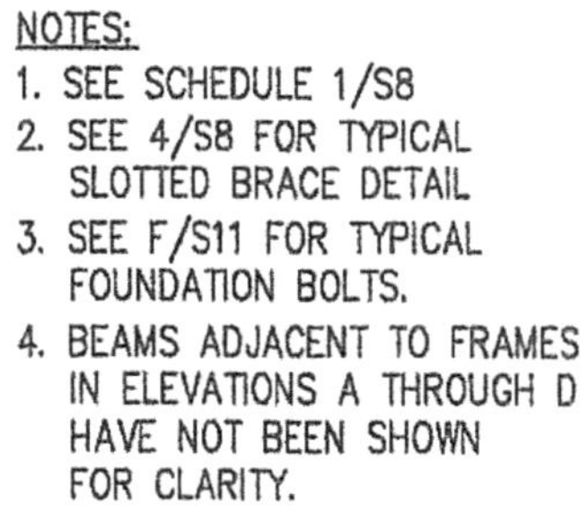

Figure 7.76
Braced frame drawing notes

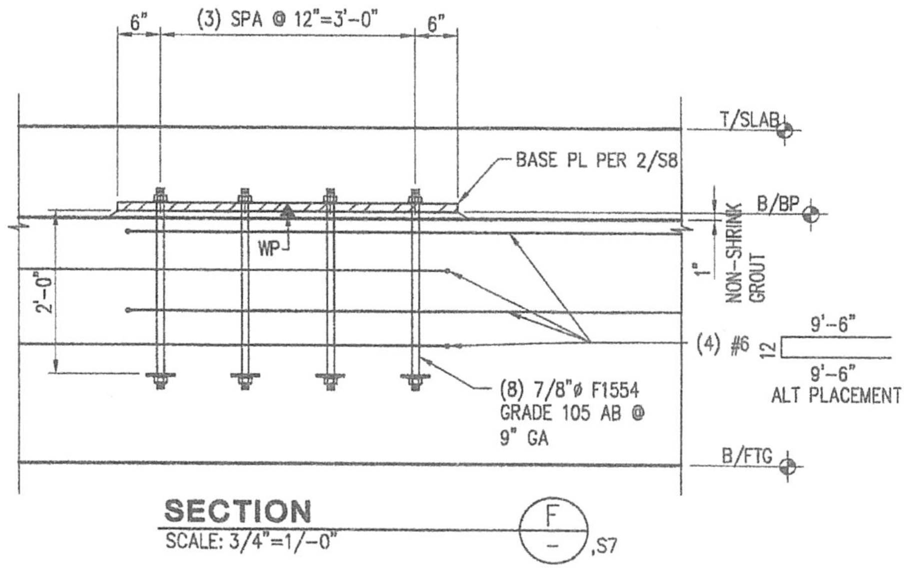

Figure 7.77 Detail F/S11, Base Plate Detail

The detail in Figure 7.77 gives us the size and quantity of the anchor bolts to be used at 3/S7, and the embed depth of 2'-0. We will need to add to that length to allow for the plate, nuts, and washers.

Looking at another braced frame (see Figure 7.78), we find some of the same details being used, although different size columns and beams will change dimensions on the gusset plates and length of bracing.

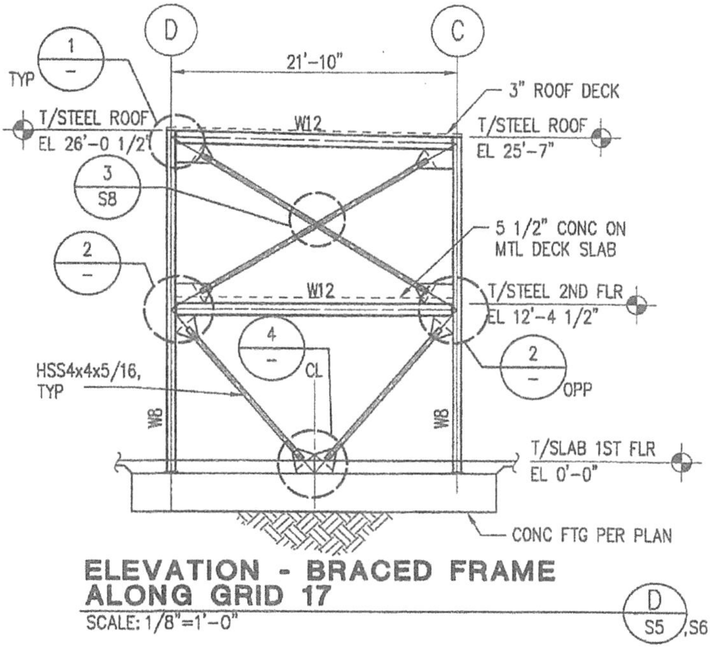

Figure 7.78
Elevation—braced frame

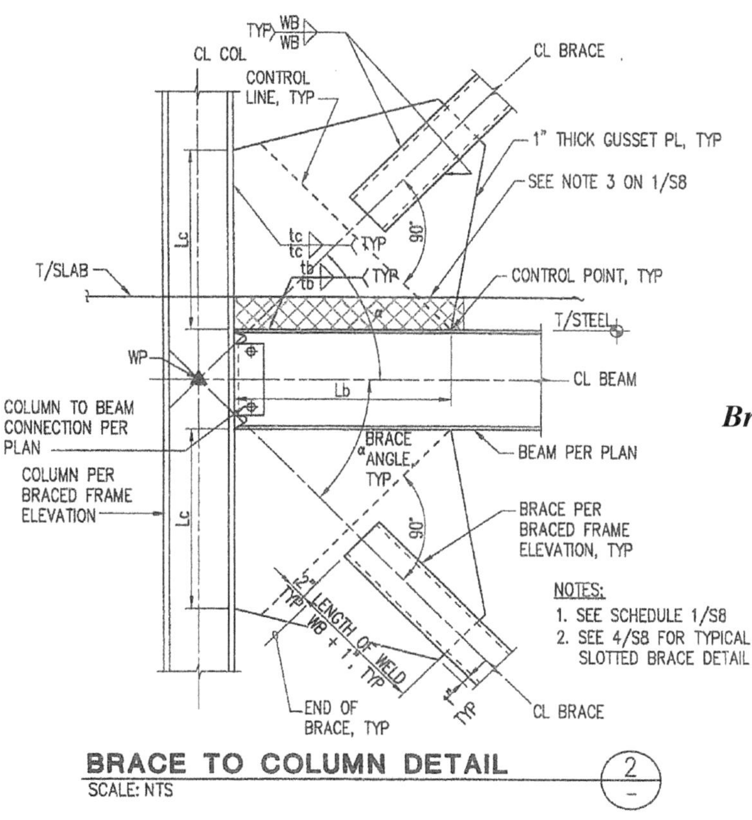

Figure 7.79
Brace to column detail

Detail 2/S7 is new with this braced frame elevation in Figure 7.78. In Figure 7.79, it shows how the bracings join at the top and bottom of the 2nd floor beam:

In Figure 7.79, the column is on the left, and the floor beam at the center with the connection to the column. We see the work point that gives us the angle of the bracing. Some of the same information is provided with regard to the size, thickness, and welding of the gusset, as well as the same references for more detail information in Figure 7.72's schedule 1/S8 and the slotted brace detail 4/S8 shown in Figure 7.80.

The detail 4/S8 in Figure 7.80 gives us a heavier weld size than is used at the gussets. This detail tells us to slot the tube steel one inch more than what is required to set the tube on the gusset (11"). It also tells us to slot the tubing at the centerline, and to slot the tube 1/16" wider than the gusset plate is thick.

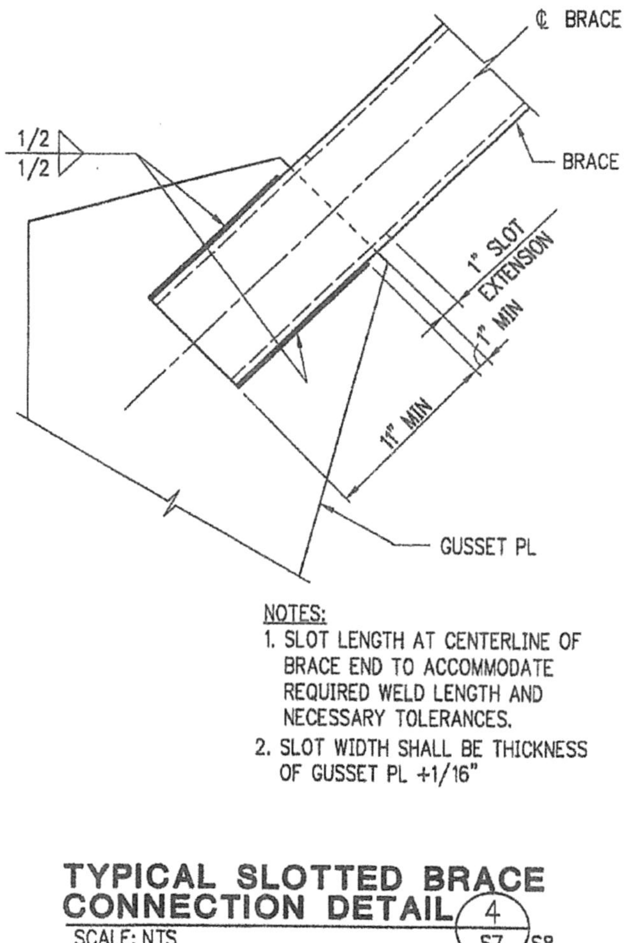

Figure 7.80
Typical slotted brace connection detail

The connection detail 3/S7 in Figure 7.81 is for the condition where the bracings cross. This detail shows that one of the braces is continuous, but the other is split in two. The notes send us back to detail 1/S8 in Figure 7.72 for our typical weld sizes, and back to 4/S8 in Figure 7.80 for the slot. What size is this plate? We will have to calculate that ourselves, building on what we know. The slots are 11" deep. We estimate that the slot in the center of the continuous tube steel brace is at least 22". We then project that the two other bracings will be 11" each. Allowing for the 4" tube brace and one inch either side for clearance makes the width about 28".

For estimating purposes, we could estimate this plate as 1" x 22" x 28", or even as much as plate 1" x 24" x 30". That measure would be good enough for an estimate. Steel detailers will worry about being exact. We are only trying to create an estimate here.

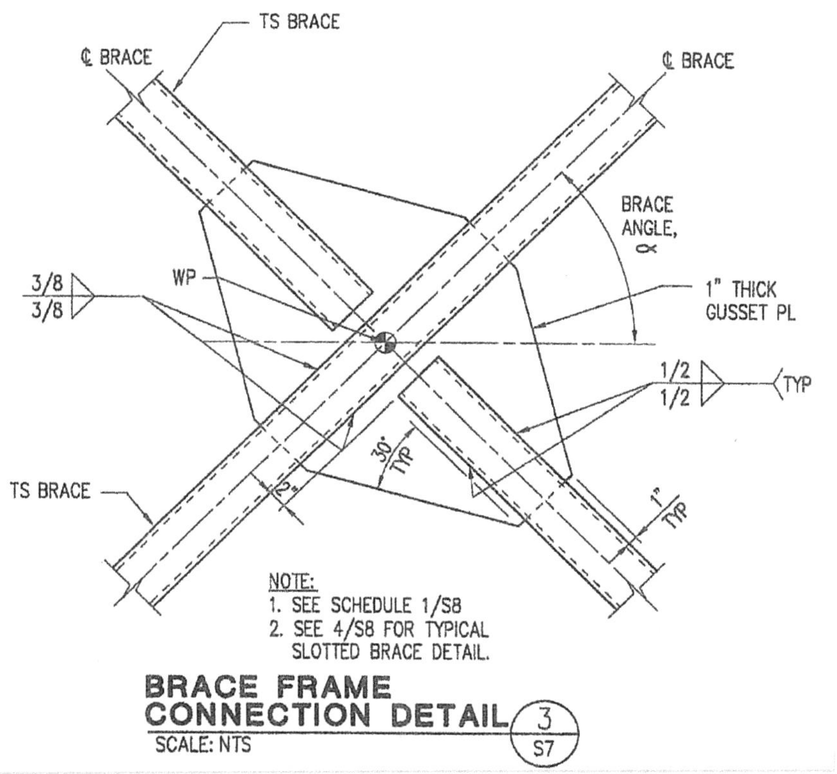

Figure 7.81 Brace frame connection detail

Going back to the first bay of the foundation plan, we see detail 4/S9 at the intersection of grid A-18 shown in Figure 7.82.

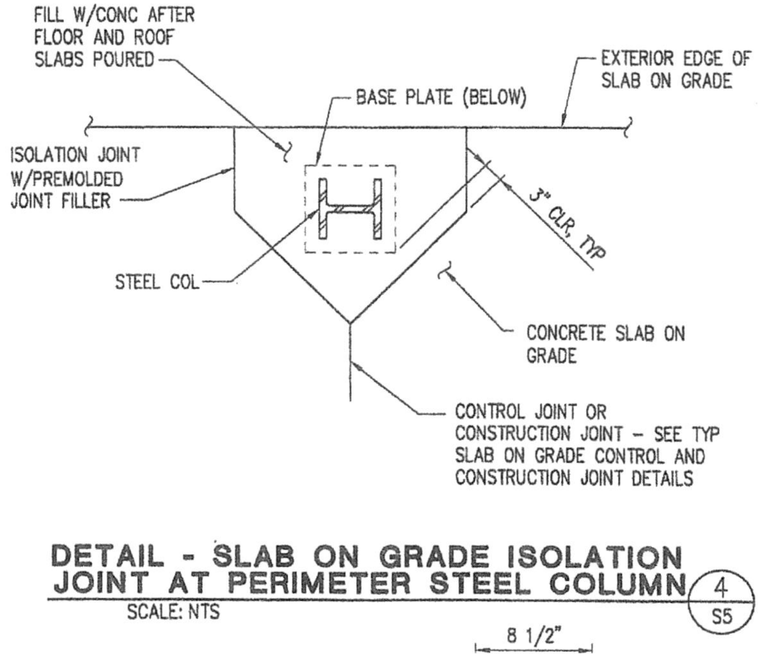

Figure 7.82 Braced column isolation slab plan

This detail in Figure 7.82 shows us the column view in plan, and how that column sits in relationship to the footing. There is nothing in particular here of interest for fabricators.

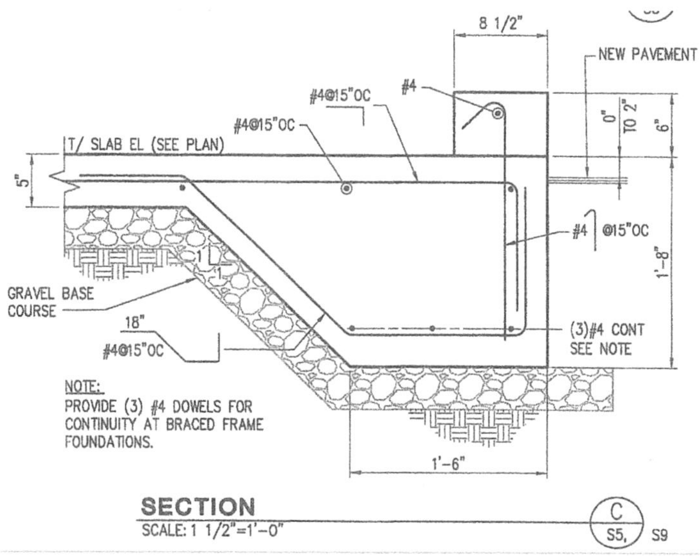

Figure 7.83 Concrete footing at braced frame

Detail C/S9 in Figure 7.83 is cut at grid 21 between grids A and B. This detail shows us the concrete foundation and curb. The embed items in the concrete are all rebar and provided by others. There is no steel here for fabricators to be concerned about.

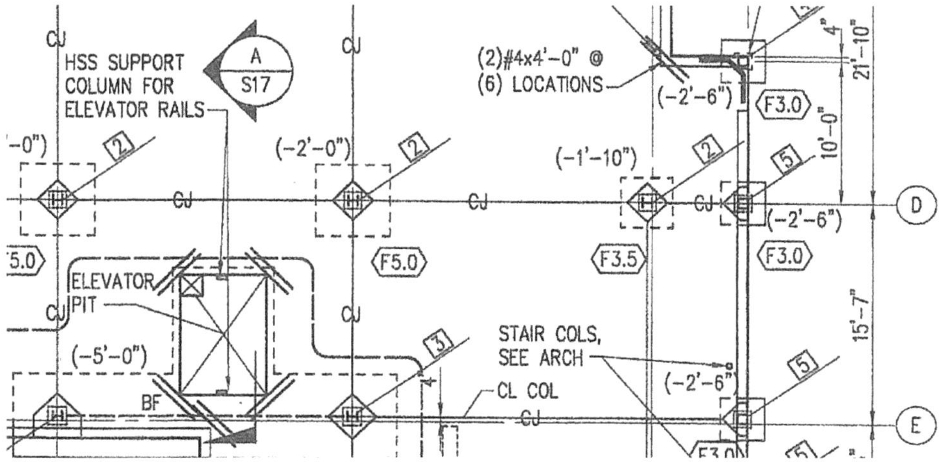

Figure 7.84 HSS support column for elevator rails

In this area of the foundation plan shown in Figure 7.84, between grids D and E, we find section A/S17 seen in Figure 7.85 cut through the middle of the elevator pit. Here are two HSS—or TS 8 x 3 x 1/4 support columns for elevator rails. The arrows from this note point to where the HSS or tube steel columns are located. These columns will support a beam at the very top that is designed to support the elevator inside that shaft.

In Figure 7.85, this elevation view A/S17 through the elevator pit shows the HSS tube that starts below the foundation level, continues up through the 2nd floor, and ends at the roof. The dimensions to the left of the picture help us know the approximate height of these tube steel columns.

Detail 2 shown in Figure 7.86 is cut at the base of the tubing to show us the connection there.

We see from the detail in Figure 7.86 that the tube goes 1'-6" below the floor and is held in place with L 6 x 4 x 1/2 x 0'-10" with anchor bolts (2) 3/4" diameter x 10" long at 7" GA (Gauge, meaning 7" apart on this 10" long piece of angle). The note for the #4 (rebar) that is 2'-0" x 10" x 2'-0" is for the concrete floor and concrete pit wall; it is not a concern for steel fabricators.

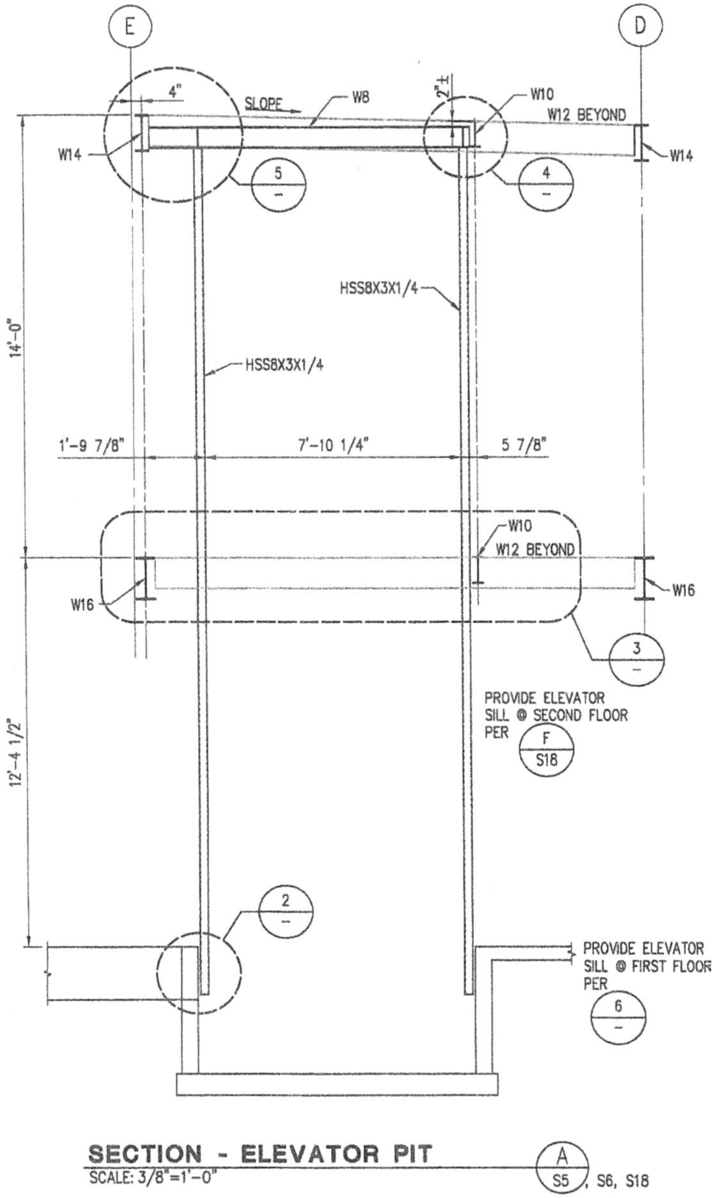

Figure 7.85 Elevation of elevator pit showing columns

Detail 3/S17 in Figure 7.87 gives us the dimensions for the location of these tube steel columns. It also shows how they connect to the steel floor framing.

Those vertical lines in Figure 7.87 with the zig-zag in them are the break lines. They give license to the person who drew the picture to allow for the

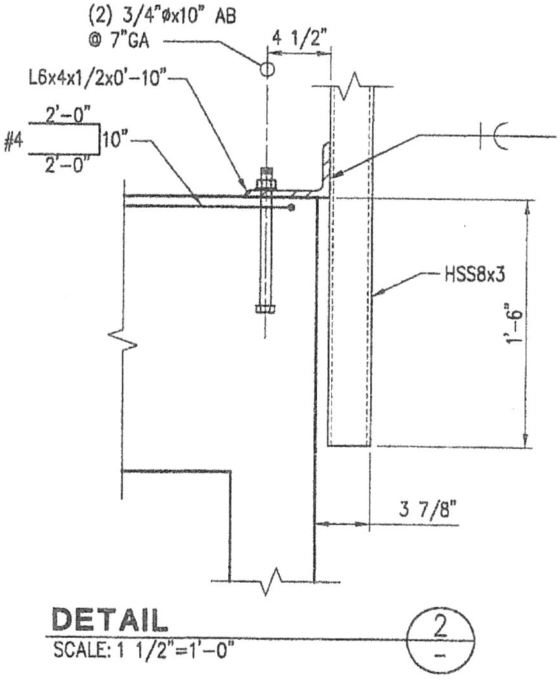

Figure 7.86
Elevator support column connection at slab

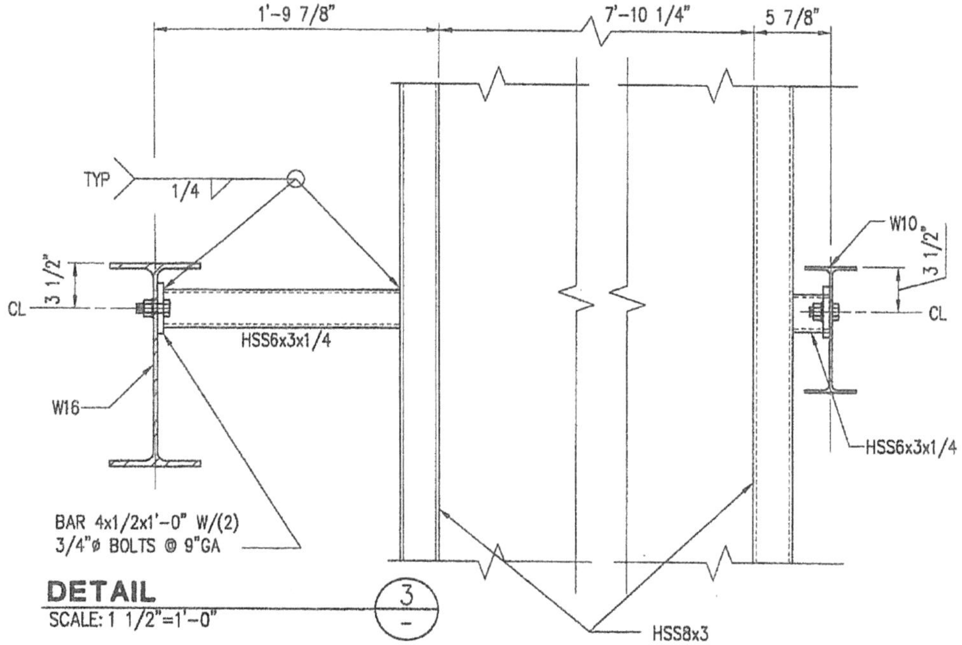

Figure 7.87 Elevator column brace to floor beam framing

actual width required for the picture without having to draw completely to scale. Notice how there is a zig-zag in the dimension line below the 7'-10 1/4" dimension as well. The dimension given should be used at the full extension of this detail.

The engineer has used some horizontal tubes TS 6 x 3 x 1/4 to connect the vertical tubes to the steel floor beams, with flat bar 1/2 x 4 x 1'-0 long, with two bolts 3/4" diameter at a 9" GA. (Note that gauge means spaced at 9" apart, leaving 1 1/2" either end of the 1'-0 piece of flat bar. Standard allowance for either side of a 3/4" diameter hole is 1 1/2" either side, or twice the diameter of the bolt being used). Also notice that the horizontal TS 6 x 3 x 1/4 is centered at 3 1/2" down from the top of the beam. This gives us the location of the connection as it is welded to the vertical tube steel column.

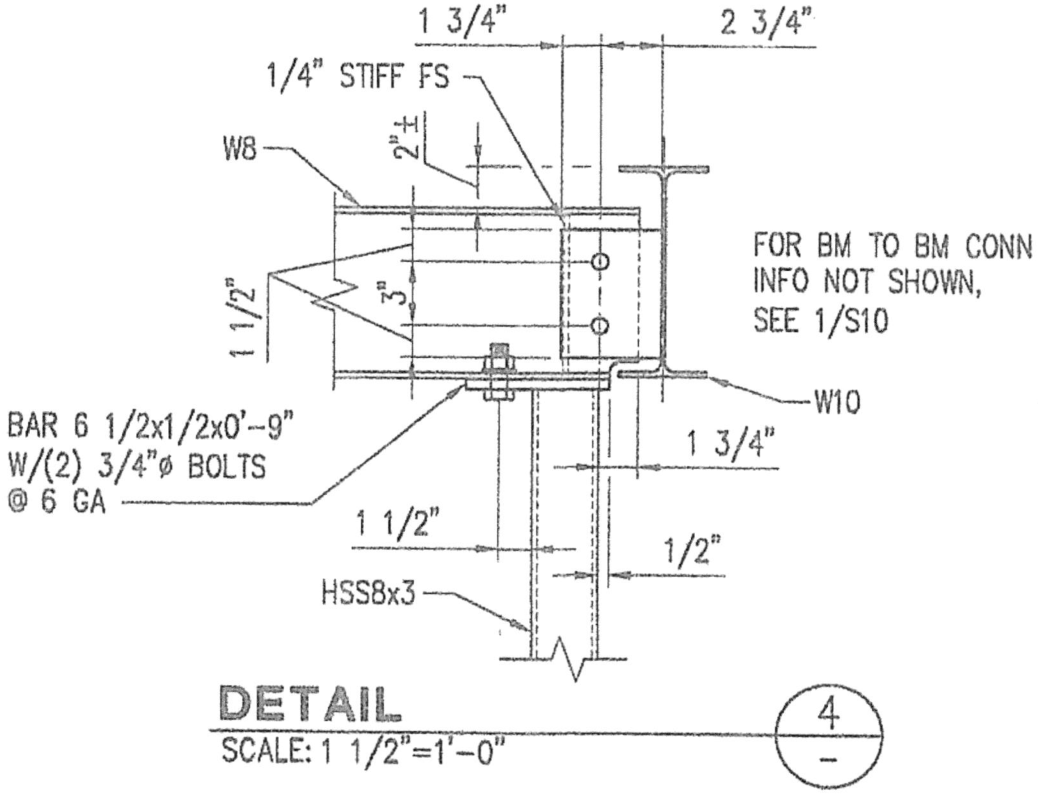

Figure 7.88 Detail at Top of Elevator column

Detail 4/S17 in Figure 7.88 is taken from the top right of the elevation view A/S17 in Figure 7.85. It shows the connection that is required at the top of the TS 8 x 3 x 1/4 columns. This is the bar 6 1/2 x 1/2 x 0'-9", which could also be a piece of cut-to-size plate, and the (2) 3/4" diameter bolts at 6" GA. This also means that the plate we see there at the top of the column is shown as 6 1/2" wide. The 9" goes into the page, with the bolts at the 6" center. These measures leave 1 1/2" each side of that plate for grip as explained in the information for detail 3/S17 shown in Figure 7.87. It shows the connecting of the elevator rail beam (W6 x 13, which we know from the roof plan, because it is located there on top of the columns) and the connection of the rail beam to the other roof framing beam.

The note calls out the location for finding the information for the typical connections, which will give us the size and thickness of the connection plate, the weld information, as well as the number and size of bolts that are required. There is also a 1/4" thick stiffener plate on the back side of the elevator rail beam behind where the connection plate is located—easy to miss if you are not looking closely. It says 1/4" stiff FS, which means they are using a plate 1/4" thick stiffener far side of the beam, and will weld in between the top and bottom flange of this beam up against the web.

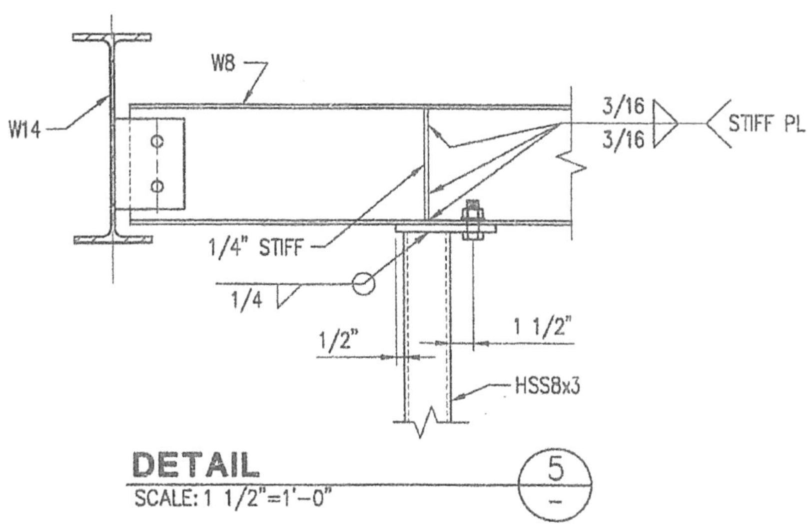

Figure 7.89 Elevator column and beam connections

Detail 5/S17 in Figure 7.89 is taken from the top left side of the elevator pit elevation in Figure 7.85. It shows the top of column connection, as well as the connection to the roof beam framing. The detail here shows a stiffener plate, 1/4" with the 3/16" fillet weld, but it doesn't clarify whether or not this plate is to be at both sides. Yet it is safe to assume that this plate will be required on both sides because the column TS 8 x 3 x 1/4 is centered under this beam, requiring that this stiffener plate is needed at both sides of this beam. At a later time, the steel detail will ask the engineer for verification. For estimating purposes, it is better to include the plate than it is to ignore it.

The detail in Figure 7.90 shows us the sill angle at the elevator from Figure 7.70. Here the designers offer the opportunity to use a 3/8" thick bent plate or a 3/8" thick angle that will be the length of the elevator shaft opening, usually about 10'-0" long. This angle or plate will be field welded to the embed angles or bent plates that have a DBA or Deformed Bar Anchor (much like rebar, but purchased from the same companies that sell the HAS or Headed Anchor Studs). In this case, each bent plate or angle that is 10" long has (2) 1/2" diameter x 2'-0" lengths of DBA welded to them at the 6" gauge.

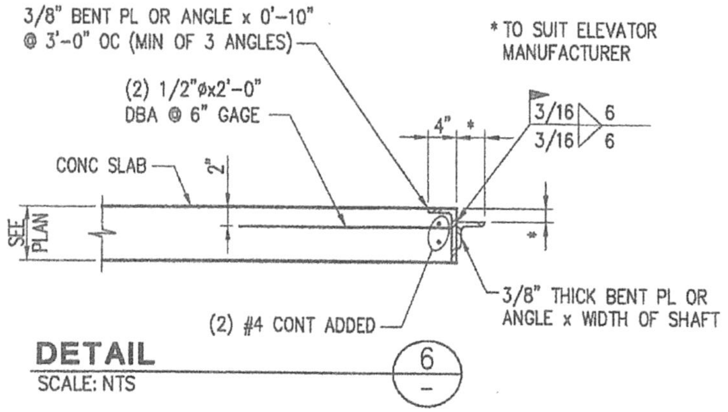

Figure 7.90 Elevator sill closure plate or angle

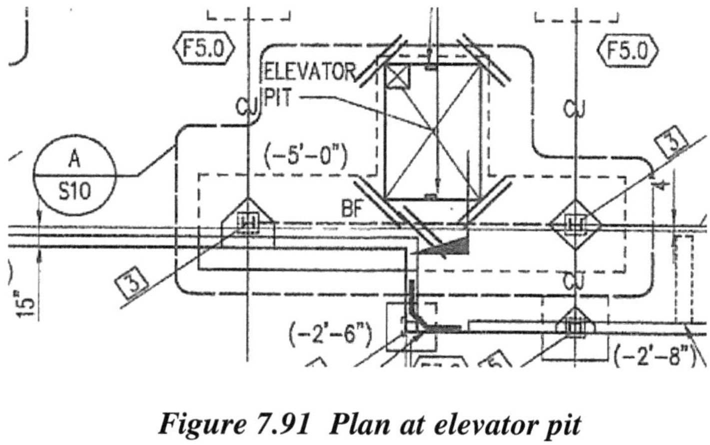

Figure 7.91 Plan at elevator pit

On the right side, just above where detail 6 is called out in Elevation A/S17 (in Figure 7.85), there is a note for detail F/S18. This is a detail we have seen before while reviewing the 2nd floor plan on S6. The enlarged view of this area can be found at B/S18—look back to Figures 7.46 and 7.48. This connection is brought to your attention because we want to be careful not to duplicate information within our own take off information.

Figure 7.91 has more information available for this elevator shaft area in the foundation plan. Detail A/S10 is in Figure 7.92.

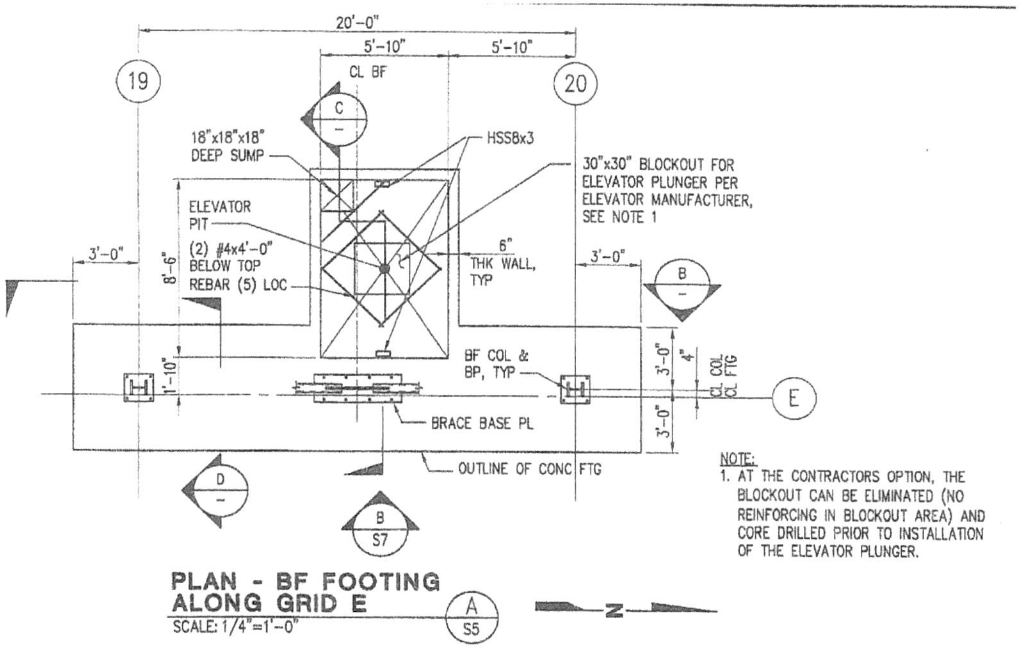

Figure 7.92 Section cuts through elevator pit

This detail in Figure 7.92 also shows the TS 8 x 3 columns. In addition, it brings our attention to details B and C. Because detail C is cut through the pit, we will look at it first.

The detail in Figure 7.93 shows us the TS 8 x 3 supports for the framing that we have already reviewed. Additionally, we are looking at the sideways view of a braced frame at grid line E, which also references details for the braced frame we have already reviewed. What we do see now that is new is the 18" x 18" sump pit—a common find for elevator pit areas. Generally there is an angle frame to support a grating to cover this open area that will be shown in the architectural drawings because it is not shown here. Additionally,

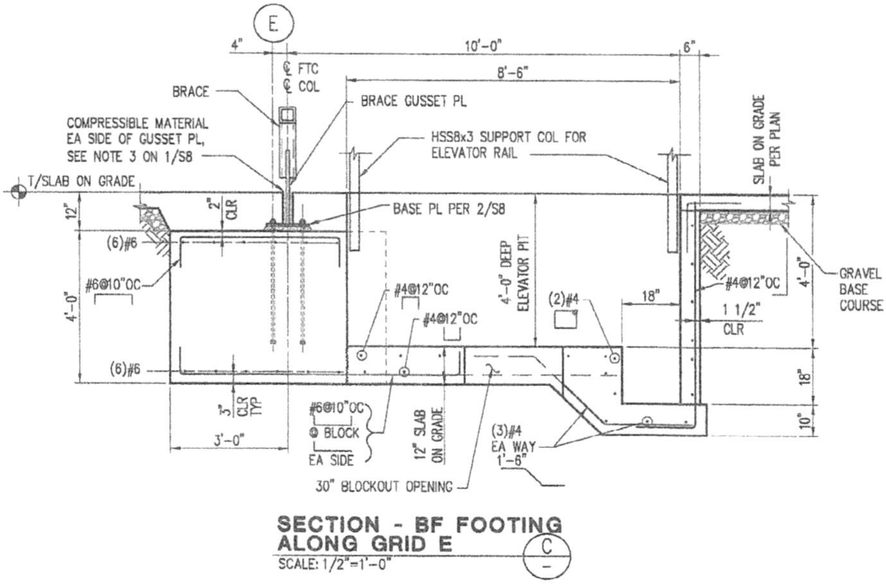

Figure 7.93 Elevation of elevator pit

there may be an elevator pit ladder shown in the architectural drawings; this ladder allows access to the bottom of this pit.

In Figure 7.94, we see that Section B is cut through the entire area of section A/S10 (Figure 7.92); it shows the braced frame B/S7 at grid line E between grids 19 and 20 (similar to Figures 7.70 and 7.78). We can see the braced frame columns that sit atop the footings below the top of slab. We can see the bracing coming into the gusset plate at the foundation level as well.

We can see the elevator pit in the middle and the TS 8 x 3 columns that support the elevator tail. There is no new steel here because we have reviewed all the other details showing us this framing from other views.

The detail at Figure 7.95 shows the standard bolted beam connections to be used on this project. The chart provides the connection plate material thickness, the weld size, and the number of bolts to be used for each size of beam. The Detail—Standard Bolted Beam Connection—shows a beam framing into a column at the floor level. Look at the connection plate; you can see that the engineers show the bolt centers are to be at 4", "typ UNO" or Typical Unless Noted Otherwise. We are to allow 2" for the plate allowance top (and bottom) of the shear plates, from the center of the hole to the edge of the plate.

The remaining dimensions of the shear plate in Figure 7.95 are shown. They are 2 3/4" from the face of the column to the centerline of the holes in the plate, and 1 3/4" from that same centerline of holes to the other side of the plate—the total is 4 1/2" wide. The Standard Bolted Connection Schedule A shows us that the thickness of the plate is going to be 1/4" for beams W8 up

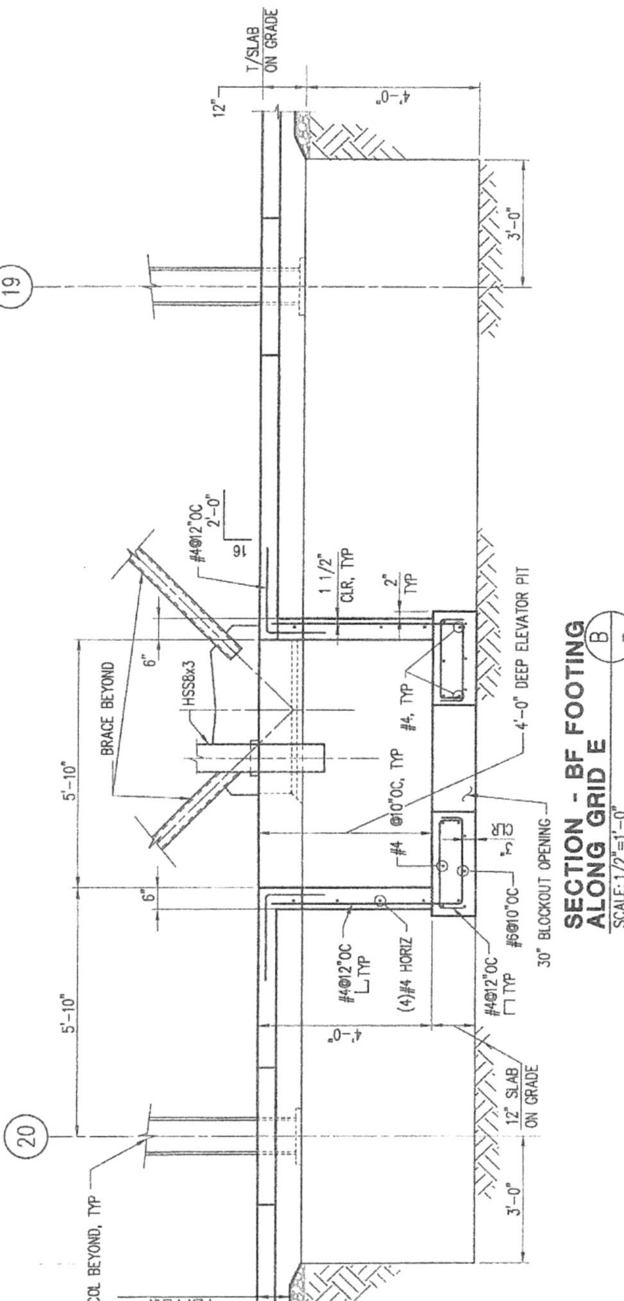

Figure 7.94 Rotated elevation view of elevator pit

STANDARD BOLTED CONNECTION SCHEDULE "A"			
BEAM SIZE	NUMBER & SIZE OF BOLTS REQ'D	PLATE THICKNESS	WELD SIZE (W)
W8	(2) 7/8"ø	1/4"	1/4"
W10	(2) 7/8"ø @ 4" GA	1/4"	1/4"
W12	(3) 7/8"ø	1/4"	1/4"
W14	(3) 7/8"ø	1/4"	1/4"
W16	(4) 7/8"ø	5/16"	1/4"
W18	(4) 7/8"ø	5/16"	1/4"
W21	(5) 7/8"ø	3/8"	5/16"
W24	(6) 7/8"ø	3/8"	5/16"

* USE HORIZONTAL SHORT SLOTTED HOLES IN THE PLATE AT COLUMN CONNECTIONS ONLY.

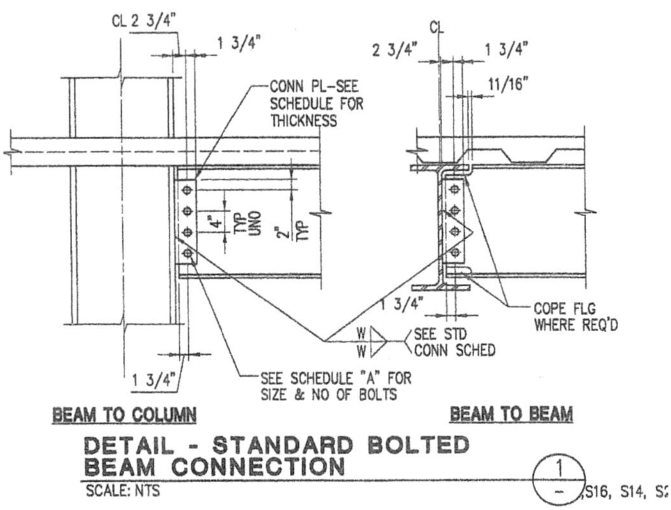

Figure 7.95 Detail: Standard bolted beam connection

to and including W14. Then we are to use 5/16" plate for W16 and W18 beams, and 3/8" thick plate for the W21 and W24s.

In using the dimensions from this detail in Figure 7.95, we have been able to determine that the plates, no matter what thickness they are, will be 4 1/2" wide for a single line of bolts. We know we are to allow 2" from the top edge of the plate to the first hole, then 4" centers for every hole after that, and then 2" again from the last hole to the bottom edge of the plate.

By understanding how to use this information to calculate the connection or shear plate length, we will be able to determine how long to make our connection plates—based on the number of holes for each connection. For example; if we want to calculate the connection for a W8 x 10 beam, we know this beam requires (2) bolts, each end. This means a total of four bolts per beam. Using the formula that this detail shows with a two-bolt connection, we allow 4" for the center of the bolts and two inches each end for edge distance. Adding those dimensions together, we have a total of 8". Therefore, the shear plates for W8 x 10 beams will be Plate 1/4" x 4 1/2" wide x 8" long with (2) each 7/8" diameter.

But how long are the bolts? We can get close to what the detailers might use by calculating the length of the bolt needed to accommodate the thickness of the shear plate (1/4") and the thickness of the web of the beam being used—3/16" for a W 8 x 10. Add these together, then add 1 1/2" for grip, which will accommodate the nut and washer as well as some thread length. You now have 1-15/16". Therefore, we would use a 2" long bolt. Some engineers will specify a length, others leave it up to the detailers to calculate. But it is good to get to a close length if you have to for pricing purposes.

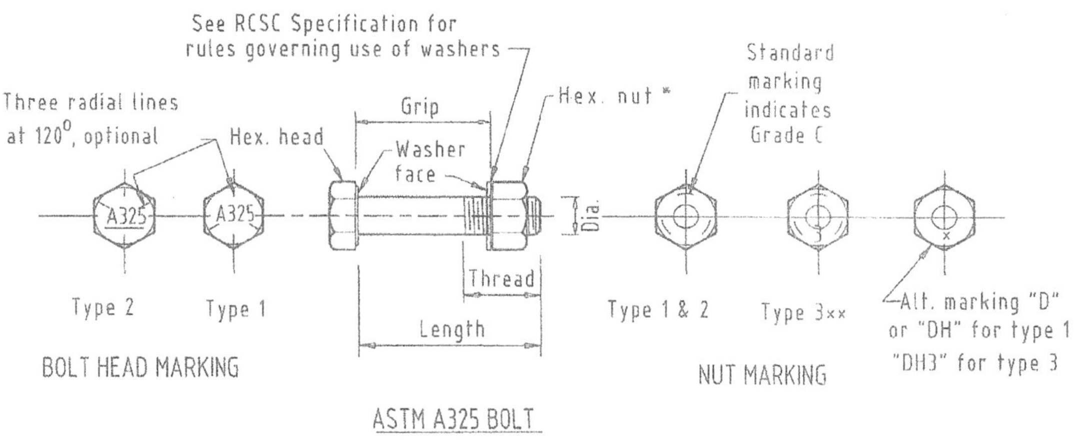

Figure 7.96 Standard ASTM A325 Bolt

Figure 7.96 is a picture of a standard ASTM A325 bolt. These bolts are available in a variety of lengths and sizes, although in some conditions they may need to be special ordered for some sizes. These can also be ordered galvanized.

Note 3 in Figure 7.97 shows the bolting and connection information. These general notes will apply to all the steel connections.

The following list explains the connection notes that were indicated in Figure 7.97.

CONNECTION NOTES:

1. ALL BOLTED CONNECTIONS TO BE TYPE N WITH FULLY PRETENSIONED ASTM A325-N BOLTS PER AISC STANDARDS EXCEPT AS DESCRIBED IN NOTES 2 AND 7.
2. BOLTS IN BEAM TO BEAM CONNECTIONS MAY BE TIGHTENED TO AISC "SNUG TIGHT" CONDITION UPON APPROVAL OF ENGINEER AND OWNER.
3. CONNECTION PLATES TO HAVE AISC STANDARD ROUND HOLES EXCEPT AS NOTED OTHERWISE.
4. ALL CONNECTION PLATE MATERIAL TO BE A36 UNLESS NOTED OTHERWISE.
5. BEAM CONNECTIONS TO BE PER THE STANDARD BOLTED BEAM CONNECTION DETAIL UNLESS NOTED OTHERWISE.
6. [5] SHOWN ON PLANS INDICATES A DIFFERENT NUMBER OF 7/8"ø BOLTS REQUIRED USING THE STANDARD CONNECTION DETAIL.
7. ALTERNATE CONNECTION DETAILS MAY BE SUBMITTED TO THE ENGINEER FOR REVIEW AND SHALL BE ACCOMPANIED BY CALCULATIONS BEARING THE SEAL AND SIGNATURE OF THE WASHINGTON STATE STRUCTURAL ENGINEER WHO IS RESPONSIBLE FOR THE DESIGN. ALTERNATE CONNECTIONS SHALL HAVE EQUAL OR GREATER CAPACITY THAN THE CONNECTIONS SHOWN ON THE DRAWINGS.

BOLTED BEAM CONNECTION NOTES 3/–

SCALE: NTS

Figure 7.97 Bolted beam connection notes

- A325-N is an ASTM grade for high strength bolts. There are other grades, and of the high strength category, but this category is the least expensive of the high grade bolts available. A490 and grade 5 bolts are examples of higher grade bolts.

- This note for tightening is one of the easiest for the field to achieve; there are some conditions where fastener tightness may be required to be calibrated.

- Standard round holes are generally 1/16" larger than the diameter of the bolt. In some conditions, the use of slotted holes is allowed and will be indicated as such, but for the most part they are not allowed.

- All connection plates may be A36. There are conditions where high strength plate?A572 and A998?may be required, and will be noted as such.

- This note directs us to use the standard bolted beam connections at all conditions unless special cuts are shown at the drawings to indicate otherwise.

- This note shows how the connections that are different than required in the Standard Bolted Connection details will be called out at Plan.

- This note lets you know that if you want to design your own connections, you will have to hire an engineer to design them, have the engineer stamp the drawing of the proposed detail drawing, and submit that proposed detail drawing for approval to the engineer of record. This cost is born by the fabricator unless previously agreed otherwise.

The picture on the right at detail (1) for the Standard Bolted Beam Connection in Figure 7.95 shows a beam framing into another beam. Figure 7.98 shows another sketch of that same condition:

In Figure 7.98, we can see the beam coming at us on the page. The beam that is framing into it will require a section cut out of the top in order to allow for clearance to make the framing work—this is called a beam "cope." In some conditions where the beams are framing together, the beams may need to be coped on top and bottom to clear the flanges of the beam they are framing into. Another condition where coping may be required is where beams frame into the column web for a column that is narrower than the beam is wide. Then all four corners of the top and bottom flanges may need to be cut back to allow for clearance.

The clearance or gap between the beams is shown as 1" in the standard connection details. We know this because the Standard Bolted Connection detail in Figure 7.95 shows the shear plate as having 1 3/4" either side of the holes, leaving 1" to the left of the 1 3/4" dimension to the point where the shear plate welds to the beam.

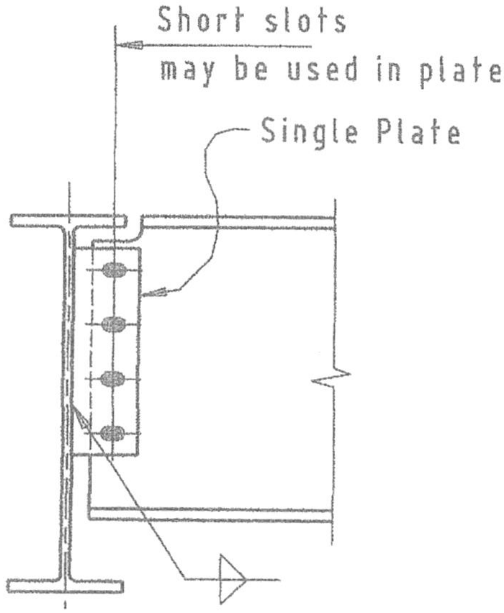

Figure 7.98
Typical beam to beam connection

The detail in Figure 7.99 shows the beams that frame into the column web. In this condition, the shear plates are longer, allowing the beam to stop at the face of the column flanges, which you can see in both the plan and elevation views here. The engineer has designated that stiffeners be added in between the column flanges; you can see them welded in there. Schedule F shows us the plate thickness that is required based on the number of bolts that the connection requires (which is based on the size of the beam at the Standard Bolted Connection Schedule). In using Schedule F together with the Standard Bolted Connection Schedule, the stiffeners at the column using 3/8" plate are for all beams sized from W8 up to and including W18. The 1/2" stiffener plates are used for all W21 and larger.

According to this detail, the weld sizes would then come from the Standard Bolted Connection Schedule *A*. In the plan view for this detail, the engineer has given the option both for clipping corners of the stiffener plate (see how the plate is drawn on the right side of the column) and for using short-width plates (that is how the plate on the left side of the column appears). However, the note for the short width plate says it can be used only at columns that have a *bf* or bottom flange of 14" wide or greater. Here the engineer recognizes that the welder will have a hard time getting in to weld if the space is too tight.

Because architects and engineers have not worked in the shop or out in the field, working space needs for the shop fabricators and field erectors are some-

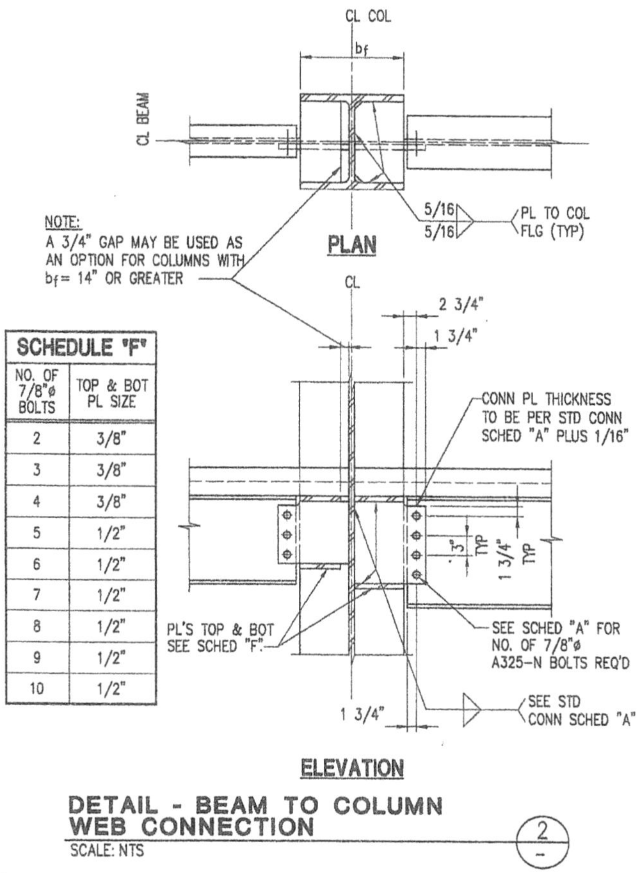

SCHEDULE "F"	
NO. OF 7/8"ø BOLTS	TOP & BOT PL SIZE
2	3/8"
3	3/8"
4	3/8"
5	1/2"
6	1/2"
7	1/2"
8	1/2"
9	1/2"
10	1/2"

Figure 7.99 Detail: Beam to column web connection

times not fully considered. On occasion, designs could occur that may be too tight or too small for a fabricator or the erector to perform the required work. Steel detailers will address these conditions at the time the project is being detailed. You may be able to see this during the time when you prepare your estimate, but more often than not experienced steel detailers will see it and send out an RFI on such conditions in order to get them modified or changed. Fabricators cannot afford to move forward with fabrication that is not practical for the shop, nor create a condition that is not workable for the field.

Gaining knowledge about working space and handling will come as you continue your work in the industry. There are so many trade-specific conditions and complications—the best way to learn them is as they come up. Keep your eyes and ears open for such things and more will be revealed as you develop your career in steel fabrication.

Figure 7.100 shows the details of the steel beam's connection to an HSS or tube steel column. The same standard connection schedule for shear plate and bolts is applicable here. The condition that is different is the use of the "knife plate" in detail 4. The fabricator has to slot the column in the center to slide that plate through and then weld in place at the shop.

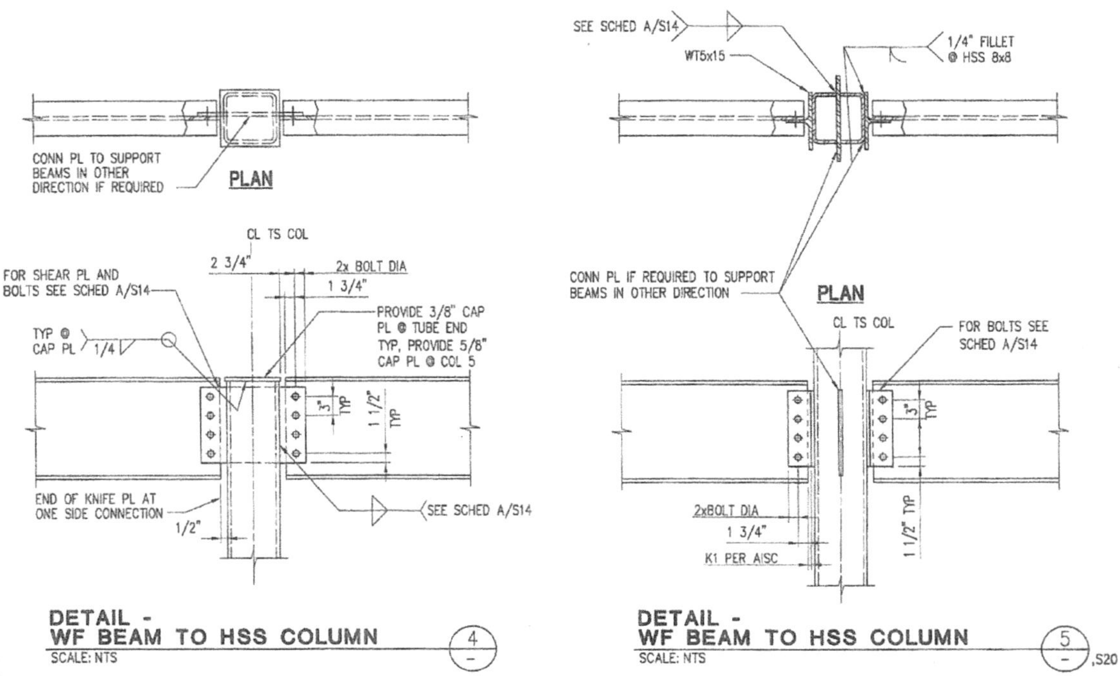

Figure 7.100 Wide flange beams to tube steel column connections

Notice here that the bolt spacing and the shear plates are a different size. This detail probably has been borrowed from a different set of plans and was not converted or corrected to include the specifics with the use of 7/8" diameter bolts (this detail is for the use of 3/4" bolts). We know this because of the dimension callouts for edge distance and the center location of the holes. The 3/4" bolts require a 1 1/2" edge distance. Therefore, the 3" center of bolts together with the dimension for the 1 1/2" edge distance is a match with requirements for the use of 3/4" bolts.

While going through these details, it will be necessary on occasion to do some detective work. Determine what is required to put this building together out of what the engineer does show. If we know information is incorrect or inaccurate, or if we are just not sure, then we can always qualify what we think will be true or used in our bid letter Inclusions, Exclusions, and Qualifications.

The detail in Figure 7.101 shows the tube steel columns and beams to be field welded. This field welding beams to columns does happen on occasion; will be called out specifically in the plan views. This condition doesn't show the need for erection aids—added angle clips or plates to the columns for the tube steel beams to sit on while the field hands weld them in place. The requirement is noted in the general notes and is shown in Figure 7.4. If the requirement is not found as used in Figure 7.4, then a reference to the note on providing erection aids should be excluded. Now that a condition has been located, the application of this note will be included with the information for this bid item in the scope letter.

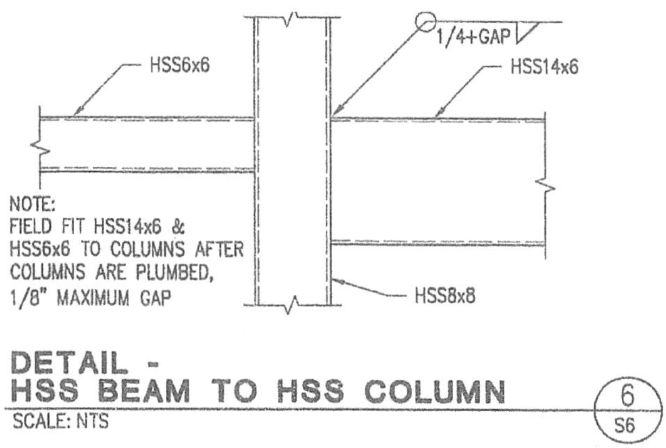

Figure 7.101 Detail: HSS beam to HSS column

The detail in Figure 7.102 on typical slab opening framing will probably not be called out in the plan views. You will find these typical details in the sheets beyond the foundation and floor plans.

You need to return to the floor and roof plans to identify and count the different size openings, and then list them to be able to determine which detail will apply for which size materials to use.

We offer further explanation of the notes in Figure 7.102:

1. This note is self explanatory: we will not need to pay attention to any openings 9" and smaller.

2. This note refers to openings between 9" and 2'-0" without edge framing, which is the edge form we have discussed in previous pages. These openings will be reinforced according to Type A

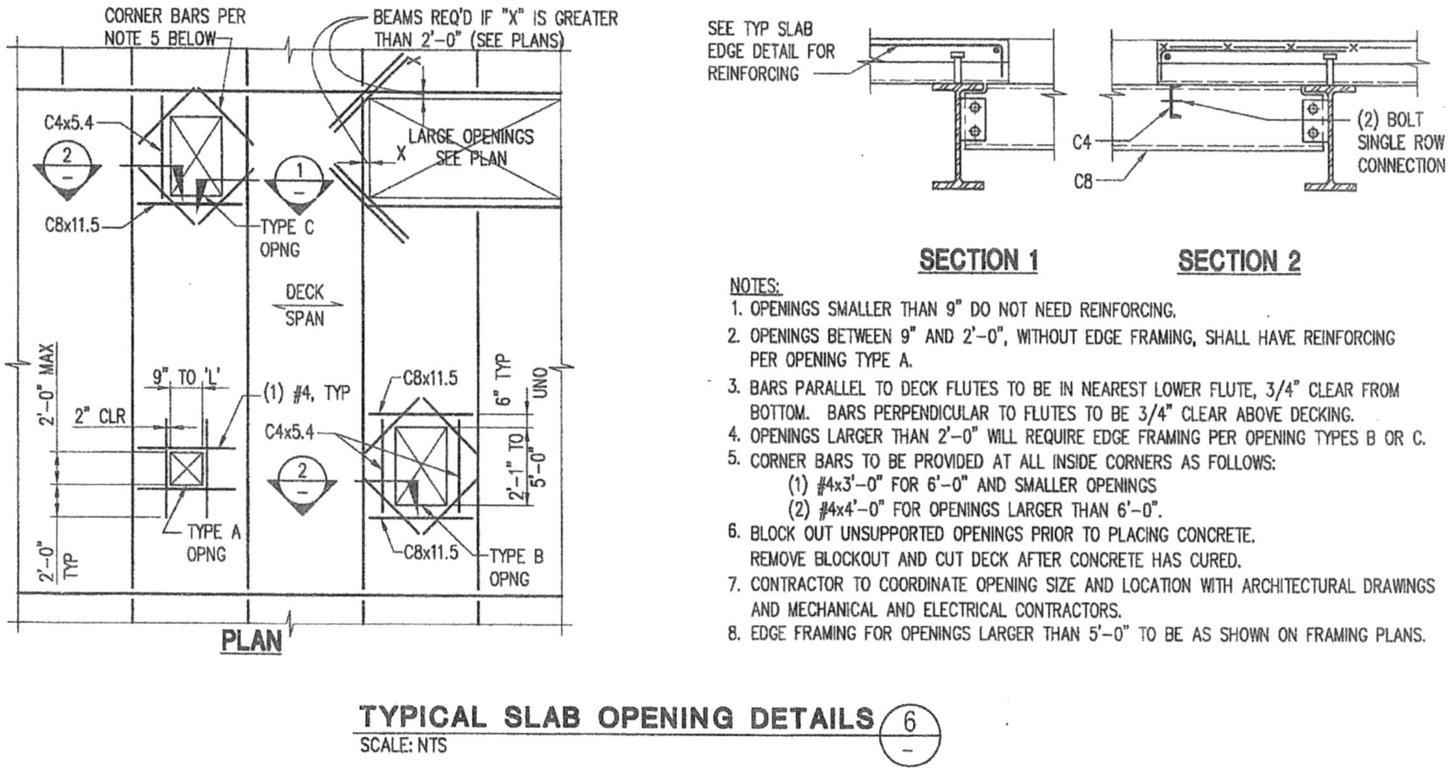

Figure 7.102 Typical slab opening details

openings, which have #4 rebar. Because #4 rebar is not supplied by the steel fabricator, it will need to be listed in your bid letter's exclusions.

3. This note talks about the placement of reinforcing bars; it is not a concern for steel fabricators.

4. This note tells us we need to use framing for opening types B or C for all openings larger than 2'-0". Type B is close to a steel beam. It only needs the channel on one side because the beam will provide the support for the opening on the other side. Type C is to be used for all mid-floor span style openings.

5. This note covers the rebar in the concrete.

6. This note tells the concrete contractor to block the opening.

7. Note 7 advises the contractor to coordinate openings with the architectural, mechanical, and electrical drawings. This note refers to you as the steel contractor. You may choose to list typical opening frames in the exclusions, or to qualify your quantity in your inclusions where you list this detail and the material sizes to be used.

8. Note 8 indicates that all openings larger than 5'-0" will be shown in the plan.

In going through the detail sheets for projects, you may often find conflicting or additional information that may need to be applied in similar conditions. We look at these drawings with the goal of creating an estimate for the steel fabrication. We are concerned with both accuracy and economy. When we do have a conflict, it may be our choice to use the less expensive method, then qualify that use in our bid letter inclusion for that specific item.

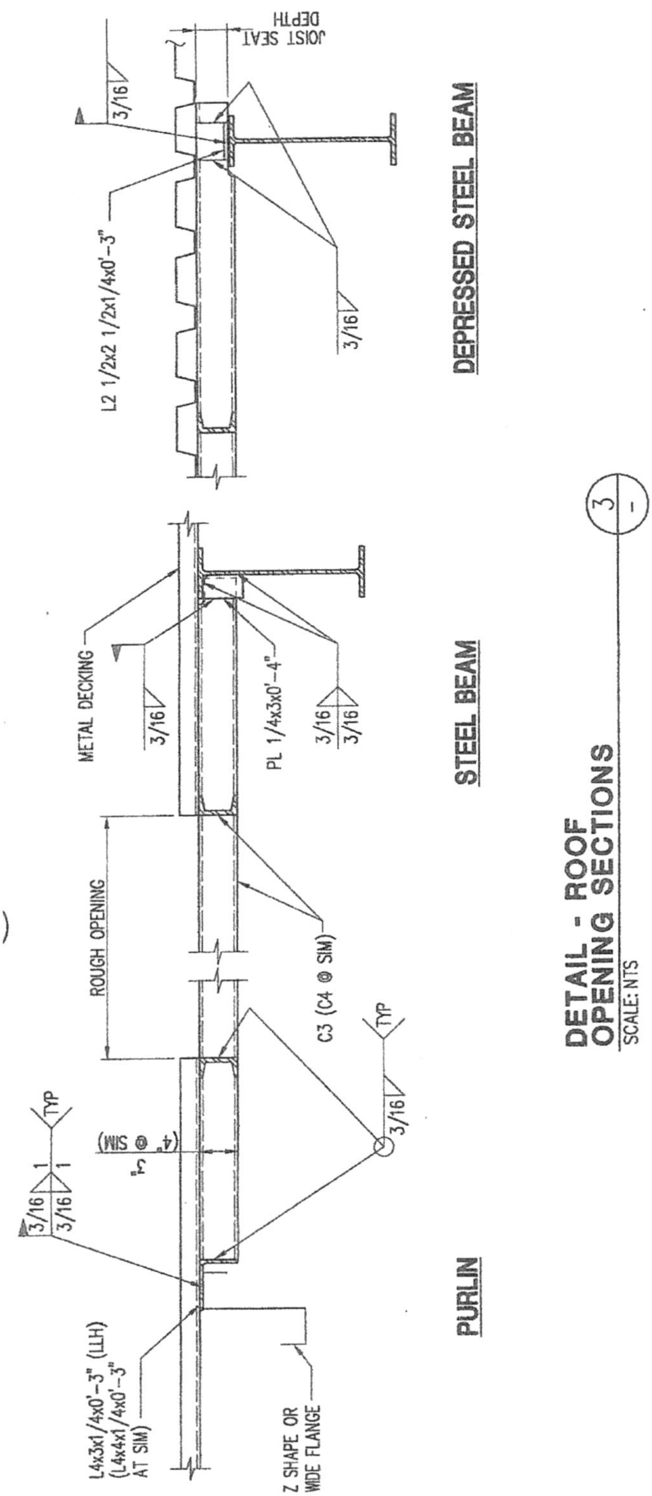

Figure 7.103 Detail: Roof opening sections

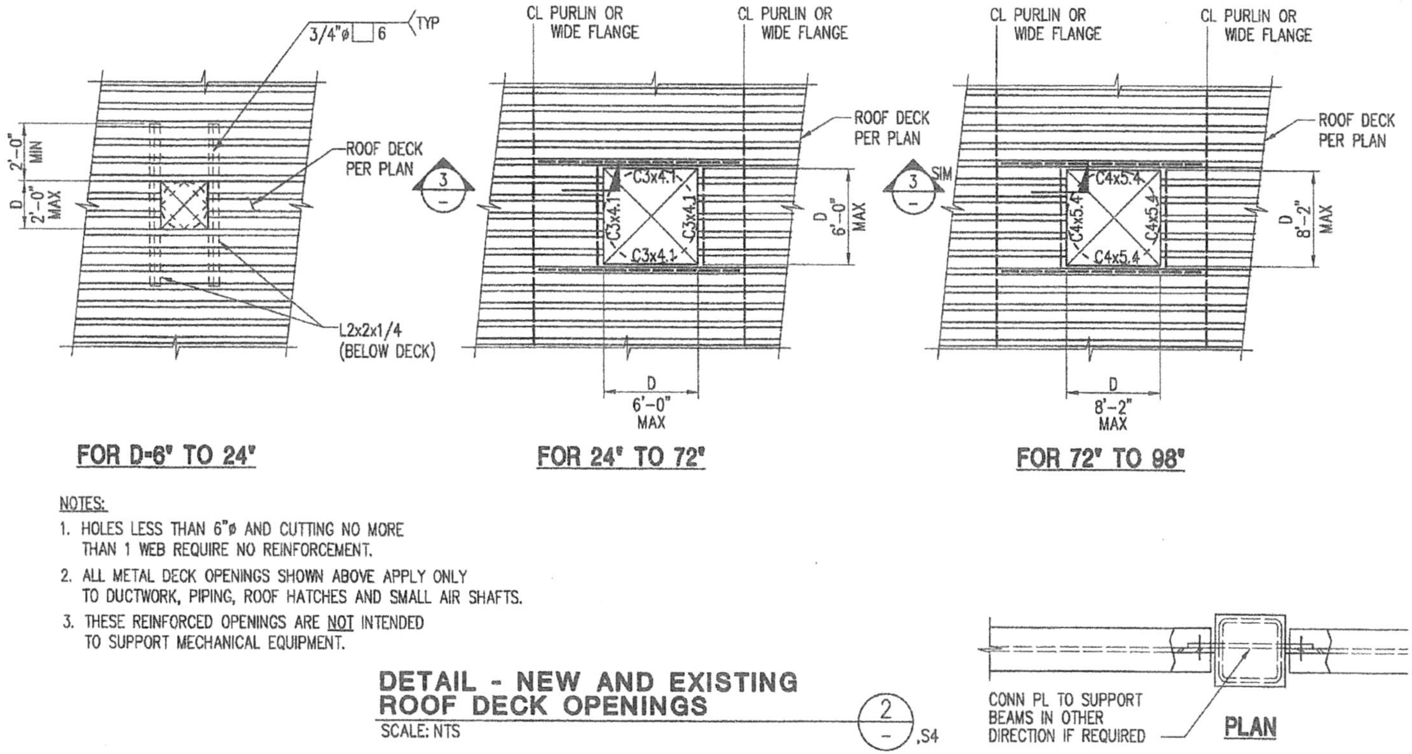

Figure 7.104 Detail: New and existing roof deck openings

This elevation view of the roof deck openings in Figure 7.103 shows framing that is not supplied by the steel fabricator. The *Z* shape shown at the left of this detail is by the stud framing suppliers that provide the "cold formed metal framing." The channels C3 and C4, the L 4 x 3 x 1/4, and the L 4 x 4 x 1/4 clips, as well as the plate 1/4 x 3 x 4" connections, will be items to be listed in the steel estimator's take off. The plan view in Figure 7.104 shows this same framing and the size of openings to which the framing applies.

Figure 7.104 shows a typical detail of a roof deck opening. It is not to be confused with the floor opening details. This area will have no concrete on top of the roof decking. Therefore, the framing material is smaller because it will not be supporting the additional weight.

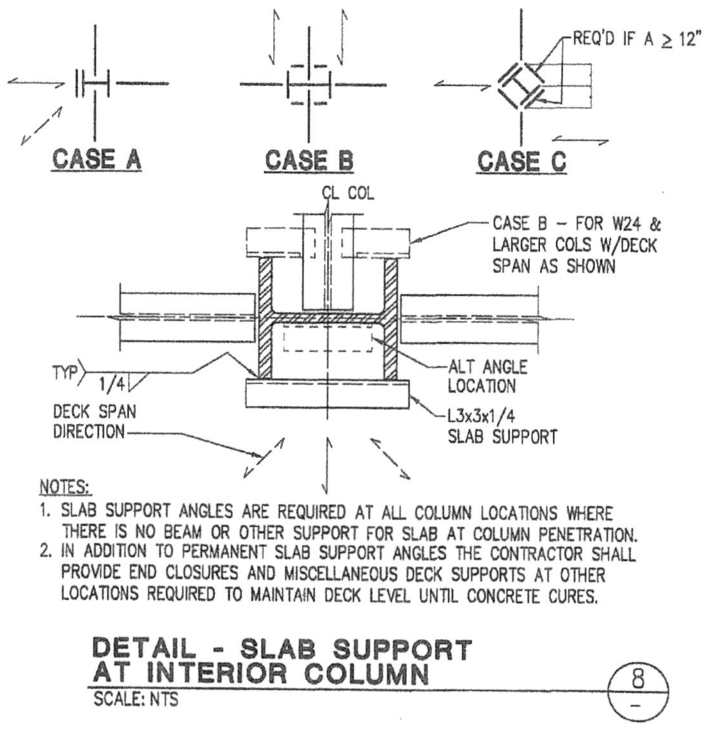

Figure 7.105 Detail: Slab support at interior column

The detail in Figure 7.105 shows how the engineer expects the slab to be supported by the columns that continue up to the next level. This material is shop welded to the column. It will need to be included at the take off or material list of the steel.

Notice that the angle is split in the condition where the beam frames into the column web. The angle spans the entire width of the column in the condition where there are no framing beams. No angle will be required on the flange sides of the beam, where the beam supports the deck.

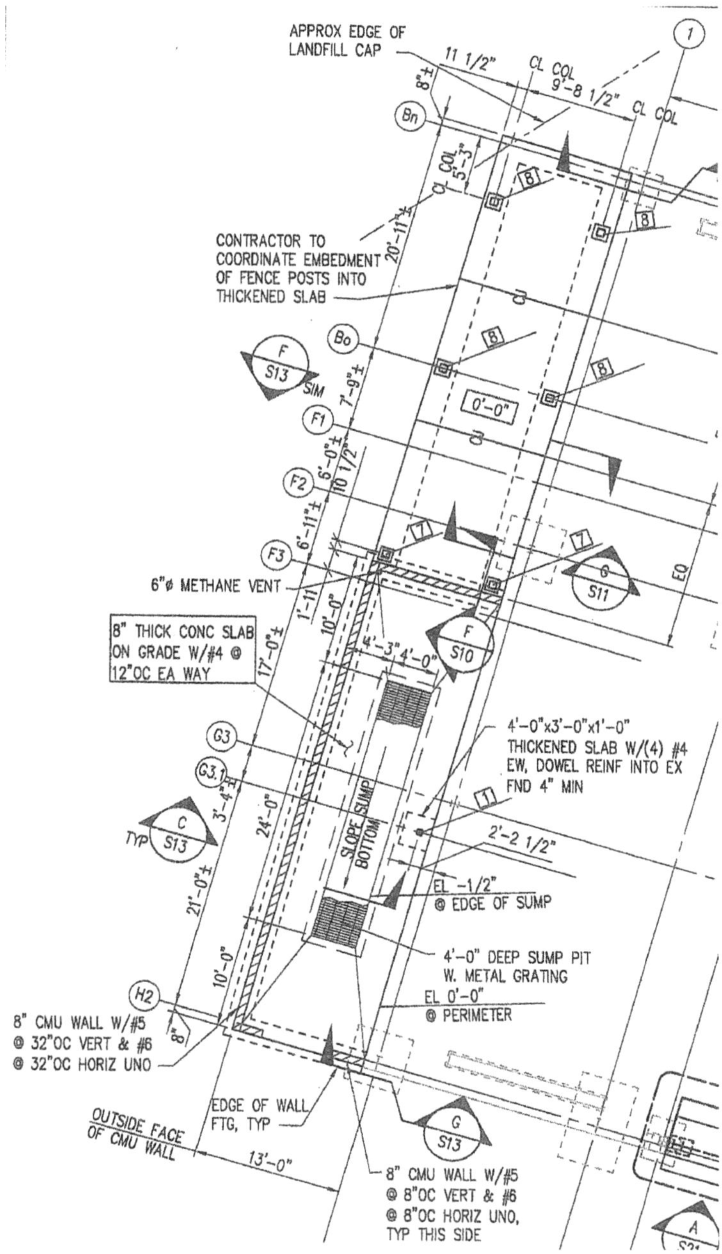

Figure 7.106 Foundation plan sump pit framing

Figure 7.106 shows the plan view of a partial building foundation area. We see a 4'-0" wide by 24'-0" long sump between grid F3 and H3 with section C/S13 cut through the middle of it. The embed steel angle and the steel grating in this location is supplied by the steel fabricator as part of the Miscellaneous Metal section 055000 scope of work.

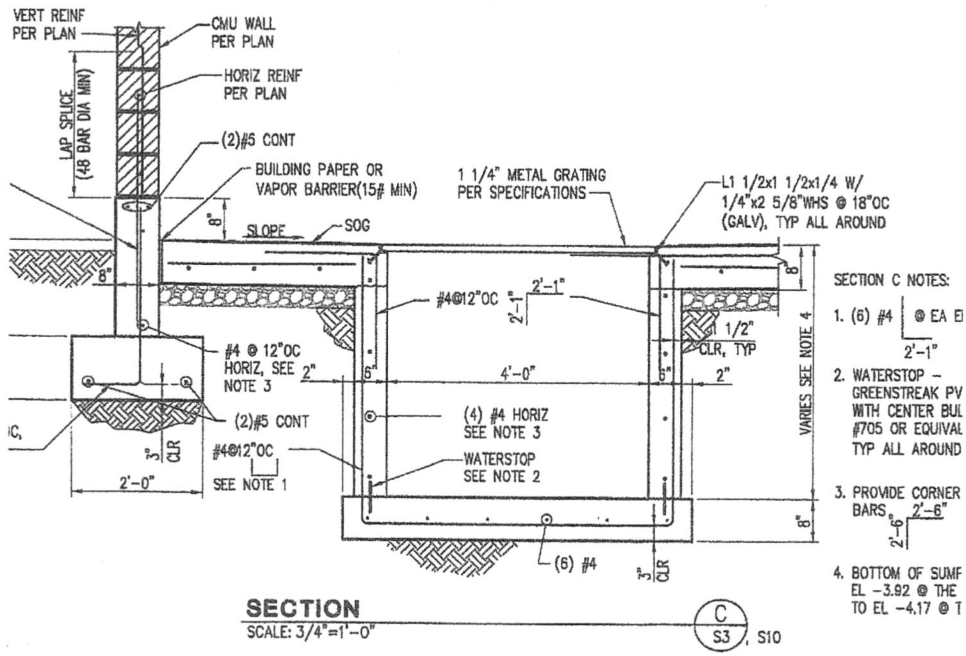

Figure 7.107 Elevation view of Sump Pit

The section C/S13 in Figure 7.107 shows the 1 1/4" metal grating per specifications shown above the pit. We find the embeds L 1 1/2 x 1 1/2 x 1/4 angle with WHS or HAS (same thing) that are 1/4" diameter x 2 5/8" long and which are welded in place at 18" cc (18" centers over the span of the 24'-0" of the angle) galv. (galvanized) typical all around.

This embed angle will be fabricated to make a frame, much like a picture frame, where the ends of the angle will actually run past the face of the opening to miter into the corner. The angle then acts as a shelf for the grating. The grating will actually be provided a bit narrower and shorter than the opening because there needs to be an allowance for clearance to set nicely on top of that angle.

The engineer does not indicate any mid-span supports, so a conversation with your grating supplier will be necessary to verify that the 4'-0" span will work with the 1 1/4" grating. Because the embed angle is galvanized, it is likely that the grating will also need to be galvanized. Verify that point in the project specifications.

There may be additional support required at 4'-0" centers over the length of the 24'-0" span. The designers don't show this support in Figure 7.107's detail. However, more information may be provided in the architectural drawings that will further clarify the required conditions.

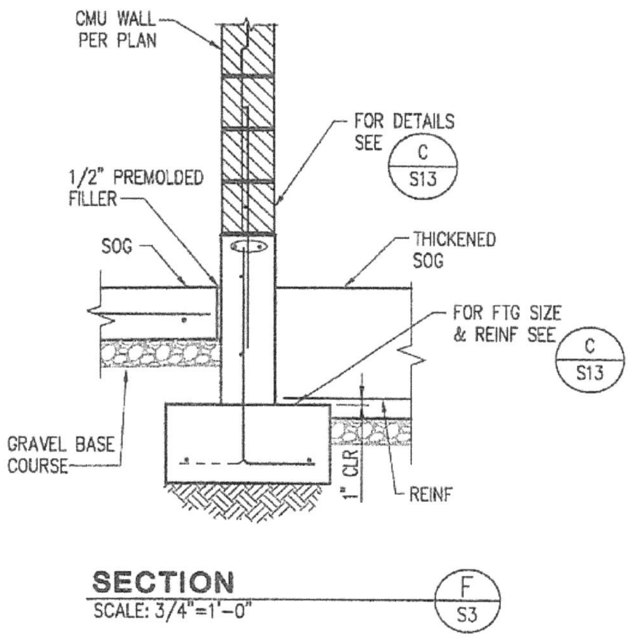

Figure 7.108 Section cut through wall

Looking back at the small foundation plan area in Figure 7.106, we see the callout for detail F/S10 in Figure 7.108 just above where the sump pit is. Although the detail shown in Figure 7.108 does not show us any new steel for the fabricator, it does show a section through the wall. You can see the block on top of the foundation walls. This detail also refers to C/S13 for more concrete information, which is the same detail where we found our sump pit (see Figure 7.107).

We see section F/S13 in Figure 1.109, cut just above grid F1 in the small foundation plan Figure 7.106. This shows the columns as they sit on top of slab. The reference for *Col & BP per sched, typ.* is intended to send us back to the typical details for column sizes and base plates. Notice how the grid line references (1) and (F). This detail is intended to show the columns in both directions—there is a break in the detail picture and in the dimension lines allowing the designer to show both directions in the same detail.

Detail G/S13 in Figure 7.110 is taken from the bottom of the foundation plan in Figure 7.103. It shows more concrete information. In the area above the foundation through the wall at grid (1) to the right, we see a reference to *EX MTL BLDG COL BEYOND*. This reference tells us *existing metal building column beyond,* identifying existing conditions that are not the concern of the steel fabricator.

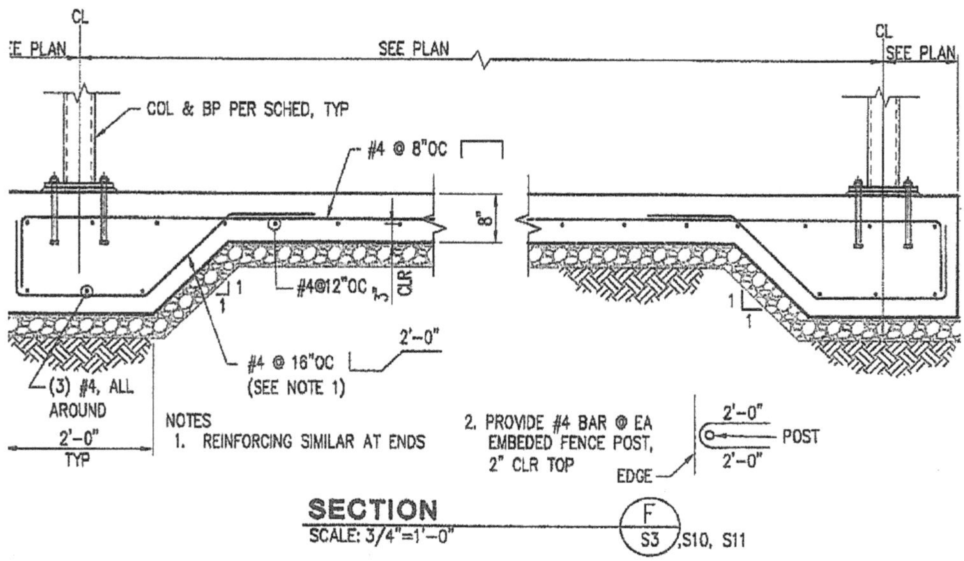

Figure 7.109 Elevation cut at foundation shows columns

Figure 7.111 shows the roof plan that is above the foundation plan that Figure 7.106 just reviewed for the sump pit. We see the beams W8 x 18, W10 x 26, and the bracing WT4 x 9 in this view.

Section A/S20 in Figure 7.112, cut through the roof in between grids G3.1 and H2 in Figure 7.111, shows the W10 x 26 coming at us on the right, and the elevation of the nearest W8 x 18 that frames into the W10 x 26. The framing above the W8 x 18 and the W10 x 26 is for the cold formed metal framing trade; it is not the concern of steel fabricators. We see that the roof framing slopes down from the top of steel elevation at the high point from the right to the left over the top of a block wall. A steel base plate with anchor bolts is at the top of the wall. We need to look at section D in Figure 7.113 to see this material.

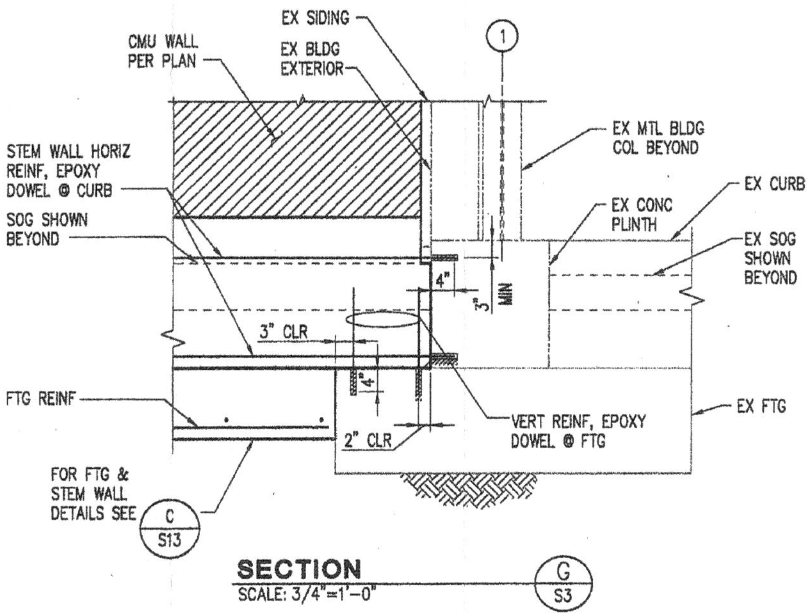

Figure 7.110 Concrete curb and CMU wall

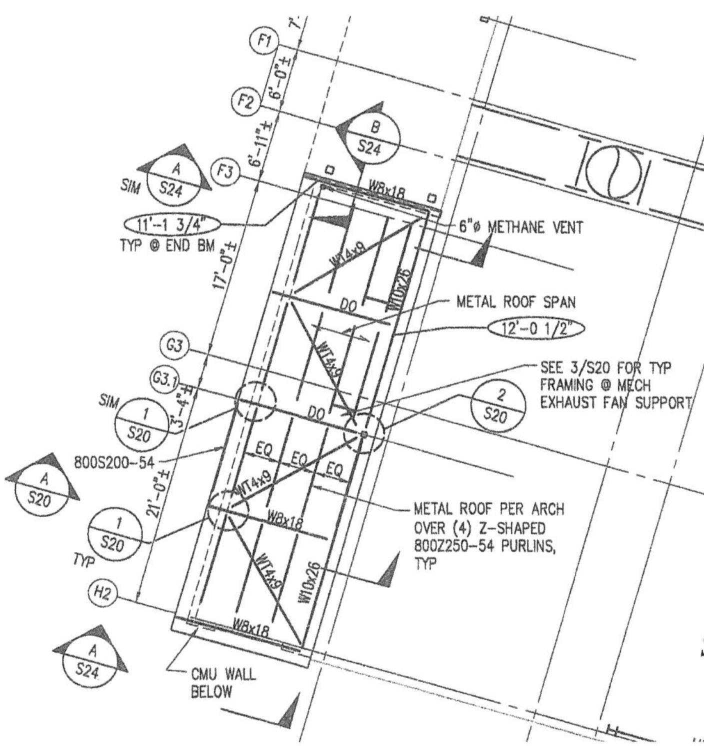

Figure 7.111 Steel roof framing over sump area

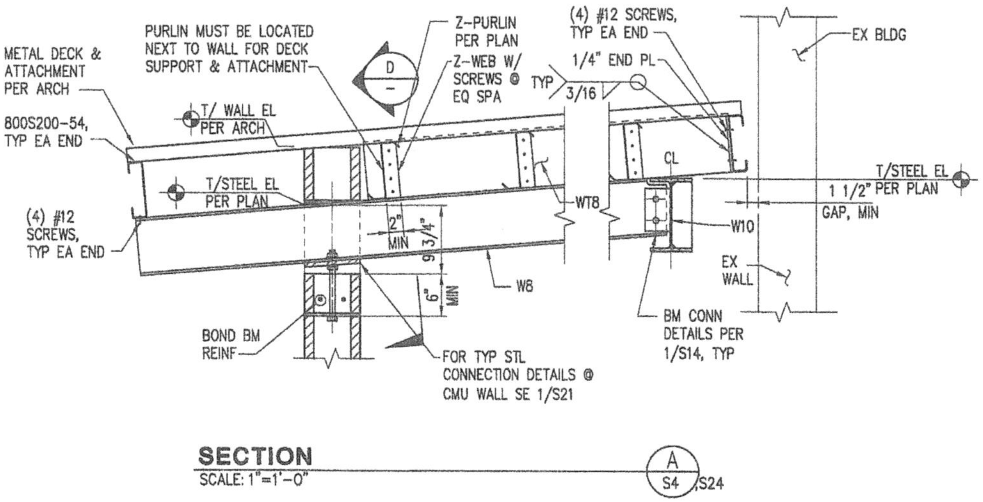

Figure 7.112 Elevation view through roof At sump area

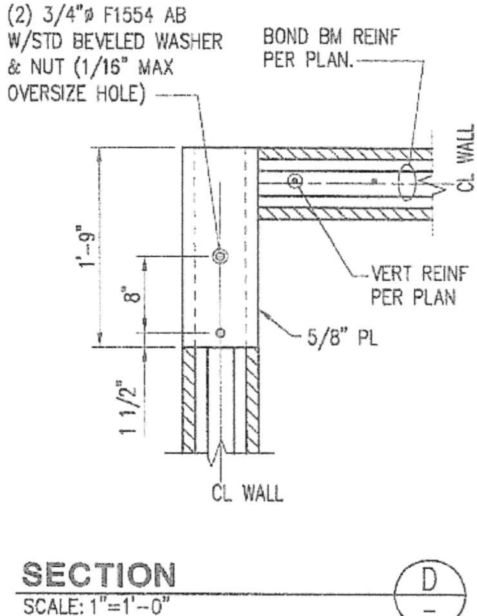

Figure 7.113 Plan view of seat plates and anchors at wall

Section D in Figure 7.113 shows the base plate 5/8" thick. The width will be the same as the block used for the wall and the length will be 1'-9". There are (2) anchor bolts, 3/4" diameter F1554 with standard bevel washers, and nuts. The holes for the anchor bolts are over size by 1/16", making them 1/8" over the diameter of the 3/4" bolt or 7/8" diameter.

This detail 1/S20 in Figure 7.114 is cut from the beam intersection at grid G3.1 in Figure 7.108, and again at midspan between grid G3.1 and H2. It is used for all conditions where the beam goes through the wall. This detail shows the connection of the bracing to the W8 x 18 beam and how the seat in the wall is to be fabricated. Section A in Figure 7.112 shows where the beam W8 x 18 goes through that wall. Centerline with the wall, we can see the 1/4" stiffener plates welded to the W8 x 18 at the point that is centerline with the wall.

We can see the seat plate that is Pl 5/8 x wall width x 11" and the anchor bolts (2) 3/4" AB w/STD bevel washers and nut set at 1 1/2" in from the edge of the 5/8" plate. We see by the note that the engineer allows for the anchor bolts to be 1/6" maximum oversize. Those holes will be 1/8" larger than the

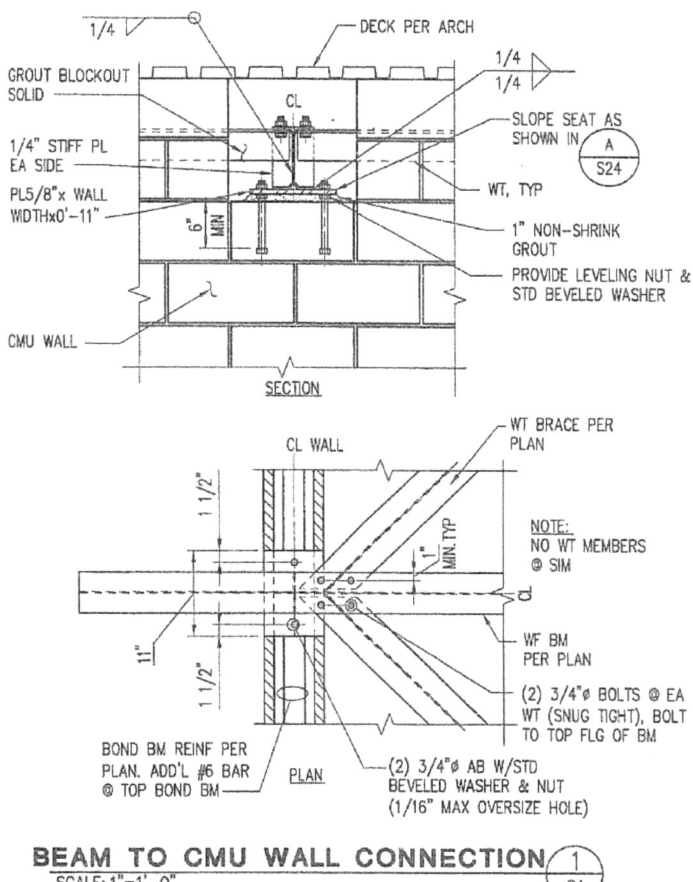

Figure 7.114 Beam to CMU wall connection

Figure 7.115 Bevel washers

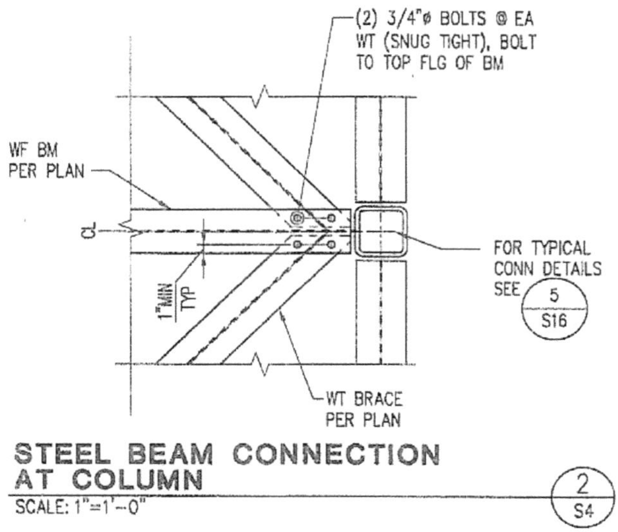

Figure 7.116 Steel beam connection at column

diameter of the bolt, which is typical for most anchor bolt holes in base plates. Bevel washers allow for the slope of the connection (see Figure 7.115).

The WT 4 x 9 bracing connection shown in Figure 7.116 will not be coped, but will bolt to the underside of the top flange of the W8 x 18 beam. There are two 3/4" bolts on each end of these braces to hold them into the W8 x 18 beams. This type of connection in Figure 7.116 is used at all locations where the bracing meets this type of intersection.

Section A/S24 in Figure 7.117 is cut from just below grid lines F3 and H2 in Figure 7.111. This detail references back to the elevation detail A/S20 in Figure 7.112 and the typical connection details that we have already reviewed. We can see that we need a stiffener plate 1/4" in the W8 x 18 at the centerline of the wall. Again, we see all the cold formed metal framing that is by others.

Section E in Figure 7.118 looks down at the top of the beams in Figure

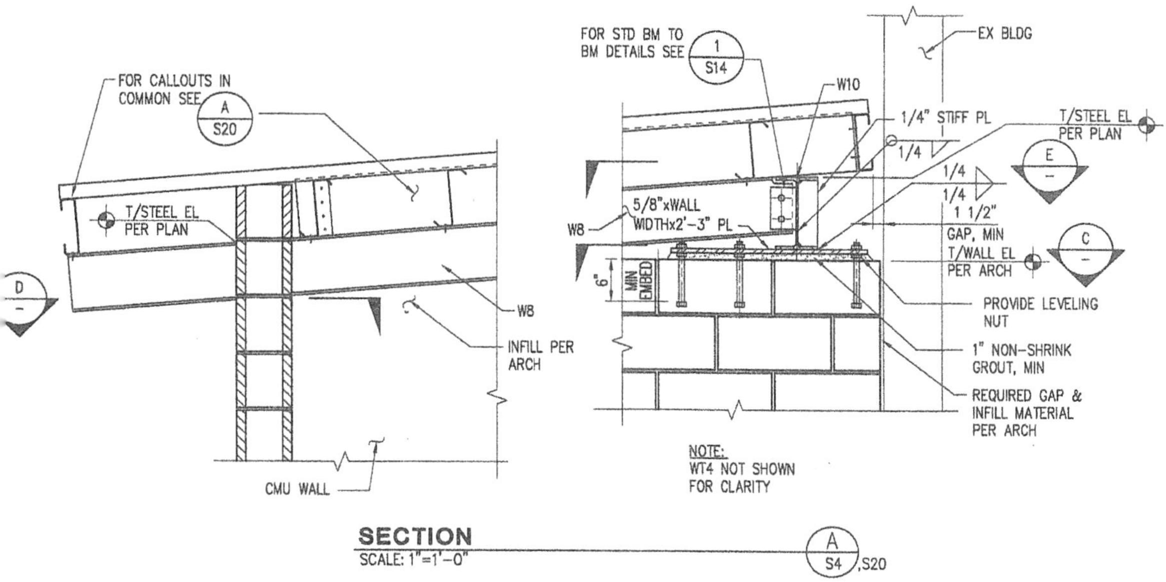

Figure 7.117 Top of wall elevation sections

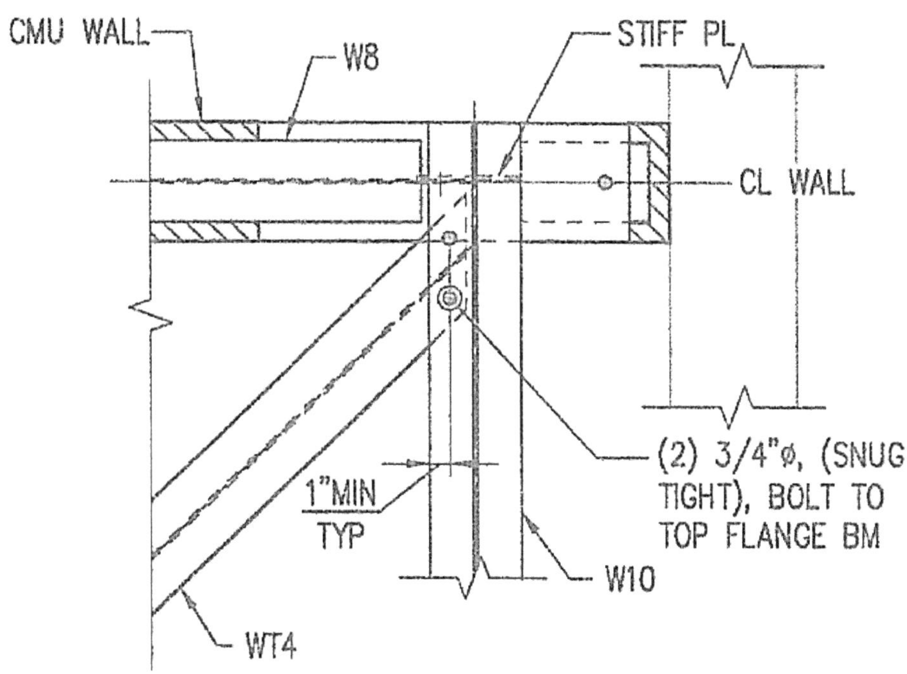

Figure 7.118 Beam and bracing intersection at wall

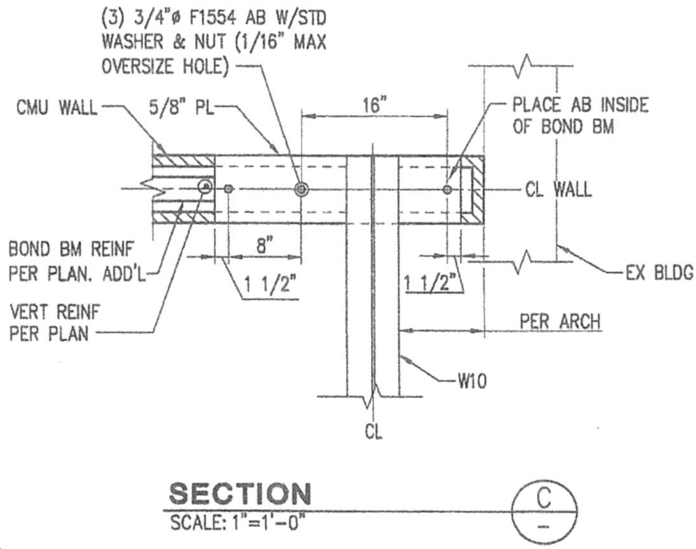

Figure 7.119 Beam seat at top of wall

7.117 that sit on top of the wall. We can see the bolted framing of the WT4 x 9 with the (2) 3/4" bolts. We need to go to detail C for the rest of the information to be shown.

At this detail C in Figure 7.119, cut just above the base plate from section A in Figure 7.117, we see the base plate at the top of the wall is Pl 5/8" thick. That measure will be the same width of the wall and a length that will accommodate all the dimensions shown to make everything fit. The 1 1/2" edge distance from either end, the 8" from the centerline of the first hole to the second one, then the 16" from the second hole to the last one—these all total 27" or 2'-3" long for this base plate. The bolts required are (3) 3/4" diameter F1554 (this is the material grade for the bolt) AB (anchor bolt) w/ STD (standard) washer and nut. Here the engineer has also noted the addition 1/16" oversize hole that is allowed. This detail should be used at all corner conditions that are similar.

This section in Figure 7.120 is cut through grid F3 at the outside wall of the roof area in Figure 7.96. This elevation view shows the block wall of the area we just reviewed. The column just to the right is the adjoining area that has a different framing requirement.

We can see the beam combination of the W8 x 18 and the WT8 x 13 on top of it. This is the style of the roof framing that covers the sump pit area.

We also see the same type of combination framing (the W8 x 18 with the

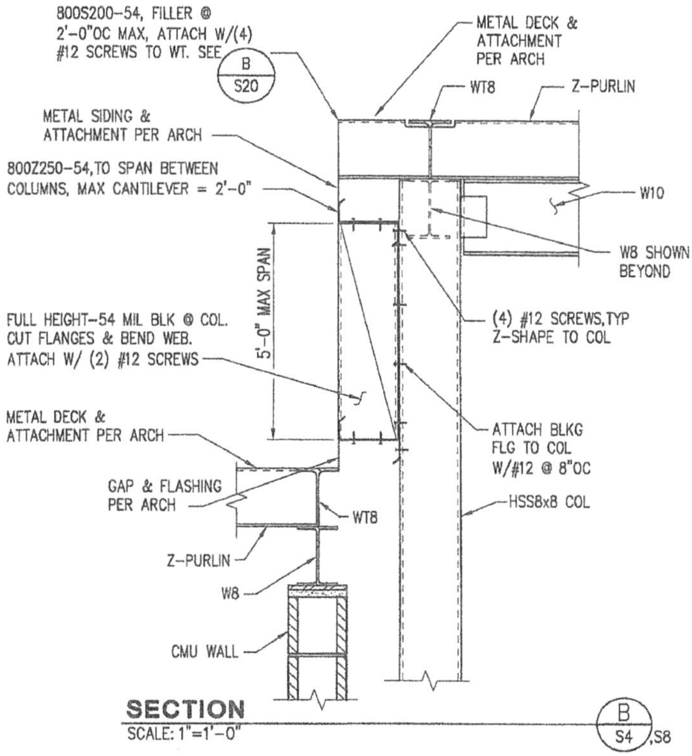

Figure 7.120 Elevation of beam to column connection

WT 8 x 13 welded on top) that connects into the TS 8 x 8 columns. The Z purlins and the light gauge metal framing are provided by others. They are not of concern to the steel fabricator.

At this partial roof plan in Figure 7.121, we see section B/S24 called out again. This detail is the same one that was called out in the roof plan that is covering the sump pit area. We see the framing, W10 x 26 and the W8 x 18.

Notice that the roof slopes down and away from grid line 1 towards the front of this roof framing area. We will look at the elevation section B/S20 to verify this slope. The W18 beams will probably be cut at a skew to accommodate this slope. The roof elevations called out in the roof plan will help steel detailers determine the degree of the slope required.

This section B in Figure 7.122, cut between grids BN and BO in Figure 7.118, shows the side view through the roof framing for this area. The WT 8 x 13 have 1/4" plate stiffeners at both ends, both sides for connection to the *Z* purlins. The W10 x 26 have stiffeners at the centerline locations of the W8 x 18 beam. The connections will be the same as required at the standard bolted

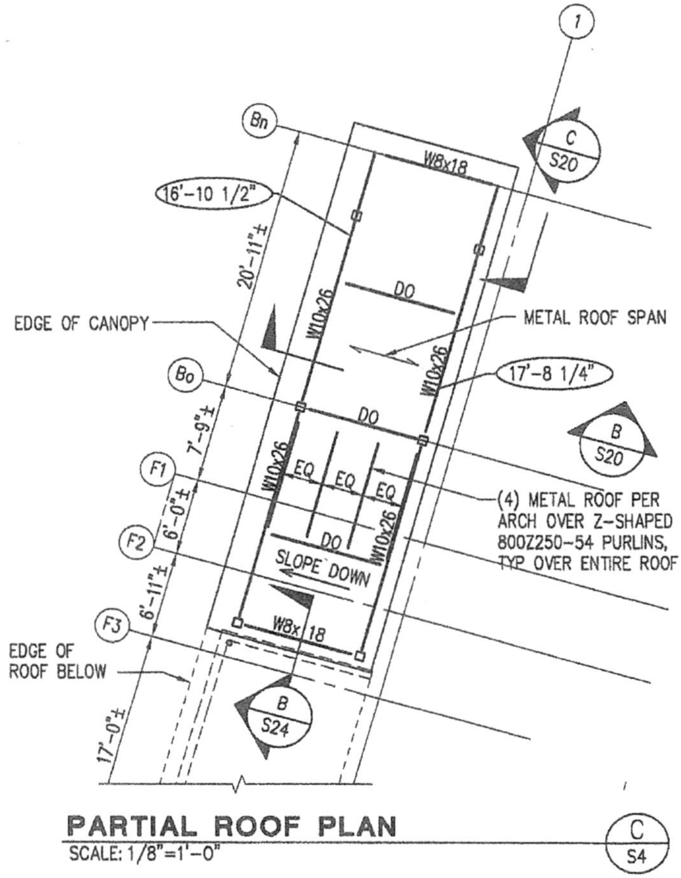

Figure 7.121 Partial roof plan

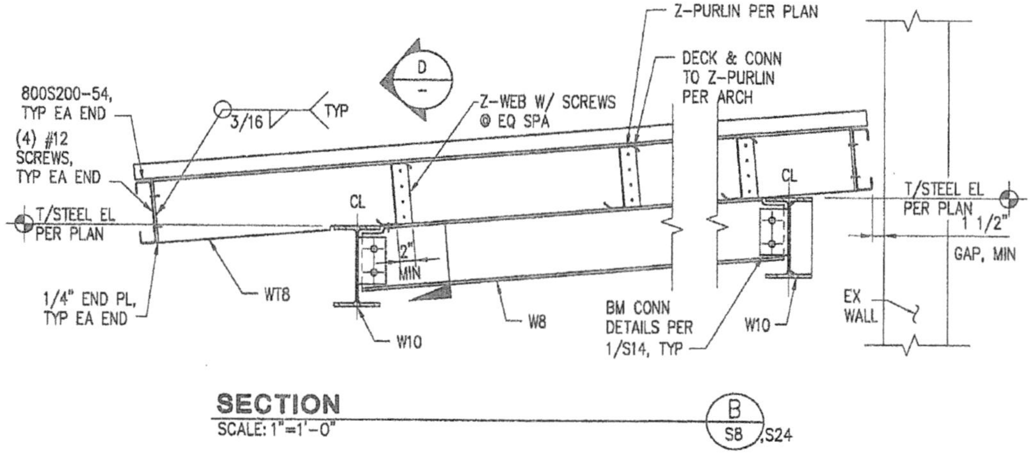

Figure 7.122 Roof elevation framing

connections. The roof slopes, with the high point being at the right of this elevation detail.

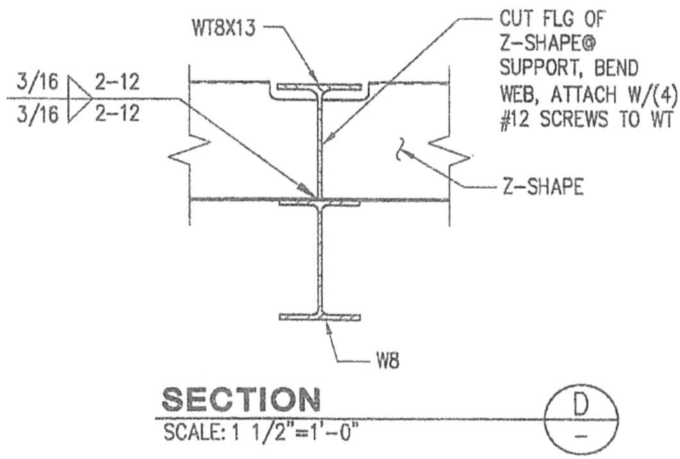

Figure 7.123 Detail condition for WT to Z shape framing

In this section D shown at Figure 7.123, we see there is a WT8 x 13 beam shop welded on top of the W8 x 18. The WT8 x 13 is not shown or called out in the plan. Therefore, this is the only detail where this material is shown and identified. The weld information, *3/16 2-12*, means that there is a 3/16" weld 2" long at 12" centers, also known as skip welding. We see that this weld should be done on both sides of the WT8 x 13. The Z framing has to cope into the steel, and those *Z* framing members will be attached in the field by others.

The section in Figure 7.124, cut from the roof plan at grid line 1 in Figure 7.121, shows us the top of the TS 8 x 8 columns with the cap plate and the connection to the beam above. There are 1/4" stiffener plates on both sides of the beam as it crosses over the top of the column at centerline; welded by using a 3/16" weld all around (this means both sides as well). The cap plate is called out here as a 5/8" bearing plate, 1/2" wider than the column or beam flange, whichever is wider. This plate will connect to the beam with four 5/8" bolts at 3" centers and located in the other direction at a location that follows the 1 1/2" edge distance rule.

When working on projects that are additions and remodels to existing buildings, it is important to carefully review the drawings for any new details that may have been added to the existing areas of the building. New steel items often may be added for seismic upgrades, new supports for equipment, any number of things that the owner may decide to add because it will be a con-

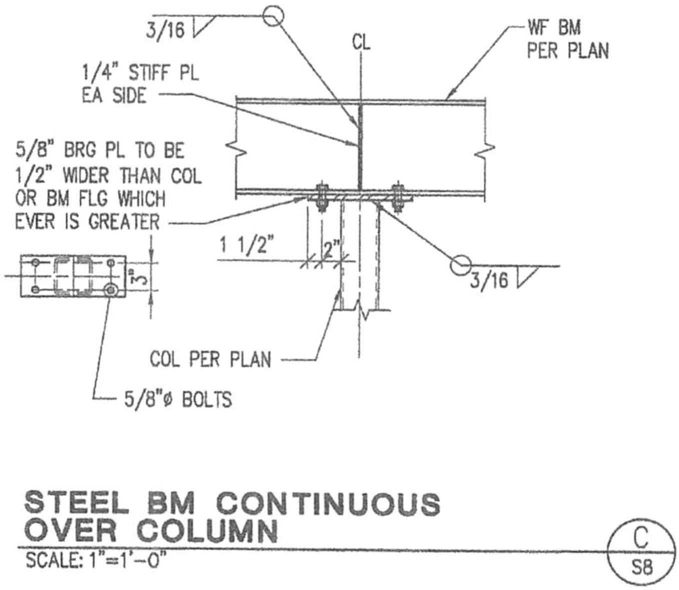

Figure 7.124 Steel Beam Continuous over column

venient time—perhaps they are mobilizing for new or additional construction anyway.

Moving forward on this premise—because the project we have been reviewing is a re-model—we do find that new framing has been added to the existing portion of the building.

While reviewing the roof plan for the existing area, we see new building section cuts shown between grids 3 and 7 in Figure 7.125.

Starting with section A/S23 in Figure 7.125, shown at the bottom of the picture along grid 6, let's look at the elevation detail that will show us what the dark lines indicate. These lines are noted as "service rack below, typ, (6) loc." See A/S20 in Figure 7.126 (The designers really mean A/S23, that A/S20 is a typo).

Here we see what the dark lines represented between grids H2 and G3, and also F1 and *Bn* in Figure 7.125. They are steel trusses and are to be provided by the steel fabricator. Material sizes are called out at the truss elevation between grids F1 and *Bn*. The columns are TS 8 x 4 x 1/4, the horizontal beams are TS 6 x 4 x 1/4, and the diagonal bracing is TS 3 x 3 x 1/4. The elevations to the right of grid line Bn provide the dimensioning we need to calculate the column height and the depth of the trusses. Detail 2 (Figure 7.127) on the left of this picture will give us the close up of the area.

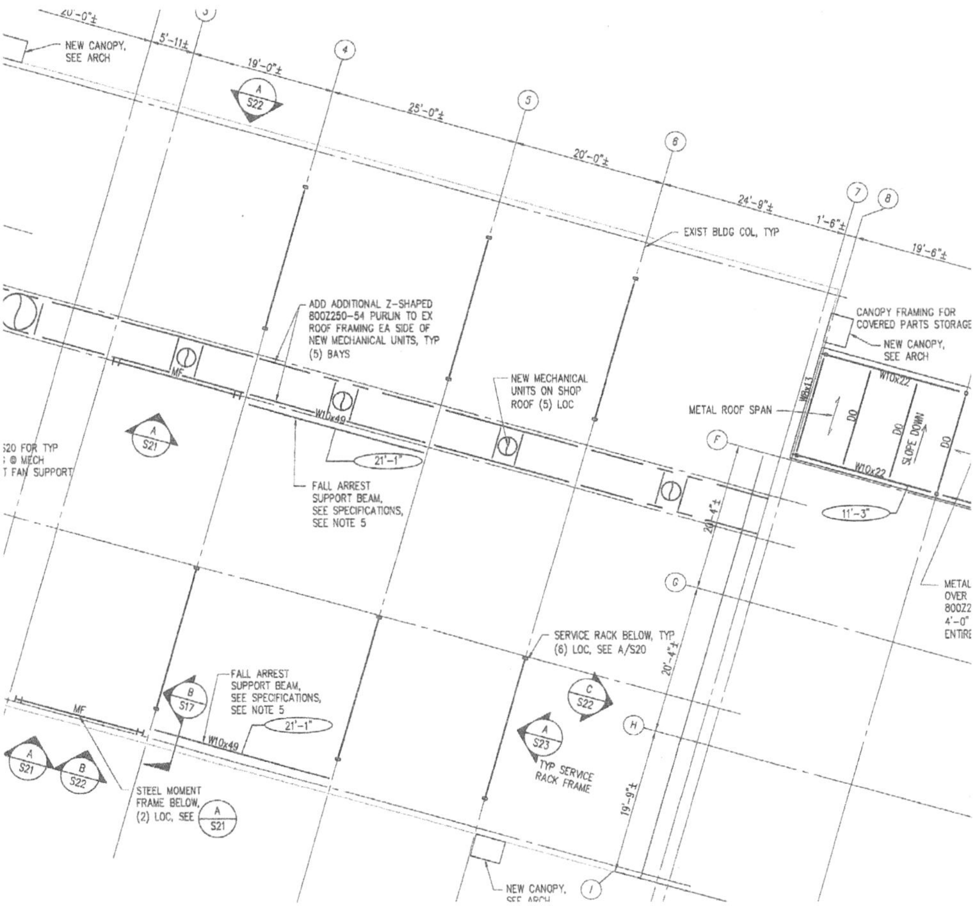

Figure 7.125 New framing at existing roof area

This elevation of the truss at detail 2 in Figure 7.127 shows us some partial material sizes and locates the C-strut channels that are to be provided by others (they are not a concern for the steel fabricator). The cap plate at the columns with the weld information, as well as the weld detail for the truss members and the spacing of the vertical bracing, is important. Detail 1 (Figure 7.128) cut from the truss on the right hand side of the drawing offers more information.

The elevation section in Figure 7.128 shows us that the welds for the diagonal bracing will be 1/4" fillet all around. The plate 1/2" x 13" x 3'-6" is set centerline of the trusses. The weld of the plate to the top of the truss is a "flare

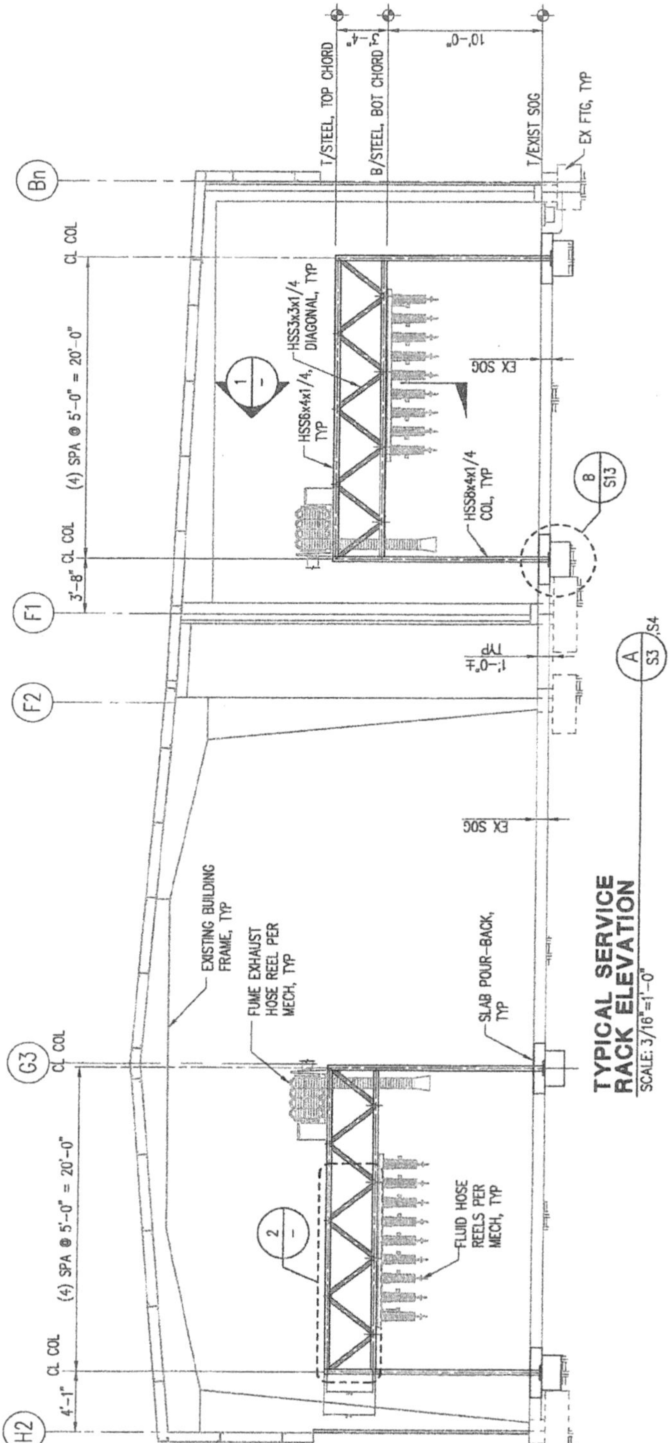

Figure 7.126 Building elevation showing typical service racks

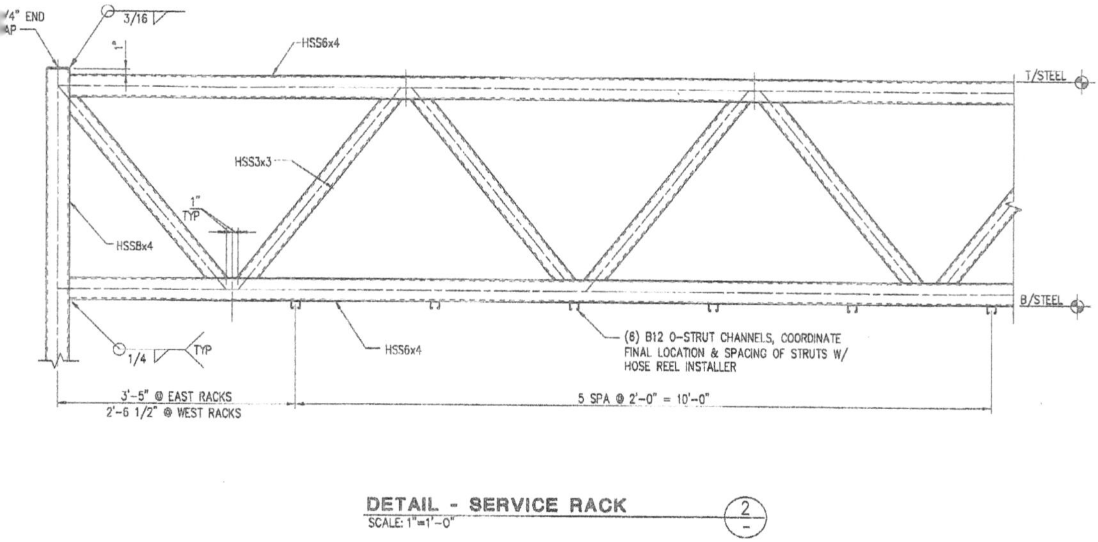

Figure 7.127 Detail: Service rack truss

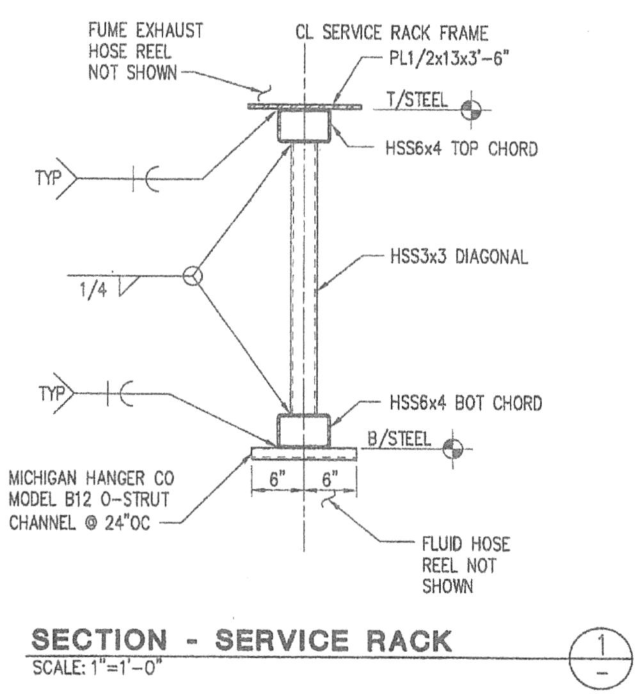

Figure 7.128 Section cut through service rack truss—end view

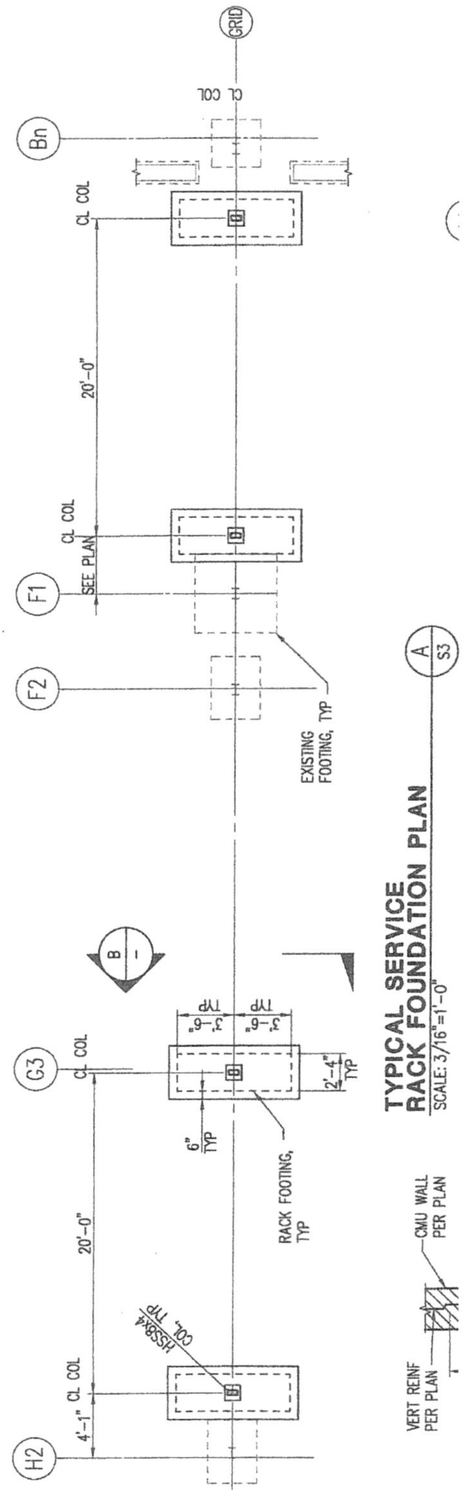

Figure 7.129 Typical service rack foundation plan

V grove" (both sides); it means they need to fill up the gap when welding it in. This information is to be used for all of the trusses indicated at the plan view at (6) locations.

Figure 7.129 is the typical foundation plan for those columns in the service rack; it shows us the size of the footings and where the columns sit in relationship to those footings.

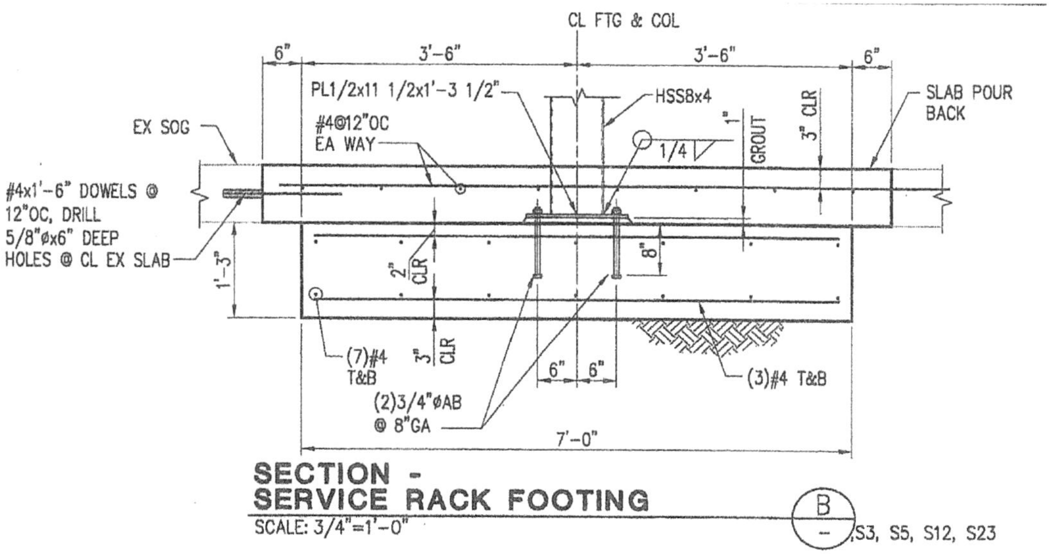

Figure 7.130 Service rack footing

Section B in Figure 7.130, taken from the plan view of the foundation and called out in the elevation view A in Figure 7.111, shows the columns sit on top of the footing, below the finished floor slab. The base plates are shown at PL 1/2 x 11 1/2 x 1'-3 1/2" and have 3/4" anchor bolts at an 8" gage. The designers are showing this "into the page" for the 11 1/2" way. We see the 1'-3 1/2" side of the plate from this elevation. Also, the elevation shows 6" from the centerline of the bolt to centerline of the column, then 6" from the column to the centerline of the next bolt. Thus, this column has four anchor bolts. One might think it could be only two bolts with the way the note is written. Sometimes the details are easy to misunderstand. We have to be very careful how we read them. When in doubt, be sure to question the designer's intent.

Summary

This chapter, *Reading Structural Drawings,* is intended to help steel estimators become familiar with a typical set of structural steel drawings. Information on how to locate conditions with the use of grid line locations, together with a review of the subsequent details, has been provided to indicate how the drawings should be read and interpreted.

Each of the notes found on the general information sheets have been explained, as well as their relationship with the foundation and floor plan. Working through this process shows how the notes apply to the building being constructed.

Through the review of the drawings and details in this chapter, steel estimators should gain an in-depth working knowledge of how to assimilate the information necessary to create a steel estimate.

8

Structural Steel Material Listing

Each and every steel column, beam, part, and piece, including all the bolts and nuts, must be accounted for in order to arrive at the right weight and price for each project. It is the only way to perform a complete and correct steel estimate.

In this chapter, you will learn how to determine the lengths and quantities of the steel items that need to be included in your material take off. You will learn more about how the steel building goes together and how to derive specific information from the steel details for application in the steel take off.

While processing the steel material listing, you will arrive at an understanding of the designer's intent for the building framing itself. You will be shown how to list the materials and how to indicate the required application of welding, holes, copes, and other functions of the fabrication that affect the labor application of the steel production.

In spite of today's technology, many steel estimators still list all their materials by hand. Spreadsheet programs are very handy and adaptable to many different types of projects or estimating needs. With computer programs, the materials can be listed, sorted, and nested for purchase. Labor hours, weights, and costs are extended and totaled; these figures are then entered into a summary sheet for the total pricing.

For purpose of instruction, this text will continue with explanations for creating a material take off by hand. This process is not as hard as it sounds. Basically we are taking notes on what the drawings are telling us. Estimators use abbreviations and small sketches that reflect much in the way of information to the reader for our take off material listing.

There are several choices for material take off sheets; a few are included here for reference. The basic designs for the take off sheets are similar. The major differences are in the level of detail that estimators want to use in listing the information. Some favor labor breakdown information. Some set aside areas for small sketches that help the reader to see specific details pertinent to the fabricated parts.

Material Take Off Sheets

PAGE# ______

PROJECT: ______ EST: KSO

LN#	MAT'L	QTY.	ITEM DESCRIPTON	REF.	MISC INFO	LO	CUT	COPE	PUN	HDL	CAM	F/ASS	WELD	STR	MISC	S/TOT	X	TOTAL	LN#
1																			1
2																			2
3																			3
4																			4
5																			5
6																			6
7																			7
8																			8
9																			9
10																			10
11																			11
12																			12
13																			13
14																			14
15																			15
16																			16
17																			17
18																			18
19																			19
20																			20
21																			21
22																			22
23																			23
24																			24
25																			25
26																			26
27																			27
28																			28
29																			29
30																			30

Figure 8.1 Sample take off sheet

Figure 8.1 shows a typical take off sheet we can use to clarify the nomenclature of the usual shop labor functions:

- **Mat'l**—Refers to the material grade. A36, A572 and A992 are some identifiers of common grades of steel. This column becomes particularly important if we are using nonferrous materials like aluminum or stainless steel. These nonferrous materials also have their special types and need to be identified in order to price the materials accurately.
- **Qty.**—Be sure to list the quantity of each item being used. We may have typical beams on the plans and we can list the steel

as only one beam, but this column is used for a multiplier as well.

- **Item Description.**—Where you list the steel item, whether it is a wide flange beam (e.g., W8 x 10) or a steel angle (e.g., L 4 x 4 x 1/4 or L 3 x 3 x 1/4), or a length of square tube steel (e.g., TS 6 x 6 x 1/4 or TS 4 x 4 x 1/4). List the item, such as TS 6 x 6 x 1/4 x 20'-10 1/2". This space can also be used to draw a picture of what we are fabricating to show the users of this take off sheet, especially if it's a complicated part.
- **Ref.**—Here you can note the detail cut reference that you used in order to comprise this list of material.
- **Misc Info.**—Enter the drawing from which the steel is being listed.

Now we move onto the columns regarding functions.

- **LO.**—Layout; enter the fabrication time to mark the intended locations, as shown on the shop drawings for the cuts, holes, added fabrications, etc.
- **Cut.**—Cut to length.
- **Cope.**—A cope or a notch may be put in the ends of the beam to allow it to fit or frame into another beam or column
- **Pun.**—Punching of holes by machine as opposed to drilling.
- **Hdl.**—Handle; required if the beam or part had to be turned or rolled over during the fabrication process, or moved to a special section of the shop
- **Cam.**—Camber; indication to put an arch in the steel so it has a rise with the high point in the center.
- **F/ASS.**—Fit and/or Assemble
- **Weld.**—The time it takes to weld. Weld times are based by the thickness (3/16, 1/4, 5/16, 3/8, 1/2, and so on) and the length of the weld. The number of passes it takes to accumulate the particular thickness of weld is important here.

- **Str.**—Straighten; parts with lots of weld may bend from the isolated heat zone. They will then need to be straightened (by heat again, usually) prior to completion of the part
- **Misc.**—Anything else not allowed for in the functions listed above.
- **S/Total.**—Total all your times here
- **X.**—Quantity again
- **Total.**—Where you enter your total labor time

Figure 8.2 provides a simpler version of a take off sheet.

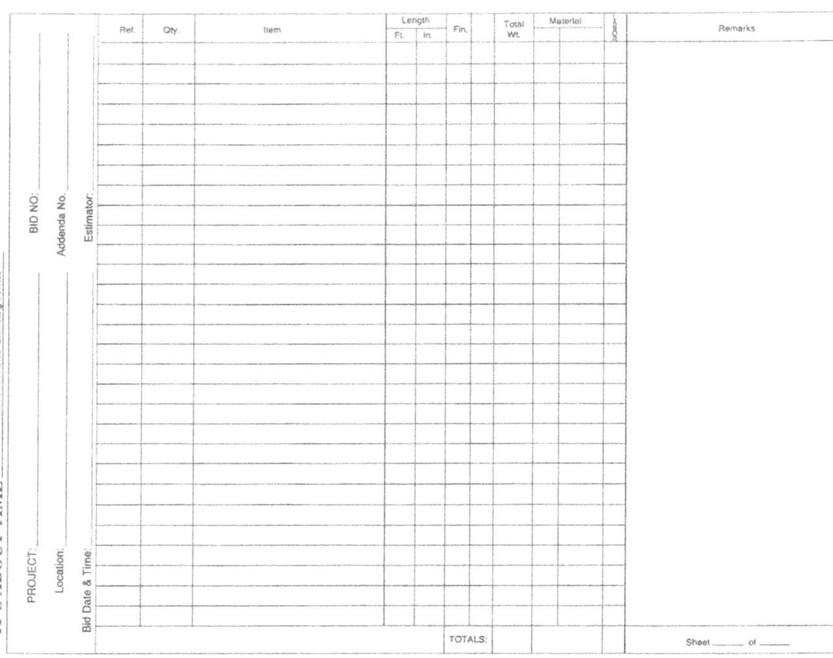
IT'S ABOUT TIME Structural & Miscellaneous Steel Estimating Service
PROJECT:
BID NO:
Location:
Addenda No.
Bid Date & Time:
Estimator:
Ref.
Qty.
Item
Length
Ft.
In.
Fin.
Total Wt.
Material
Remarks
TOTALS:
Sheet ___ of ___

Figure 8.2 Simpler take off sheet

Figure 8.2 consolidates labor into one line item. Some steel estimators reduce their estimating functions to a simple time per part as they become more familiar with the time their shop takes to work on them, also saving time on their take off. The remarks column might be used to include small sketches of the fabricated parts or specific information to labor applications that might be needed.

Many steel estimators use a typical spreadsheet program to prepare their take off. These programs allow much in flexibility for design and application of the calculations. Listing and adding labor hours to the steel materials to be fabricated is easily done, and the format is easily modified to fit any need for your listing.

Because Figures 8.1 and 8.2 are prepared on Excel spreadsheets, the standard Excel formulas are used. You can adjust your spreadsheet to meet your needs. Then, once you have a format that you like to use, you just have to save the file as a different name for your next job and continue on with your changes and additions. The automatic application of the extensions of the weights, materials pricing, and labor time saves much time in take off preparation.

Figure 8.3 shows a sample spreadsheet developed for material take off.

Materials			QTY	Mat. Thkns	Plate Width in inches	Plate Length	Ext Wt	Cost Per lb.	Exten. Mat.cost	Exten Hdwre
S2.11 - Main Level Fdn West										
Labor		Detail	35.85	13	2.8	per ea.	Lbs			
4.8	TS 4 x 4 x 5/16	at grid A2.5 t	6			13.25	1177	0.48	$ 565	
0.8	TS 4 x 4 x 5/16	6/S6.04	1			12	178	0.48	$ 85	
1.6	TS 4 x 4 x 5/16	6/S6.04	2			8	237	0.48	$ 114	
0.5	Plate 3/8	6/S6.04	1	0.375	12	12	15	0.487	$ 7	
6	3/4 x 4 DICA		24					2.8		$ 67
0.5	Plate 5/16	6/S6.04	1	0.313	6	12	6	0.487	$ 3	
1.5	7/8 x 8 machine bolts w/n&w		6					2.8		$ 17
		2.87857143								
	Column Grid 1A. 1B	S4.06								
7	W12 x 40	Col 1B	2			39.25	3140	0.41	$ 1,287	
7	W12 x 40	14/S6.01	2			13.25	1060	0.41	$ 435	
0.5	Cap Plate 1"	guess	1	1.000	12	16	54	0.487	$ 26	
1	Splice Web Plate 5/8"	15/S34.06	2	0.625	9	12	38	0.487	$ 19	
0.15	A325X Bolts 7/8" w/N & W	30/S4.06	3					2.8		$ 8
2	Stiffeners Pl 1/2	2/S4.06	4	0.500	3.5	12	24	0.487	$ 12	
1	Connection Plate 1/2 x	20/S4.06	2	0.500	8	12	27	0.487	$ 13	
0.1	A325X Bolts 7/8" w/N & W	30/S4.06	2					2.8		$ 6
1	Base Plate 1"	3/S4.06	2	1.000	9.5	13	70	0.487	$ 34	
0.4	AB 3/4 x 1'-2	3/S4.06	8					2.8		$ 22
36							6026		$ 2,600	$ 120

Figure 8.3 Sample spreadsheet for material take off

Figure 8.4 illustrates the summary or Pricing Breakdown for this list from Figure 8.3.

After the take off is completed and the material pricing has been installed, steel estimators are done with the quote. No double entry into another program is required. All the information is in one file and can be utilized as needed by the company. The spreadsheet programs are inexpensive, easy to use, and can accommodate a multitude of needs with sorting, pricing, and breakouts for specific fabricated items all in one file.

Summary	Hours Weight	Cost lb/hr	Cost Extensions		Sale Amount	
Detailing	4	$ 70	$ 280	$0.05	$ 322	
Engineering	0	$ 40	$ -	$0.00	$ -	
Labor	36	$ 40	$ 1,434	$0.24	$ 1,434	
Materials	6026	0.4	$ 2,600		$ 2,990	
Hardware		$ 120	$ -		$ -	
Grating/Forming		$ 1.80	$ -		$ -	
Paint Material	4	25	$ 103	1,136 sq ft	$ 118	
Paint Labor	6	40	$ 246	1	$ 246	
Galvanizing	0	0.42	$ -		$ -	
Handling	1	40	$ 22		$ 25	
Shipping	0.30	650	$ 196	$0.0325	$ 225	
			$ 4,880	$ 0.81 Per lb	$ 5,360	$ 0.89 Per Lb

Figure 8.4 Pricing breakdown

Building Columns

We read these structural drawings as if we were beginning our take off listing with the columns. To calculate the length of the columns, we find in the foundation plan drawings where they start and where they end. The top of the columns can be only to the next floor above, or through to the roof level.

In one- or two-story buildings, columns will be a single piece of material going up to the roof. For a multi-story building, columns that go to the roof may require splicing. Building columns may change sizes as they go up to the roof in a multi-story building. The column details will indicate when it happens.

To calculate column lengths where there is no information available, use a general "rule of thumb" allowance. With typical building heights, one may allow 12'-6" for the foundation up to the 2nd floor. (This measure consists of 6" down to the top of the footing under the concrete floor slab, 8'-0" for ceiling height and 4'-0" from above the ceiling to the next floor, allowing for duct work and electrical work to run through the building). Subsequent floors are 12'-0" because they do not have the minus 6" to the top of the footing as the first floor does. However, the 8'-0" ceiling height and the 4'-0" allowance for the mechanical stays the same.

To establish the column lengths, we must locate both the top footing elevations and the top of steel elevation where the columns stop. To determine the top of footing elevations, find them on the foundation plan. The top of footing will be the same elevation as the underside of the column base plate. The top of the column will be equal to the top of steel level for the roof of the building. Ultimately, when talking about columns lengths, we think about this length as being from the underside of the base plate to the top of the cap plate.

Top of steel elevations are usually shown on the floor and roof plans in the structural drawings. They may also be indicated in the notes on the plan view sheets. When these elevations are not shown or indicated at the structural drawings, you may have to refer to the architectural drawings to find them. In the architectural drawings, these floor level elevations are usually located on the sheets showing the floor plans, or in the notes. If not located there, they may be found with the elevation cuts through the building. To get to the top of steel, the thickness of the floor and roof materials may need to be deducted from those elevations.

Once you have found the top of steel elevation and the underside of base plate elevations at the particular location of the column, you will have established that column length. To list the material for the column with the cap and base plate information as a fabricated unit, it is necessary to locate all the specific details applicable to the column. Looking through the plans and details to locate this information requires quite a bit of work. Once you have made this effort for the first column, however, the remaining columns will use much of that same information and will be easier to list.

Using the sample drawings we will work through this exercise in more detail.

Figure 8.5 shows a foundation plan. All the notes on the drawings need to be read. These notes give you more information about the designers' intent while they were creating these drawings. Even notes that seem unrelated may have an effect on the fabrication of our steel. Specific notes on footings and concrete may have an effect on column heights.

In Figure 8.6, Notes 1 through 6 provide necessary information for us to calculate the column lengths. These notes were reviewed in the chapter on Chapter 7, Reading Structural Drawings. Here we will go back to these notes in terms of how they apply to our search for specific information.

- **Note #1** in Figure 8.6 indicates that the spread footing types are shown on sheet S9.
- **Note #2** in Figure 8.6 shows us how to find the elevation to the bottom of the footing (a minus dimension shown at –X'–XX"). We know that the columns sit on top of the spread footings. Therefore, we have to find out how far below the finished floor the top of these footings will be. The spread footing types, together with the indicator of the footing elevation and the note

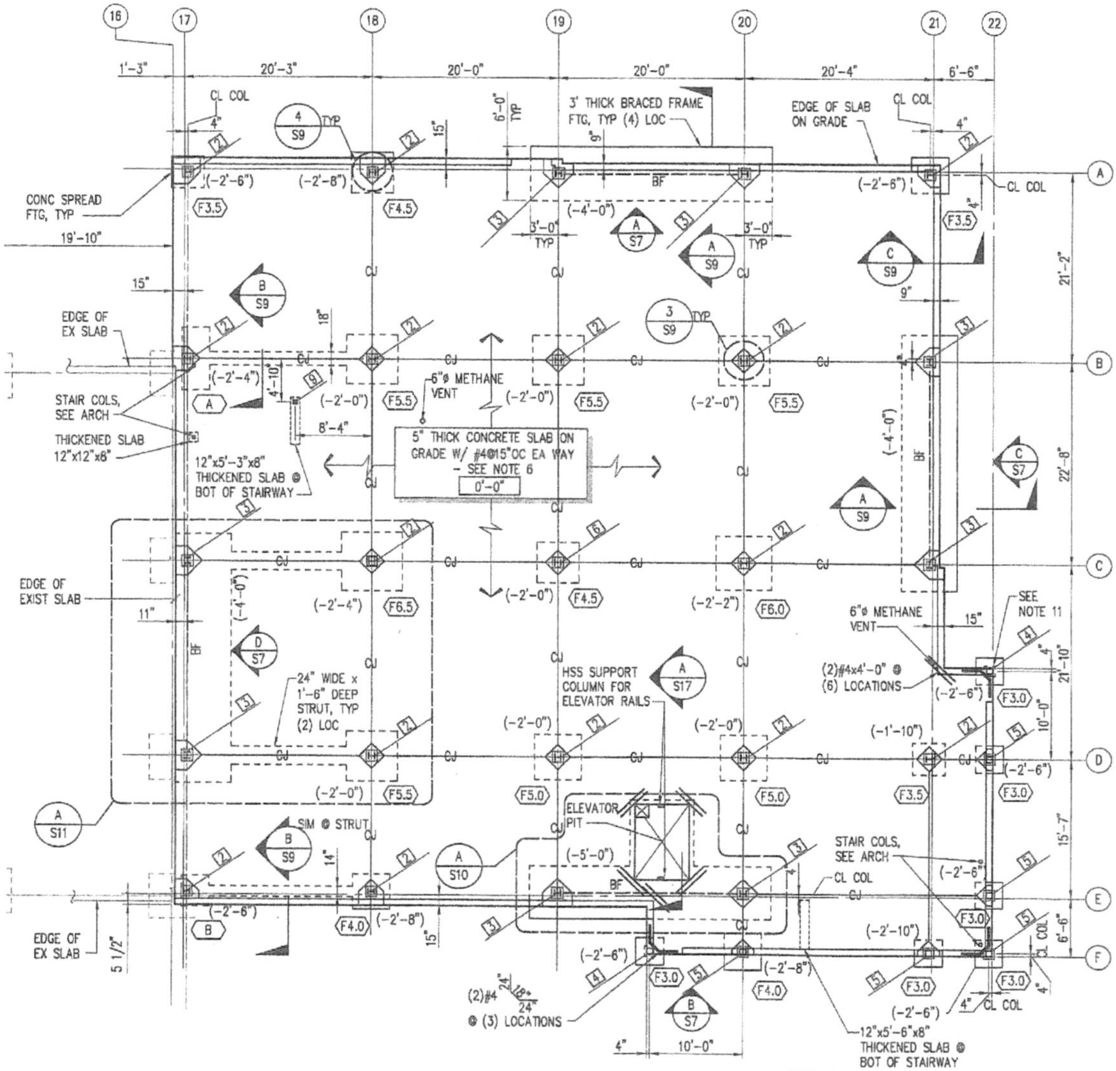

Figure 8.5 Sample foundation plan

that directs you to the typical foundation and slab on grade details, will assist us in determining the column base elevation.

The spread footing schedule gives us the thickness of the concrete footing. The depth to the underside of this footing is given in the foundation plan. To find the top of the footing, add the thickness dimension of the footing to the depth; this sum is the underside of column elevation.

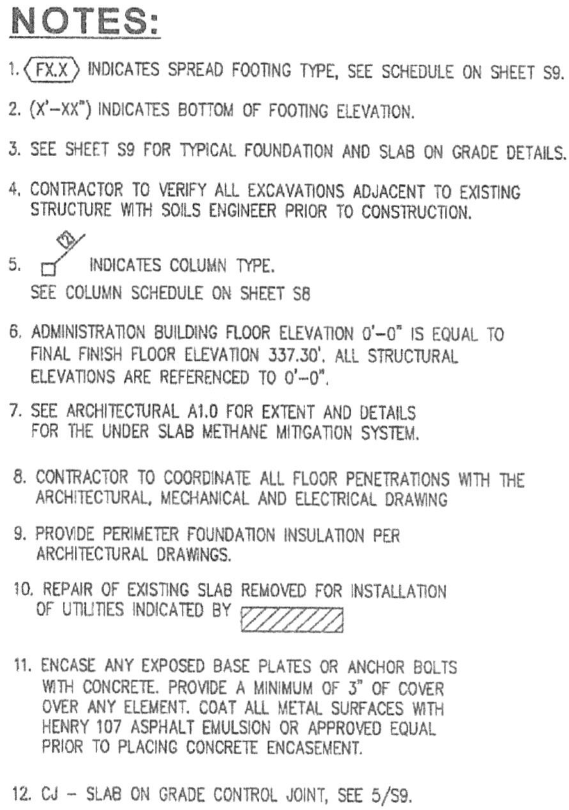

NOTES:

1. ⟨FX.X⟩ INDICATES SPREAD FOOTING TYPE, SEE SCHEDULE ON SHEET S9.
2. (X'–XX") INDICATES BOTTOM OF FOOTING ELEVATION.
3. SEE SHEET S9 FOR TYPICAL FOUNDATION AND SLAB ON GRADE DETAILS.
4. CONTRACTOR TO VERIFY ALL EXCAVATIONS ADJACENT TO EXISTING STRUCTURE WITH SOILS ENGINEER PRIOR TO CONSTRUCTION.
5. INDICATES COLUMN TYPE. SEE COLUMN SCHEDULE ON SHEET S8
6. ADMINISTRATION BUILDING FLOOR ELEVATION 0'–0" IS EQUAL TO FINAL FINISH FLOOR ELEVATION 337.30'. ALL STRUCTURAL ELEVATIONS ARE REFERENCED TO 0'–0".
7. SEE ARCHITECTURAL A1.0 FOR EXTENT AND DETAILS FOR THE UNDER SLAB METHANE MITIGATION SYSTEM.
8. CONTRACTOR TO COORDINATE ALL FLOOR PENETRATIONS WITH THE ARCHITECTURAL, MECHANICAL AND ELECTRICAL DRAWING
9. PROVIDE PERIMETER FOUNDATION INSULATION PER ARCHITECTURAL DRAWINGS.
10. REPAIR OF EXISTING SLAB REMOVED FOR INSTALLATION OF UTILITIES INDICATED BY
11. ENCASE ANY EXPOSED BASE PLATES OR ANCHOR BOLTS WITH CONCRETE. PROVIDE A MINIMUM OF 3" OF COVER OVER ANY ELEMENT. COAT ALL METAL SURFACES WITH HENRY 107 ASPHALT EMULSION OR APPROVED EQUAL PRIOR TO PLACING CONCRETE ENCASEMENT.
12. CJ – SLAB ON GRADE CONTROL JOINT, SEE 5/S9.

Figure 8.6 Foundation plan notes (covering all notes)

If the footing elevation shown in the foundation plan is minus 2'-0", then you add 10" for the thickness of that footing, arriving at minus 1'-2" for the top of footing elevation, which is minus 1'-2" below the finished floor elevation of 0'-0".

- **Note #3** tells us that all the rest of the concrete foundation information will be at sheet S9. This sheet will need to be researched thoroughly for us to determine if there are any other steel items (including embeds in the concrete) that should be of concern to us.
- **Note #4** does not concern any steel fabrication. However, it does inform the contractor about moving dirt near existing

buildings. (This information helps to prevent any unanticipated structural impact being built or added to the existing building).

- **Note #5** in Figure 8.6 tells us where to find the column types and sizes on sheet S8. It also directs the reader to the steel column schedule sample in Figure 8.7.
- **Note #6** gives us the finished floor elevation at 0'-0". We use this elevation together with the top of steel elevations found in the roof plan to help us determine the column lengths.

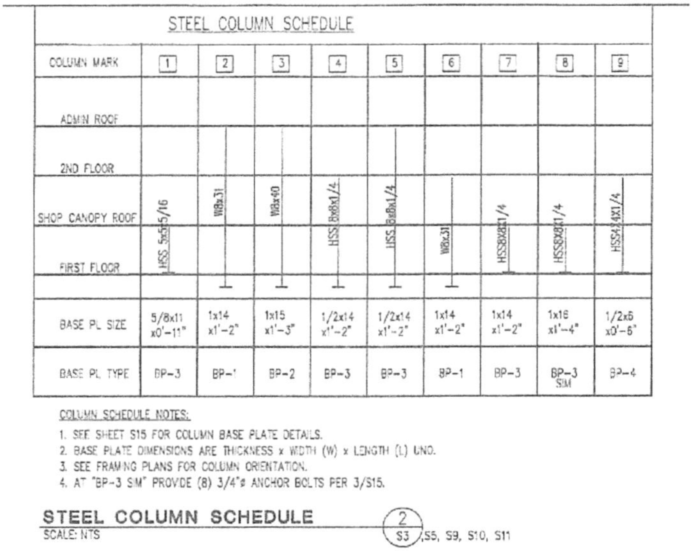

Figure 8.7 Steel column schedule

Not all projects provide a column schedule like this one in Figure 8.7, but it is handy when they do. This schedule gives you the column marks, shows you where they start and stop, as well as the base plate type and size. What we do have to verify is how many of each column type there are, counting the column types as they are located on the foundation plan.

Notice also the reference back to the floor plans where the columns are found. See where it reads "Steel Column Schedule" with the detail bubble 2/S3 at the end, but then also lists *S5, S9, S10,* and *S11*. This note lets you know the other places where this detail is referenced, saving you a lot of time. Always look for additional notes by the detail bubbles to see if the back references are indicated.

Note #6 in Figure 8.6 gives the building floor elevation as 0'-0", that figure being equivalent to the (sea level) elevation of 337.30 that is noted in other details. Knowing that the finish floor is at 0'-0" helps us further to resolve the elevation questions we are having with determining the columns. That number also gives us our starting point when using the bottom of footing and top of steel elevations in calculating our column lengths.

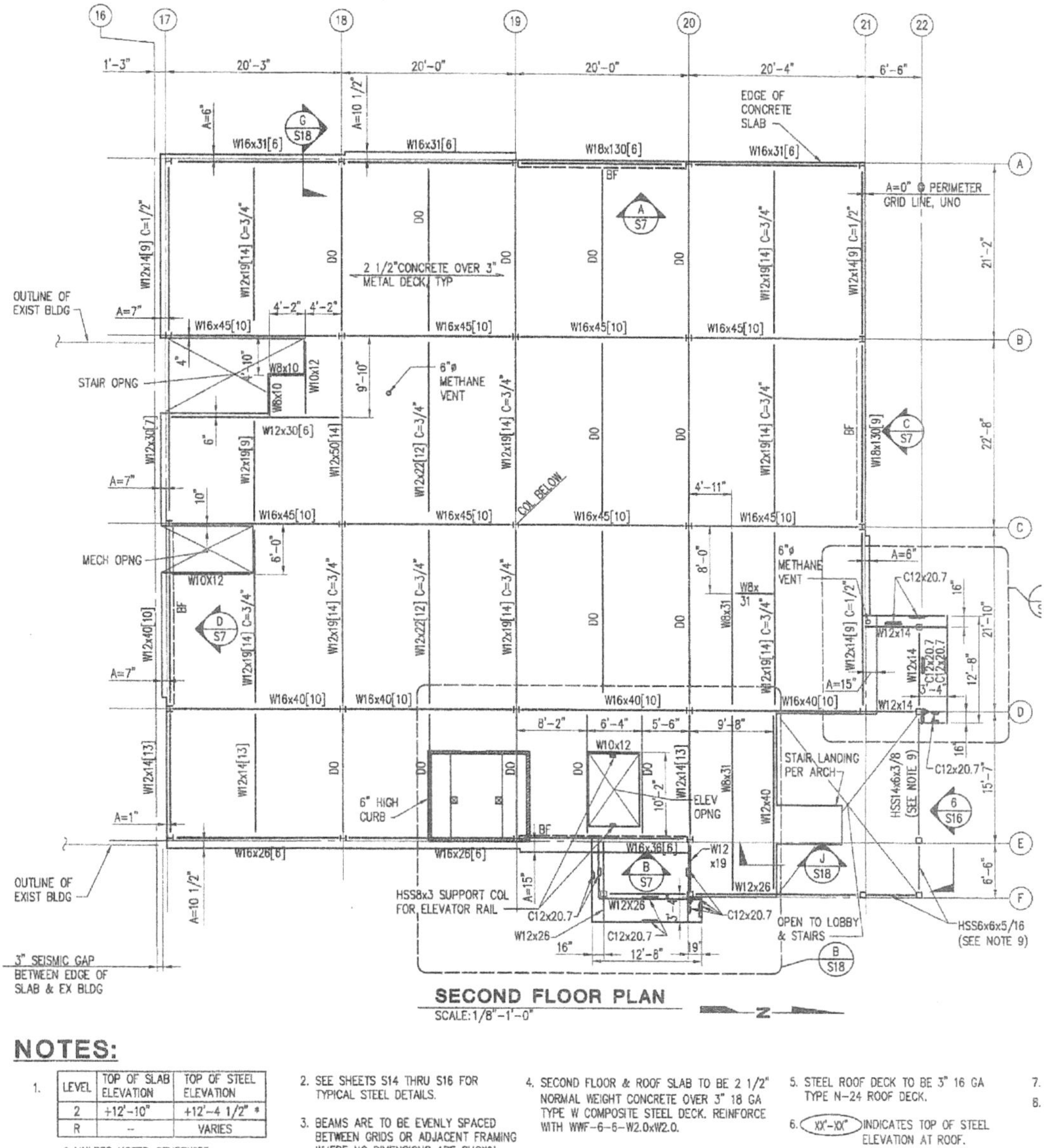

Figure 8.8 Second floor plan

When looking for columns in the foundation plans of the drawings, you will find most at the grid line intersections of the drawings. However, be on the lookout for columns in other areas of the building. Special framing or extra heavy loading of the building may add columns at locations that are off the grid centers.

Columns may be added between floors where you don't expect to find them. Special framing for large entries, landings, or balconies; supports for overhead doors; window walls; special signs, and overhead equipment are a few examples where additional columns may be added.

Steel columns are usually easy to list once you locate the elevations necessary to establish the column heights. From there, it is just a matter of finding out how thick and what size the base and cap plates are. Also check for connection plates for beams or any other materials that may be attached to them.

Looking at the 2nd floor plan in Figure 8.8 and the roof plan in Figure 8.9, we find most of these columns are the type 2 that go up through to the roof. The column located at grid intersection C-19 stops at the 2nd floor level in Figure 8.8, noted *Col on previous page.*

The notes in Figure 8.8 give you the top of steel elevation at the 2nd floor. They tell you where to find the information for typical steel details, beam spacing, concrete, and decking, as well as how to find the top of steel elevation at the roof.

The notes in Figure 8.9 give you more information on a moveable partition. They provide a directive to refer to the architectural and mechanical drawings for openings in the floor and storefront beam elevations. The notes also indicate where to locate the edge form for openings, and information on addition support framing.

We have reviewed the foundation plan in Figure 8.5, the 2nd floor plan in Figure 8.8, and the roof plan in Figure 8.9. We now know that with the exception of a column that stops at the 2nd floor, all the rest of the columns go on through to the roof. At the roof plan we see the callout for the top of steel elevation. The top of steel elevation is also the top of the steel column (25'-1 1/2"). You could use the column schedule in Figure 8.7, putting this elevation at the Admin Roof Level. The notes for the 2nd floor plan in Figure 8.8 show you the top of steel elevation at 12'-4 1/2". You can also put this information in your column schedule at sheet S8 or Figure 8.7 for your 2nd floor. Now we need to find out where the bottoms of the columns are.

Go back to the foundation/first floor plan in Figure 8.5; look at the column at grid B-20 where it calls out the detail 3/S9. This section at the Figure

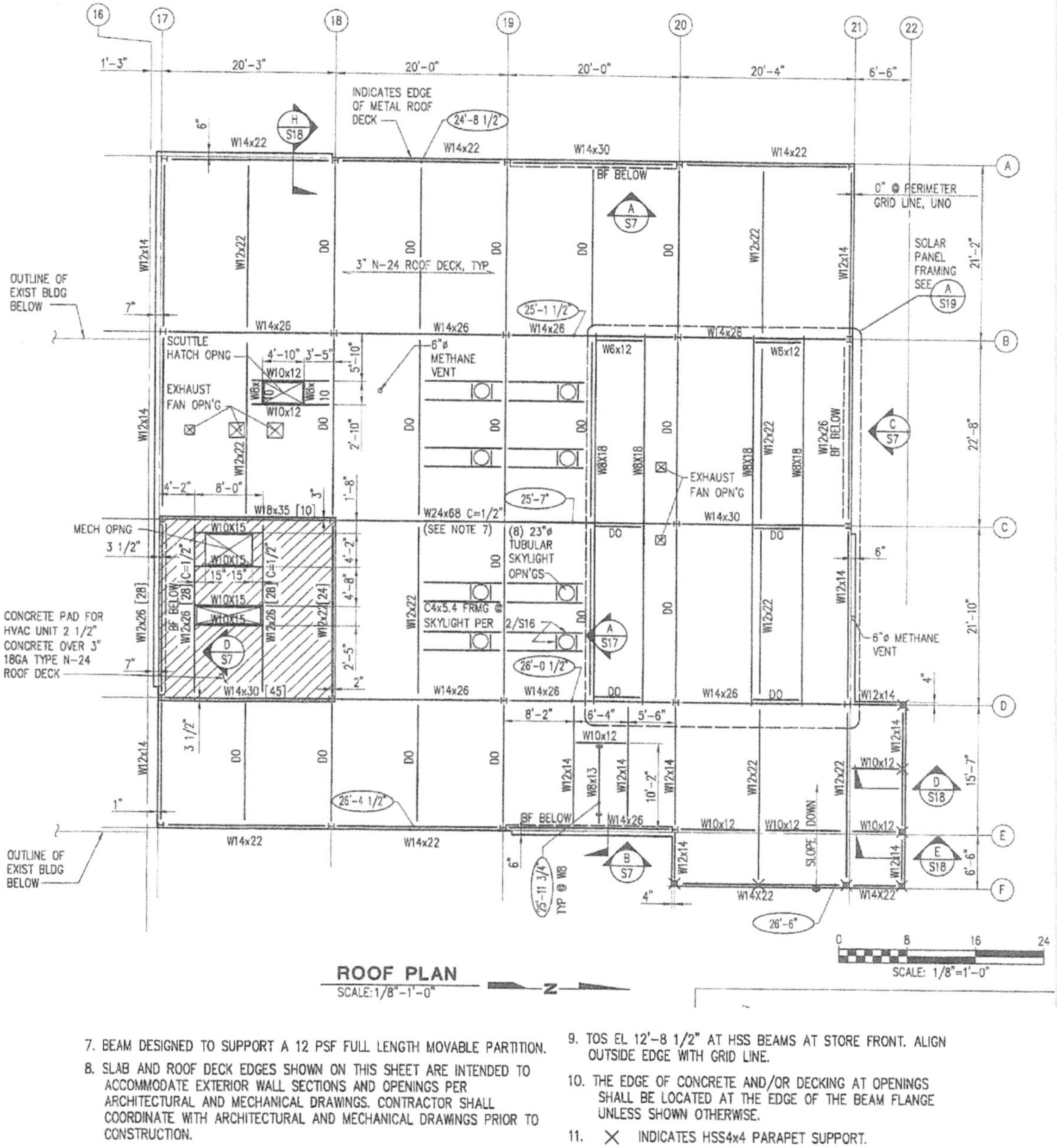

Figure 8.9 Roof plan

8.10 shows us the plan view (or looking down on the column as it sits on the concrete slab) of this condition. Notice also this is a column type 2, which is a W8 x 31 beam.

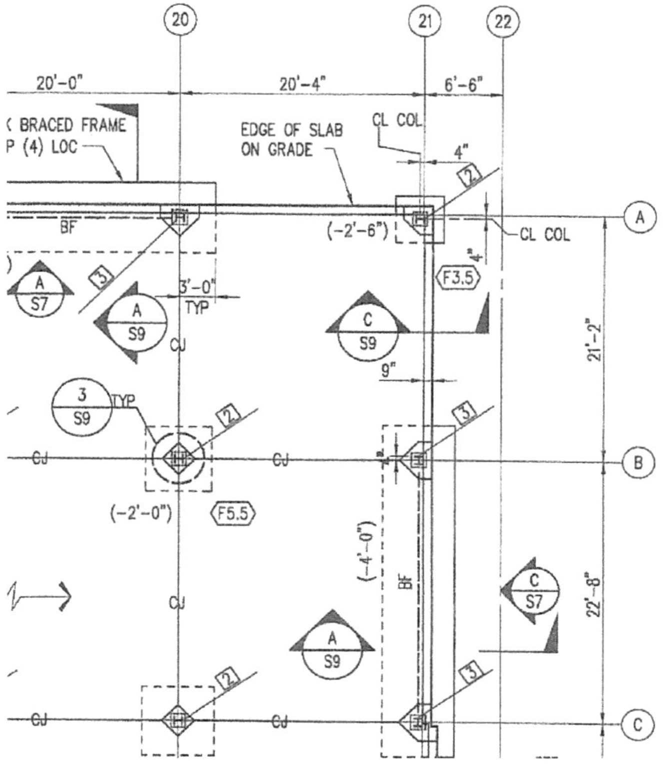

Figure 8.10 Close up of the foundation plan from Figure 8.5

We see more foundation details in Figure 8.10. Note that detail 2 shows us the elevation view of the column as it sits on top of the footing, which is below the concrete slab. Because this is a typical detail, there is no specific callout in the foundation plan—this detail applies to all columns in this area. Together with the other information we have found on these sheets, we will calculate the length of the columns.

The two pieces of information in Figure 8.11 that are particularly helpful for calculating the column length are the column spread footing schedule and the thickness of that footing. In going back to the foundation plan in Figure 8.5, you can see dimensions by the footings shown as (-2'-0") and (-2'-8") or (-4'-0") and so on. Our particular column at grid B-20 shows (-2'-0") as the bottom of footing elevation. This is 2'-0" below the finished floor elevation of 0'-0". Further, this column has footing type F5.5. Look at the details below

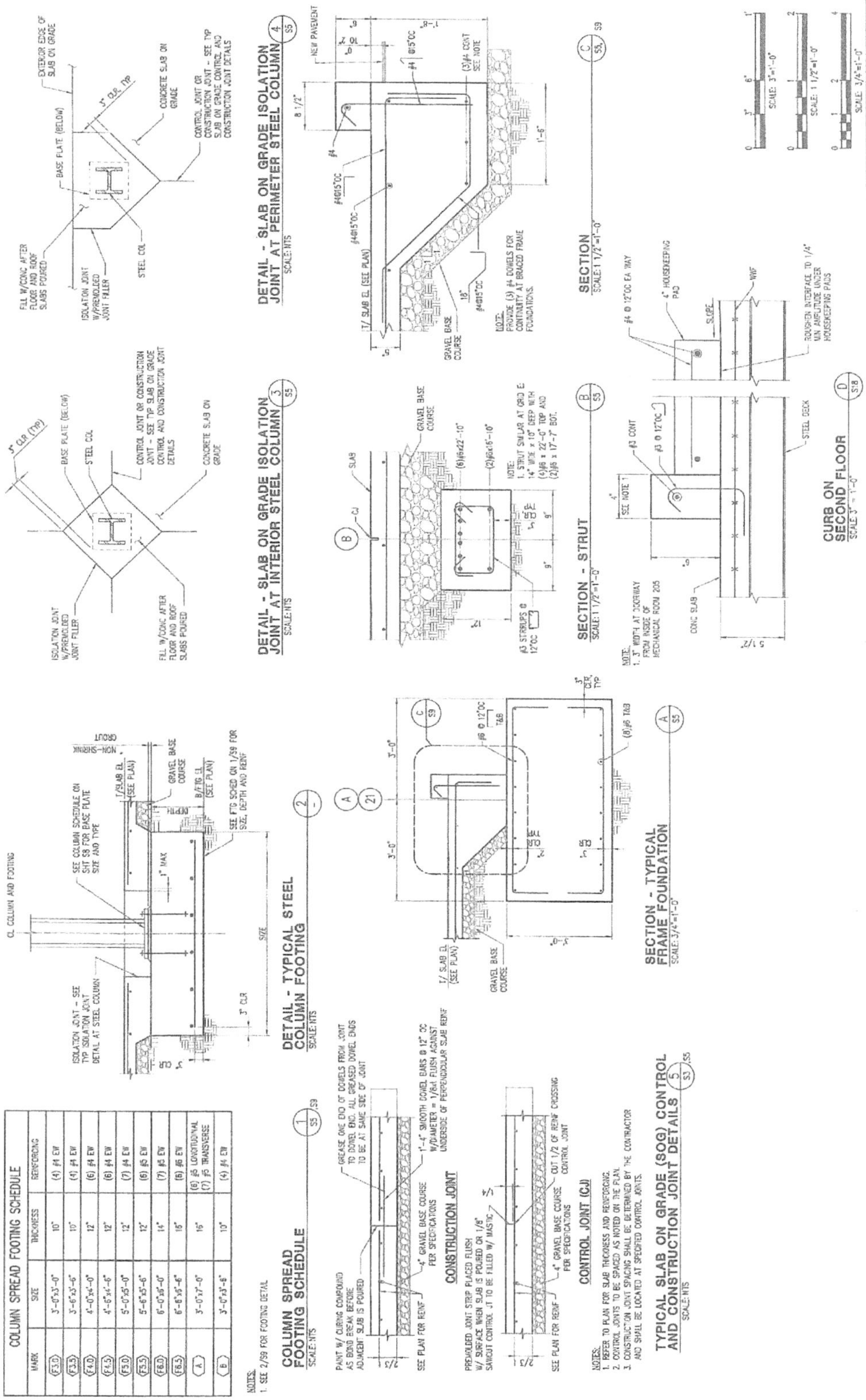

Figure 8.11 Concrete footing details

COLUMN SPREAD FOOTING SCHEDULE			
MARK	SIZE	THICKNESS	REINFORCING
F3.0	3'-0"x3'-0"	10"	(4) #4 EW
F3.5	3'-6"x3'-6"	10"	(4) #4 EW
F4.0	4'-0"x4'-0"	12"	(6) #4 EW
F4.5	4'-6"x4'-6"	12"	(6) #4 EW
F5.0	5'-0"x5'-0"	12"	(7) #4 EW
F5.5	5'-6"x5'-6"	12"	(6) #5 EW
F6.0	6'-0"x6'-0"	14"	(7) #5 EW
F6.5	6'-6"x6'-6"	16"	(6) #6 EW
A	3'-0"x7'-0"	16"	(6) #5 LONGITUDINAL (7) #5 TRANSVERSE
B	3'-0"x3'-6"	10"	(4) #4 EW

NOTES:
1. SEE 2/S9 FOR FOOTING DETAIL

COLUMN SPREAD FOOTING SCHEDULE
SCALE: NTS
1 S5 ,S9

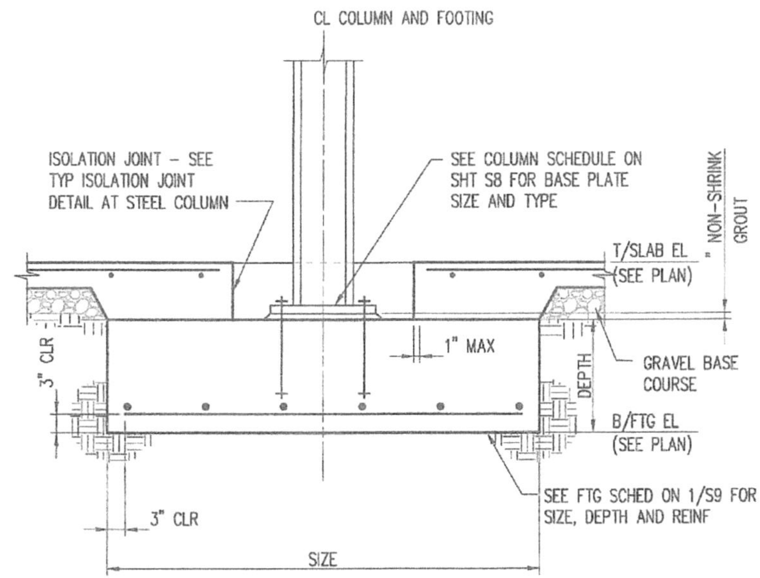

DETAIL - TYPICAL STEEL COLUMN FOOTING
SCALE: NTS
2 –

Figure 8.12 Details at footing supports for columns

and notice that footing F5.5 is 12" thick. The grout that goes underneath the column for leveling that column has a typical thickness of 1 1/2".

To calculate the length of this column at grid B-20, we begin by locating the top elevation 25'-1 1/2" for the columns at this grid line B. The elevation is found in Figure 8.9 showing the roof plan. Add 1'-0" to the 25'-1 1/2" for the drop under the slab to the top of the footing and we have 26'-1 1/2". (The bottom of the footing F5.5 at the "Column Spread Footing Schedule" is 2'-0"; the thickness of that footing is 1'-0". Therefore, we drop only 1'-0" below the slab.) Subtract another 1 1/2" for the thickness of leveling grout. The total height of the column is 26'-0".

Discovering all of the remaining column lengths happens in this same manner. Now that you know how to find the necessary information, calculating the subsequent column lengths will become easier.

To list the steel columns we have to take our research a step further. We must calculate the actual use length of the steel beam that makes this column. For Column Mark 2, we have a 1" thick base plate as shown in the Steel Column Schedule at 2/S8 in Figure 8.7. We next have to subtract that thickness from the total column height.

This column has a cap plate, which is found in the details showing the top of the column. This cap plate thickness (we are making an assumption here) of 1/2" needs to be subtracted from the total column height. The result of subtracting this 1 1/2" from the 26'-0" is 25'-10 1/2". This is the actual use length of the beam that makes up the column.

Another way to think of this is to add the combined thickness of the base plate, the cap plate, and the grout depth. Your total deduction is then 3".

For example, the column is 1'-0 below the floor. Therefore, you would add 1'-0" to the top of steel elevation of 25'-1 1/2", the result of which is 26'-1 1/2". Subtract the 3" (the sum of the total thickness items to deduct) and you have 25'-10 1/2". That is (1) W8 x 31 x 25'-10 1/2".

You go through this process for all the columns to arrive at all the correct lengths.

Looking back at the foundation plan in Figure 8.5, two columns are located at gridline locations A-17 and A-21 that are the same: they have the same type of footing with the same footing elevation. We can group these columns as (2) W8 x 31. The top elevation per the roof plan in Figure 8.9 is 24'-8 1/2". The bottom of the footing is minus 2'-6" and the thickness of the footing is 10" per Figure 8.12, at footing mark F3.5 (minus 2'-6 plus 10" is minus 1'-8"). Add this 1'-8" to the 24'-8 1/2", then subtract 3" for the base plate, cap plate, and 1 1/2" of grout. The total column length then becomes 26'-1 1/2"

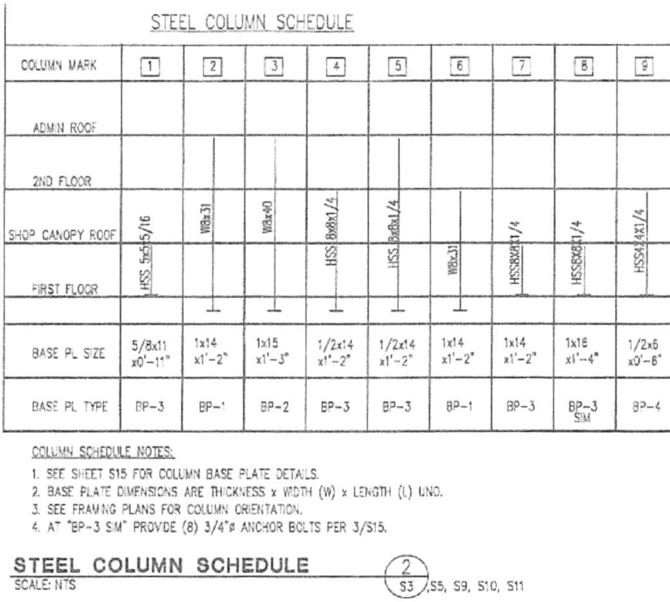

Figure 8.13 Steel column schedule

for both of these columns. Thus we have (2) W8 x 31 x 26'-1 1/2" to be listed in your take off.

The column at A-18 in Figure 8.5 is a type 2 column of W8 x 31 (see Figure 8.13) with a top of steel elevation of 24'-8 1/2 per the roof plan in Figure 8.9. The type of footing is an F4.5, which is 12" thick, and the bottom of footing elevation is minus 2'-8" per Figure 8.9. The minus 2'-8 plus 12" is minus 1'-8". Add 1'-8" to 24'-8 1/2", minus 3" is 26'-1 1/2". This is then (1) W8 x 31 x 26'-1 1/2".

The result of this exercise is that there are (3) columns W8 x 31 x 26'-1 1/2". These columns are the same even though the footing at A18 is lower by 2". The reason is the footing is thicker by 2", making them the same length. Figure 8.14 shows the footing callouts for these column locations.

From this base plate detail in Figure 8.15, we see the weld size required, the anchor bolt size and length, and the size of the holes required in the base plates for the anchor bolts.

The engineers indicate that the anchor bolt holes in the base plate are to be the bolt diameter plus oversize. The anchor bolts are shown as 3/4" diameter, and the allowance for oversize holes in base plates is 1/8". Therefore, the finished hole size is 7/8" diameter.

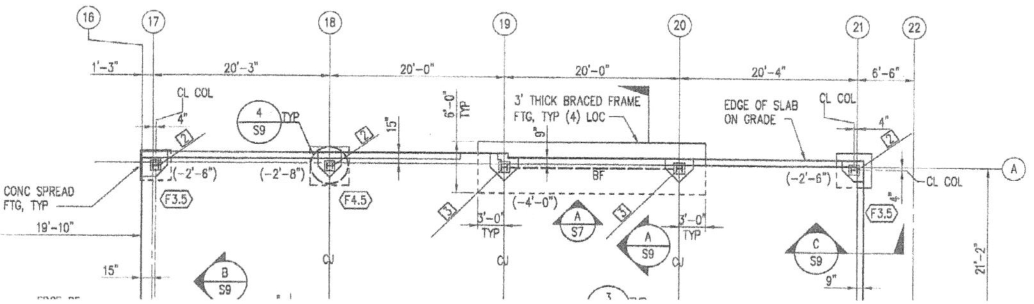

Figure 8.14 Foundation plan close up along grid line A

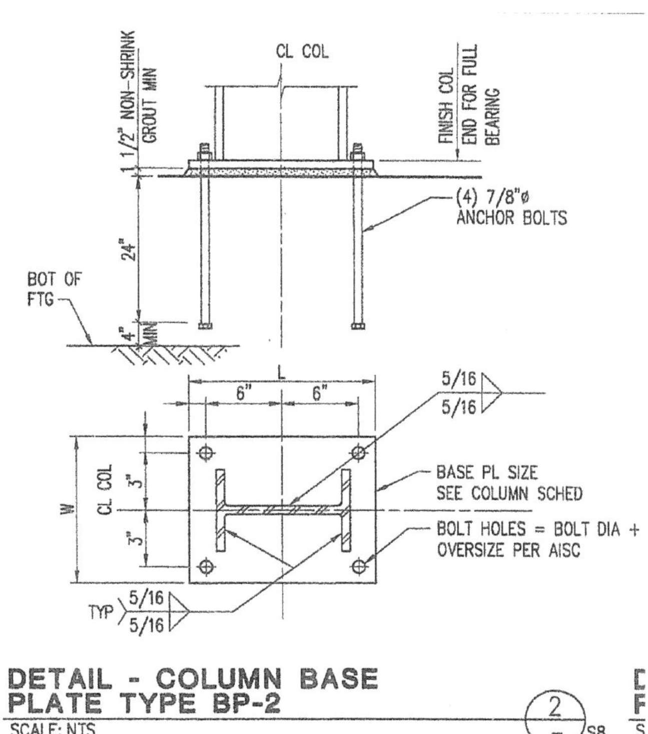

Figure 8.15 Detail—column base plate type BP-1

To calculate the length of the anchor bolts, the engineers measure 14" from the underside of the bolt head to the underside of base plate. Then we allow for the 1 1/2" grout, the 1" base plate, the nuts, and the washer, which is about another 1" including thread grip. These anchors then will be a minimum of 18" long.

The length of the weld that connects the base plate to the column will need to be listed on the take off sheet under the base plate listing itself. For this column, the base plate has a 5/16" weld. This weld will need to go all around the column base. The engineer shows the 5/16" weld to be on both sides of the flange, as well as both sides of the web. The total length of the weld is the sum of all these dimensions for both sides of the flanges and the web.

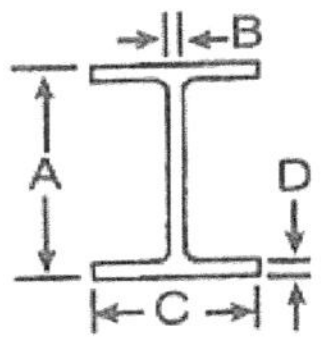

Figure 8.16 Sample of a wide flange beam section

For example, use the steel products chart for actual beam dimensions, with Figure 8.16 for the W8 x 31 beam. The A distance x 2 and the C distance x 4 will suffice for the total weld length of 48", or 4'-0". This is listed on the takeoff sheet as *wfl* (weld–fillet) 5/16 x 4'-0" under the part; the person adding the labor hours will know how much time to allow for this welding.

While reviewing the structural drawings, a typical detail on the standard detail sheets is discovered. It refers to Beam to Column Web Connections (see Figure 8.17). On this typical detail, we see the requirement for stiffeners with the columns that are not specifically called out in the drawing plan views. These stiffener plates need to be listed for all of the columns. The correct application of the labor requirements relative to the braced frame columns must be listed with those columns.

Refer to the Beam to Column Web schedule, chart F, in Figure 8.17. The stiffener plate thickness required is 3/8". This stiffener plate changes to a heavier thickness because there are more bolts used at the connections—notice the change from 3/8" to 1/2" when the connection condition changes from a 4-bolt connection to a 5-bolt connection. To calculate the width and length of these stiffener plates, we need to find out the space allowance in the column from flange to flange and from face of flange to face web:

What the detail in Figure 8.18 indicates is that the columns have stiffeners both above and below the connection plates for the beams that frame into the columns. These stiffener plates have to be sized just right to fit in between the beam flanges; they have to be flush to the column web and the face of the flanges. The plan view above looks down on the column and shows the beams as they frame into that column.

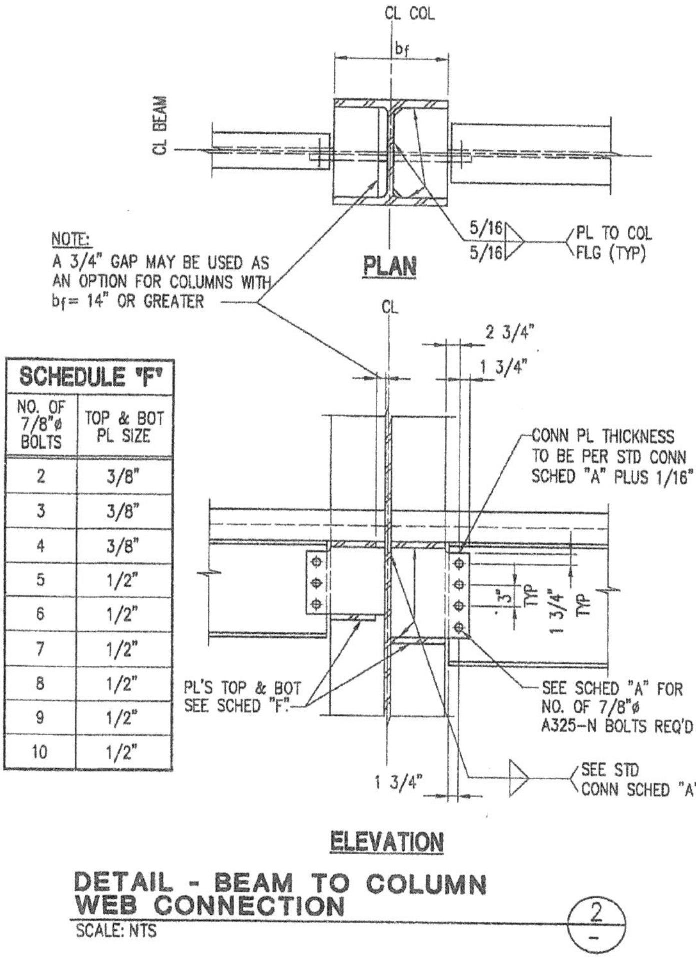

SCHEDULE "F"

NO. OF 7/8"ø BOLTS	TOP & BOT PL SIZE
2	3/8"
3	3/8"
4	3/8"
5	1/2"
6	1/2"
7	1/2"
8	1/2"
9	1/2"
10	1/2"

Figure 8.17 Detail: beam to column web connection

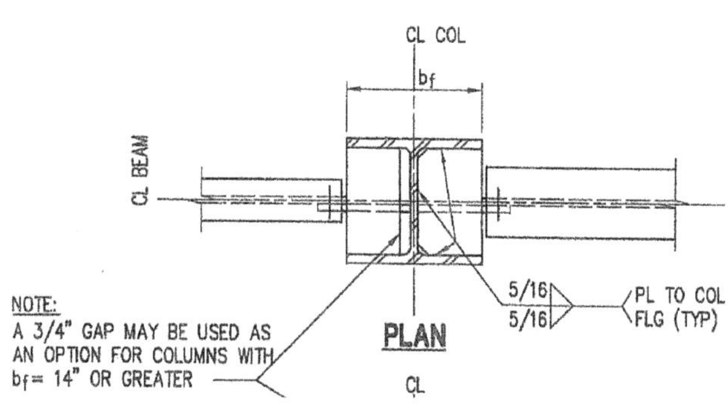

Figure 8.18 Plan view of beam to column web connection

The inside dimensions of the column flange and web distances must be determined in order for us to size this plate. To find this information, you can reference a beam chart from a steel supplier's product catalog for the overall depth of beam, half of the bottom flange, and half of the *tw* or thickness of web. These dimensions will help in calculating the size of the stiffener plates. See Figure 8.19 for the required section dimensions needed to support the calculations.

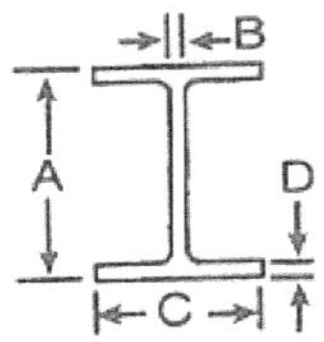

Figure 8.19 Typical beam elevation view for dimensions

To find the length of the stiffener plate, we review a beam chart from a steel supplier's product catalog, or from the product information found in the AISC manual. In either of these manuals, we will find that the W8 x 31 has a beam depth of 8" (see to the left of the end view of the beam). From that *A* dimension, we subtract twice the 7/16" thickness of the flange to get the actual space between the flanges. To enable the plate to slide in between the flanges, subtract twice another 1/16", or a total of 1/8". Thus,

[8" – (7/16" x 2) – 1/8"] = 7"

The result is 7". Therefore, our 3/8" plate is 7" long.

To determine the width of the plate, we need to use half of the bottom flange dimension of 8", or 4". Then subtract half of the B for the thickness of the web, or 3/16" for this size beam. Then subtract 1/16" clearance. Thus,

4" – 3/16" – 1/16" = 3 3/4"

The result is 3 3/4". Now we have determined the plate for the stiffeners to be Pl 3/8" x 3 3/4" x 7". It is a bit of work to find all of this, but this is as accurate as it gets without having to detail the project.

If all of the building columns and braced frame columns are the same size beam, you have to find this size only once. Because these stiffeners are at both the roof and 2nd floor level, each column will have (8) of them. According to the detail *Typical Beam to Column Web Connection* in Figure 8.17, that shear plate goes all the way back to the web of the column.

Apply Standard Bolted Connection Schedule A, like that shown in Figure 8.23. The shear plate for the W16 x 31 beam at the 2nd floor level Figure 8.8, will be Pl 5/16" thick x 4 1/2" wide x 1'-4" long. The shear plate for the W14 x 22 beam at the roof level in Figure 8.9 will be Pl 1/4 x 4 1/2 x 1'-0". The shear plate for the beam going the north-south direction will need to be calculated going to the column web. We see from the 2nd floor plan in Figure 8.8 and the roof plan in Figure 8.9 that the beams that frame into these columns are W12 x 14 beams. These beams require 1/4" thick plates that are 1'-0" long (three holes, 4" centers on the holes, with 2" edge distance top and bottom).

These plates must be welded to the column at the same time the stiffeners are installed on the column. We have calculated a dimension on the column from face of web to face of the flanges at 3 3/4". We need to add that to the width of the plate that we normally measure for the connection at 4 1/2". The connection plates will now be sized at 8 5/16" wide to accommodate all these dimensions. The connection plate for the W14 will be 1/4" x 8 5/16" x 12" deep (allowing for 4" centers at holes times two, plus 4" edge distance).

Figure 8.20 provides a simple take off sheet that lists the material we have identified so far.

Labor	Materials	Detail	QTY	Mat. Thkns	Plate shown in inches Width	Length (shapes in ft/dec of the ft)	Ext Wt	Cost Per lb.	Exten. Mat.cost	Exten Hdwre
	Column at grid intersection B-20									
Labor		Detail	9.51024	1	9.5	per ea.	Lbs			
0.8	W8 x 31 x 25'-10 1/2"	at Grid B-20	1			25.875	802	0.48	$ 385	
0.5	Plate 1 x 14	Base - 4 H	1	1.000	14	14	55	0.487	$ 27	
0.5	Plate 1/2 x 8	Cap - 4 H	1	0.500	8	8	9	0.487	$ 4	
1.067	WFL 5/16 x 4'-0"		2			4				
1	AB 3/4 x 1'-6 w/nut & washer		4					2.8		$ 11
2	Plate 3/8	stiffener	4	0.375	3.75	7	11	0.487	$ 5	
1.333	WFL 1/4 x 1'-2 1/2		8			1.25				
0.4	Plate 1/4 x 8 1/4"	Conn. Pl/3H	2	0.250	8.25	12	14	0.487	$ 7	
0.2	Plate 5/16 x 4 1/2"	Conn. Pl/4H	1	0.313	4.5	16	6	0.487	$ 3	
0.2	Plate 1/4 x 4 1/2"	Conn. Pl/3H	1	0.250	4.5	12	4	0.487	$ 2	
0.889	WFL 1/4 x 1'-7 1/2		4			1.6667				
0.355	WFL 1/4 x 1'-4		2			1.33				
0.267	WFL 1/4 x 1'-0		2			1				
10							902		$ 434	$ 11

4H = 4 holes

Conn. Pl/3H = Connection Plate, 3 holes

Figure 8.20 Simple take off sheet

This listing for the material for the first column at grid B-20 in Figure 8.20 shows the column with the attached base and cap plates, the stiffeners required at the floor and roof levels, the shear plates to column web for both the floor and roof, and the stiffeners to the column flanges for both levels of framing. This listing represents all of the steel materials, bolts, holes to be punched or drilled, and welding required to build this one column.

Braced Frames

The columns at the braced frame at grid A between grids 19 and 20 can be taken off as the same as regular columns. However, due to the nature of the labor and the complexity of the details, this sample will have the braced frame columns and beams all together.

Figure 8.21 shows the elevation of braced frame A.

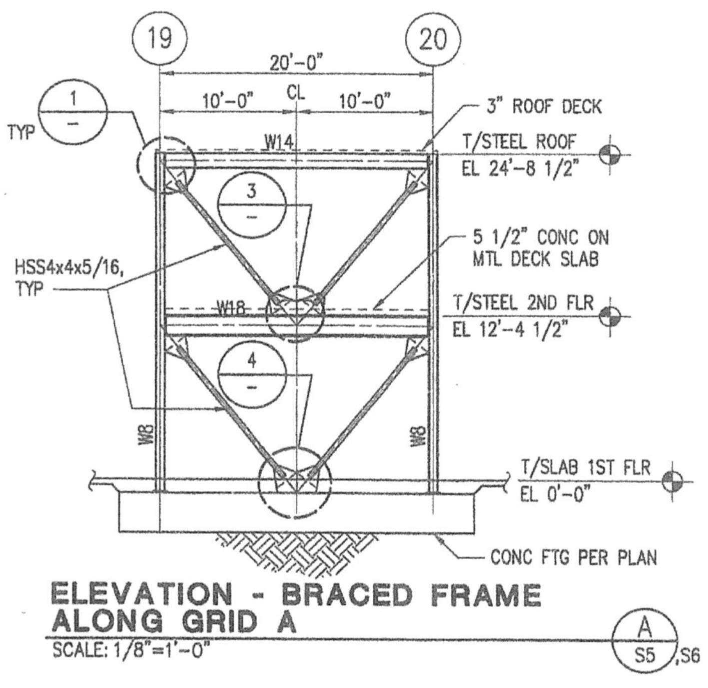

Figure 8.21 Elevation: Braced frame along grid A

These columns are type 3, W8 x 40. The bottom of the footings is 4'-0" below grade, as found in the notes on the foundation plan in Figure 8.5, and is 3'-0" thick. The 1'-0" will be added to the elevation 24'-8 1/2" shown at the roof plan in Figure 8.9, for the sum of 25'-8 1/2". From that 25'-8 1/2" we subtract 3", which is the total of 1 1/2" grout, 1" base plate, and 1/2" cap plate. Therefore, these two columns will be 25'-5 1/2".

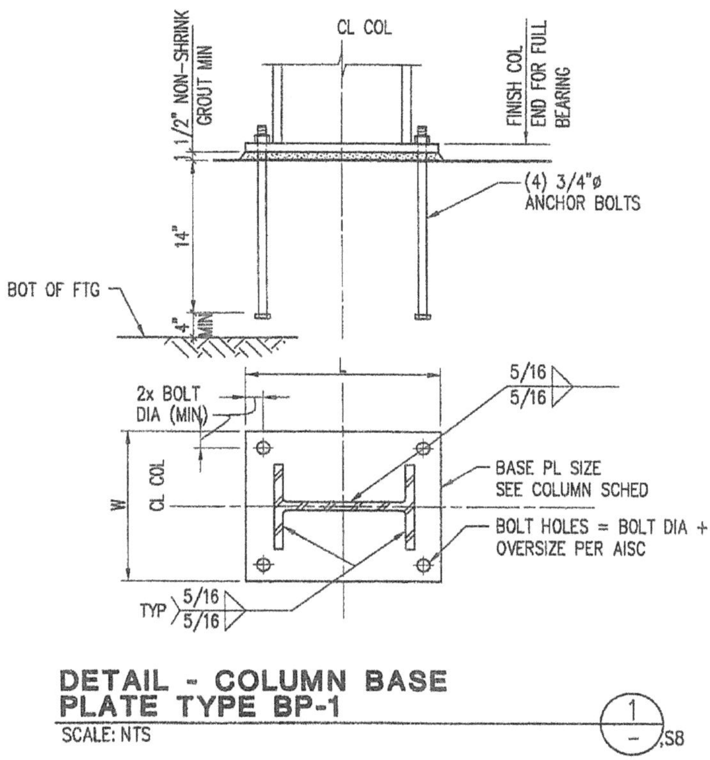

Figure 8.22 Detail: Column base plate type BP-2

The base plate detail for BP-2 at these columns in Figure 8.22 is close to the same as BP-1, although the anchor bolts are 10" longer. To figure the connections for the beams to these columns, we need to go to the Standard Bolted Connection Schedule A in Figure 8.23. From this standard bolted connection schedule, see that the W14 beams need 3 bolts each end whereas the W18 beams need 4 bolts each end.

If you ever find the need to estimate a project where connection details are not shown, the Standard Bolted Connection Schedule shown in Figure 8.23 is what is typically required for most projects that do not have any special moment or seismic conditions where more bolts and heavier welds are required. A general rule of thumb is one bolt for every 3" web depth of beam minus 1 1/2" edge distance. If a W12 beam has a 12" depth, then you would have a three-hole connection, allowing for 19”. The remaining 3” would be the 1 1/2” edge distance top and bottom of the connection beam.

STANDARD BOLTED CONNECTION SCHEDULE "A"			
BEAM SIZE	NUMBER & SIZE OF BOLTS REQ'D	PLATE THICKNESS	WELD SIZE (W)
W8	(2) 7/8"ø	1/4"	1/4"
W10	(2) 7/8"ø @ 4" GA	1/4"	1/4"
W12	(3) 7/8"ø	1/4"	1/4"
W14	(3) 7/8"ø	1/4"	• 1/4"
W16	(4) 7/8"ø	5/16"	1/4"
W18	(4) 7/8"ø	5/16"	1/4"
W21	(5) 7/8"ø	3/8"	5/16"
W24	(6) 7/8"ø	3/8"	5/16"

* USE HORIZONTAL SHORT SLOTTED HOLES IN THE PLATE AT COLUMN CONNECTIONS ONLY.

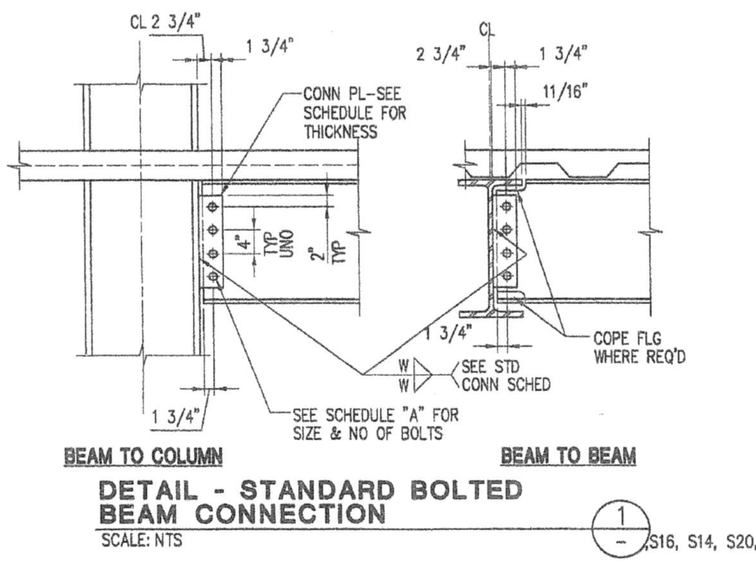

Figure 8.23 Detail: standard bolted beam connection

The typical standard details refers to Beam to Column Web Connections, shown in Figure 8.23. Figures 8.23 and 8.24 both apply to the take off for the braced framed columns. Refer back to the Beam to Column Web Schedule chart F in Figure 8.17; you’ll find that the stiffener plate thickness required is 3/8”. To calculate the width and length of these stiffener plates, we need to find out the space allowance at the column from flange to flange and face of flange to face web:

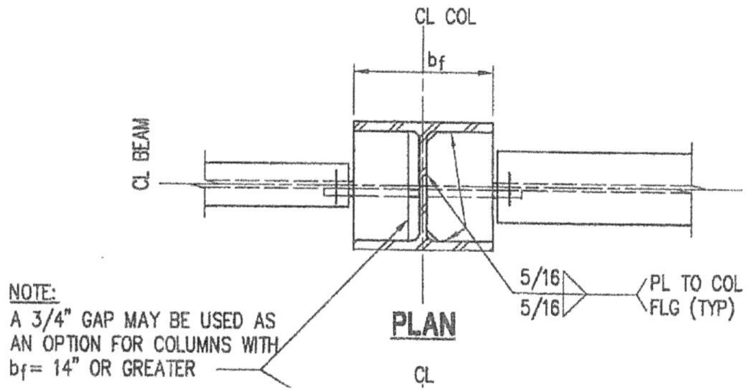

Figure 8.24 Plan view of braced frame beam to column web

In calculating the plate width, it is discovered in Figure 8.25 that the stiffener plates at the columns are a little bit wider:

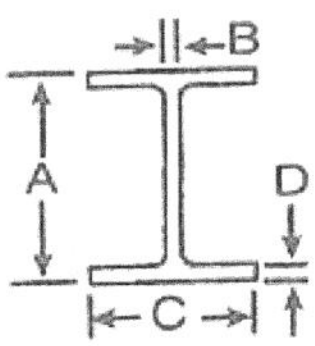

Figure 8.25 Column plan view for dimensioning

In looking at a product catalog beam chart, find the dimensions indicated in Figure 8.25 for the W8 x 40. The *A* distance (top to bottom of the beam flanges) for the W8 x 40 beam is 8 1/4". From that dimension, we will need to subtract twice the flange thickness or *B* dimension of 9/16 (allowing for top and bottom flange thicknesses) to determine the actual space between the flanges.

Once that calculation has been completed, subtract twice 1/16" clearance to enable this plate to slide in between the flange, or a total of 1/8".

[8 1/4" – (2 x 9/16") – 1/8"] = 7"

The result is 7". We now know that our 3/8" plate is 7" long.

To find the width of the plate, we need to use half of the bottom flange or C dimension which, for this beam, is 8 1/8". Subtract half of the B or thickness of the web, which is 3/16". Half of the bottom flange width C is 4 1/16"; then subtract 1/16" clearance.

(8 1/8" – 4 1/16" – 3/16" – 1/16") = 3 13/16"

The result is 3 13/16". Now we have calculated the plate for the stiffeners to be Pl 3/8 x 3 13/16 x 7". If all of the braced frame columns are the same size beam, you only have to find this size once. These stiffeners are at both the roof and 2nd floor level—each column will have (8) of them.

At the plan view for the braced frame columns in Figure 8.24, the columns are turned so that the beam actually frames into the flange of the column. Therefore, shear plates as shown at the Standard Bolted Connection Schedule A per Figure 8.23 will be Pl 5/16" thick x 4 1/2" wide x 1'-4" long.

The shear plate for the beam along the north-south direction will need to be calculated going to the column web. We then need to locate the beams that frame into the braced frame columns at the 2nd floor and roof. We see from the 2nd floor plan in Figure 8.8 and the roof plan in Figure 8.9 that the beams framing into these columns are W12 x 14 beams. These beams require 1/4" thick plates that are 1'-0 long (three holes, 4" centers on the holes, and 2" edge distance top and bottom).

These plates will need to be welded to the column at the same time the stiffeners are installed on the column. We have calculated the dimension on the column from face of web to face of the flanges at 3 13/16". We will need to

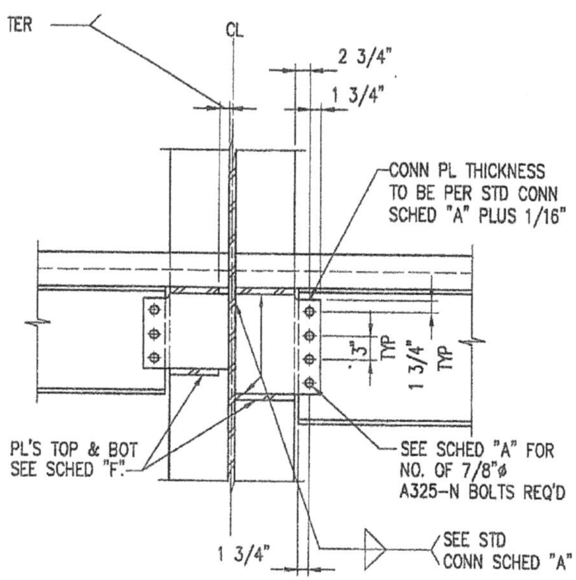

Figure 8.26 Braced Frame Beam to Column Web

add that to the width of the plate. We normally calculate the width for the connection to be 4 1/2". Now your connection plates will be sized 8 5/16" wide to accommodate all those dimensions. The connection plate for the W14 will be 1/4" x 8 5/16" x 12" deep (allowing for 4" centers at holes times two, and 4" edge distance). See Figure 8.26 to see how this connection will appear.

Figure 8.26 shows the framing condition of the braced frame beam connection into the web of the braced frame column. For the connections of the W18 beams at the roof, the plate from the chart in Figure 8.17 is 5/16" thick. The width of the plate will be the same at 8-5/16", but the length will be 4" longer for the additional hole at 16". Now we have two more plates (2) Pl 5/16 x 8 5/16 x 16".

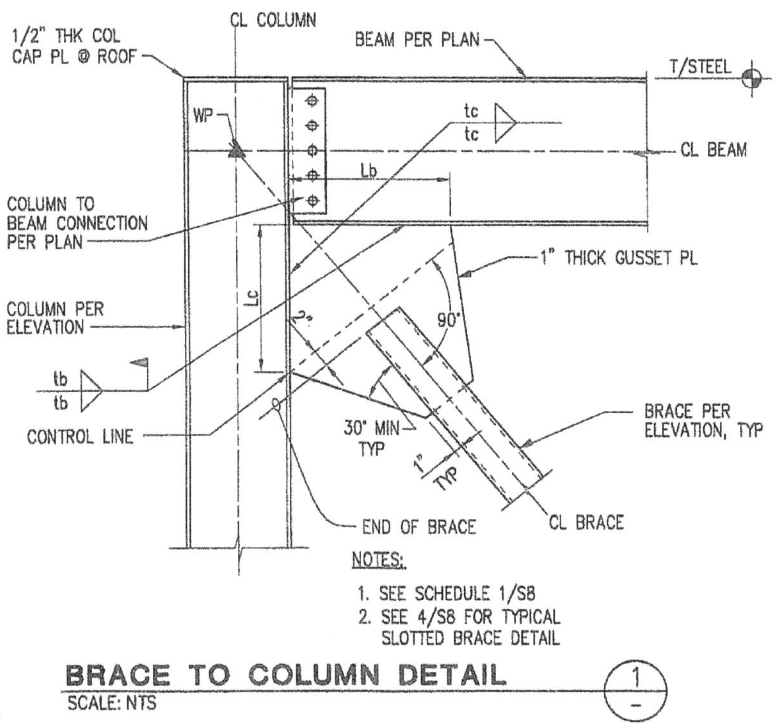

Figure 8.27 Brace to column detail

These gusset plates in Figure 8.27 are welded to the column in the shop, and then field welded to the beams after the beams are installed. This set of structural drawings has a chart for determining the size of the gusset plates, with specific information allowed for the individual braced frames. The width and length of the widest points of the gussets are designated by *Lb* and *Lc*.

How these dimensions are applied will be seen by the referenced detail number for that specific location for each braced frame. The thicknesses of the welds are shown at *tb* and *tc*, with *tc* being the thickness of the shop weld required.

BRACED FRAME ELEVATION	DETAIL	LOCATION	Lb	tb	Lc	tc
A/S8	1/S8	ROOF	17"	3/8	16.5"	3/8
	3/S8	ABOVE 2ND FLOOR	22"	3/8	12.5"	3/8
	1/S8	BELOW 2ND FLOOR	18"	3/8	14"	3/8
	4/S8	FOUNDATION	17"	3/8	18"	3/8
B/S8	1/S8	ROOF	25.5"	3/8	17.5"	3/8
	2/S8	ABOVE 2ND FLOOR	27"	3/8	17"	3/8
	2/S8	BELOW 2ND FLOOR	17.5"	3/8	15.5"	3/8
	4/S8	FOUNDATION	17"	3/8	18"	3/8
C/S8	1/S8	ROOF	17"	3/8	16.5"	3/8
	3/S8	ABOVE 2ND FLOOR	23.5"	3/8	12.5"	3/8
	1/S8	BELOW 2ND FLOOR	19.5"	3/8	14"	3/8
	4/S8	FOUNDATION	16.5"	3/8	17"	3/8
D/S8	1/S8	ROOF	26.5"	3/8	18.5"	3/8
	2/S8	ABOVE 2ND FLOOR	26.5"	3/8	18.5"	3/8
	2/S8	BELOW 2ND FLOOR	17"	3/8	16.5"	3/8
	4/S8	FOUNDATION	16.5"	3/8	17.5"	3/8

NOTES:
1. ALL GUSSET PLATES AND BASE PLATES SHALL BE ASTM A572 GRADE 50.
2. ALL BRACED FRAME WELDING IS SEISMIC CRITICAL. CONTINUOUS VISUAL INSPECTION IS REQUIRED FOR ALL BRACE TO GUSSET PLATE WELDS AND FOR ALL GUSSET PLATE TO BEAM COLUMN WELDS.
3. PROVIDE 1" COMPRESSIBLE JOINT FILLER AROUND BOTH SIDES OF GUSSET PLATE WHERE NOTED ON DETAIL.
4. SEE 4/S8 FOR TYPICAL SLOTTED BRACE DETAIL.
5. FOR DETAILS 3/S7 AND 4/S7, Lc REFERS TO THE HEIGHT OF THE SHARED EDGE.

BRACED FRAME CONNECTION SCHEDULE
SCALE: NTS
1 / S7, S8, S10, S11

Figure 8.28 Braced frame connection schedule.

In Figure 8.28, braced Frame *A* requires that the gusset at the underside of the roof beam use the *Lb* dimension 17" and the *Lc* dimension 16.5". But this doesn't give us the actual size of the plate.

The *Lb* and *Lc* dimensions are to the widest point of the plate. The remainder of the plate will have to be dimensioned, but we see that this detail is NTS (not to scale). Now we have to do some detective work.

The slotted brace detail 4/S8 in Figure 8.29 indicates that the slotted area on the brace needs to be 12" to the gusset. From the brace to column detail, we see we need to allow 2" from the widest point of the gusset plate to the point where the bracing starts.

This additional area is 90 degrees to the widest part of the gusset. The allowance for this additional plate is most easily established by making the plate a rectangle large enough to accommodate the shape being cut. A plate 1" x 20 x 24 will be large enough. The best way to calculate pricing is to make a copy of the detail, then send it to the shop, a subcontractor, or a supplier to price these plates cut to shape. Include that pricing in your estimate.

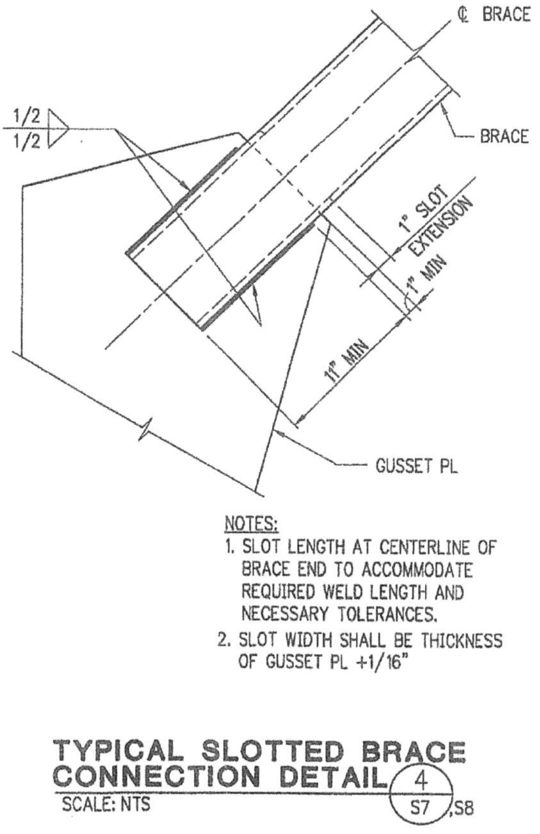

Figure 8.29
Typical slotted brace connection detail

Because we are just creating an estimate, we can calculate this plate to be sized close to what we have to buy to make it. Regardless of the shape that is to be cut, we need a plate large enough to cut this shape out of it. Because this is 1" thick plate, be careful the shape we are pricing isn't *too* large—otherwise, it could sway our pricing in the wrong direction.

Calculating the Length of Braced Frame Beams

Several steps are involved with accurately calculating the lengths of bracing beams. The following example is complicated, so work through these steps slowly. We want to find the length that the braced frame beam needs to be in order to fit between the braced frame columns.

For this example, we will use the braced frame shown in Figure 8.21. To calculate the length of the beam, start by finding the dimension to the centerline of the columns. The centerline of column dimension is the 20'0" located between grid lines 19 and 20.

Next, we need to determine what the dimension is for the width of the braced frame columns. We can then deduct half the width of each column from the centerline dimension we have just found. For this, we need to locate the column sizes.

The braced frame in Figure 8.21 is the section A/S7 shown in the foundation plan Figure 8.5, and again in a close-up view in Figure 8.14. Both figures show us that the column sizes for this braced frame are both noted as number 3. The column size number 3 found in the steel column schedule Figure 8.7 show the size as W8 x 40.

Now we need to find the physical dimensions of these W8 x 40 columns. We can refer either to a steel products guide provided by the steel suppliers or to the AISC manual for the dimensions we need of the W8 x 40, as indicated by Figure 8.19. For the W8 x 40 beams, the A dimension or beam depth is 8 1/4".

We can now proceed with calculating the braced by beam length using this information. We will calculate the length of this beam so that it will fit between the two W8 x 40 columns. After we have calculated the length from column face to column face, we will need to deduct 1" clearance allowance at the end of our calculations to allow the beam to fit between the columns.

From the 20'-0" dimension for the centerline of the braced frame columns, we will deduct half the depth of the W8 x 40 columns that we found from Figure 8.7, twice from the 20'-0" dimension. The column depth is 8-1/4", so half of that is 4-1/8". The calculation is:

(20'-0" – 4 1/8" – 4 1/8" – 1") = 19'-2 3/4"

In this example, the braced frame columns are the same size. Therefore, we could have simply deducted the depth of the column once from the 20'-0" dimension, then subtracted 1" for clearance.

Braced frame column sizes will not always be in pairs, so it is important that we remember to check the column sizes and dimensions for those columns when calculating braced frame beams in this way.

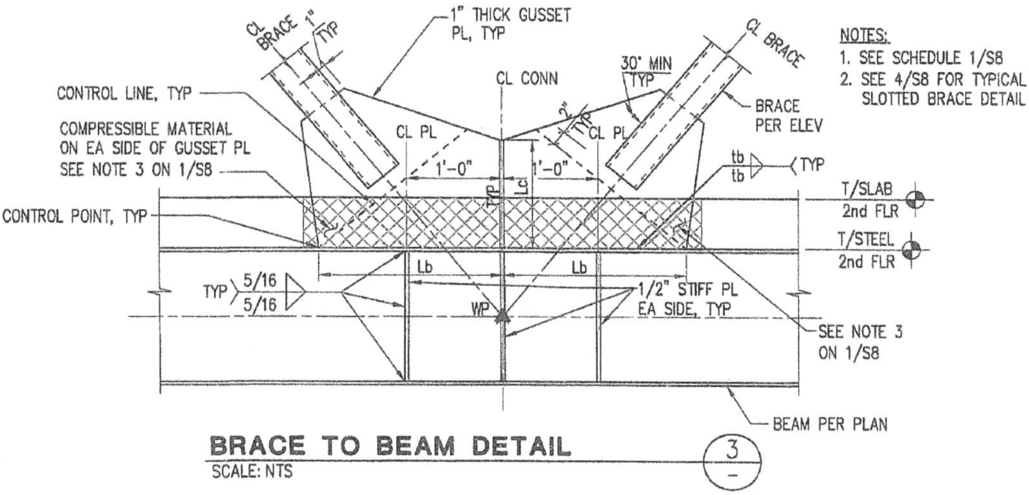

Figure 8.30 Brace to beam detail

Figure 8.30, is a section taken from the braced frame elevation Figure 8.21. This section is called out from the middle of the 2nd floor beam.

To find the size of this beam, we refer back to the 2nd floor plan shown in Figure 8.8, along grid line A between grid lines 19 and 20, where the braced frame section A/S7 is shown. The beam is W18 x 130.

This close up of the W18 x 130 beam in Figure 8.30 shows plate stiffeners at the beam web area between the flanges at three locations. The locations are indicated by the vertical lines at three points of the beam directly below the bracing. See the note to the right of the six vertical lines that says *1/2" stiff pl. ea. side, typ.* Generally, if stiffeners are shown at one side of a beam, it is industry standard that matching stiffeners will be on the other side. Based on the drawing note for this detail, there are beam stiffeners required at (6) locations, (3) on each side of the beam.

Look more closely at the plates and bracing located above the W18 x 130 beam in Figure 8.30. See the plate directly above the middle stiffener in the center of the beam with the gusset plates on either side. We will now work through the calculations required to size all of these stiffener and gusset plates shown at Figure 8.30.

To calculate the size of the stiffeners, we go to the beam chart information, using Figure 8.31.

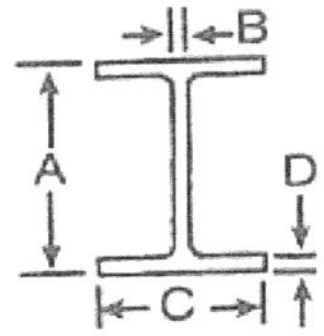

Figure 8.31 Beam section

Calculating the size of beam stiffeners

Referring to Figure 8.31, the A distance is 19 1/4" for the W18 x 130 beam. Subtract the *D* thickness of flange twice as 1 3/16", then the 1/16" clearance on both sides (as 1/8").

[19 1/4" – (1 3/16" x 2) – (1/16" x 2)] = 16 3/4"

The result is 16 3/4".

For the width of the plate, we use half of the *C* or bottom flange measure of 11 1/8", or 5 -9/16". Subtract half of B or the thickness of web (half is 3/8") and the result is 5 3/16". We will also need to deduct 1/16" clearance.

[(1/2 x 11 1/8") – 3/8" – 1/16"] = 5 1/8"

Our stiffener plates are Pl 1/2 x 5 1/8 x 1'-4 3/4" (we always list plate with the thickness first, the width next, and the length in feet and inches).

Calculating the bracing plate on top of the beam

Referring back to Figure 8.30 we look now to calculate the size of the bracing plate that is located at the top center of the beam, between the gusset plates that secure the braces. We know the plate is 1/2" thick for this bracing spacer because it is drawing to match the thickness of the stiffener below. The width of this plate will match the flange width of the beam W18 x 130 at 11 1/8". Thus far we have Pl 1/2 x 11-1/8".

To calculate the length of this bracing plate we refer to the note *Lc* shown at the dimension line running vertical to the right of the plate. This indicator *Lc* takes us back to the chart in Figure 8.27.

Using this chart in Figure 8.27, we are working with the braced frame elevation A/S8, detail 3/S8—above 2nd floor. In the *Lc* column, we see the 12.5" dimension required for the plate at this location. This will be the length required for this particular *Lc* plate. The result of the calculations for the spacer plates is (1) Pl 1/2 x 11 1/8 x 1'-0 1/2".

For the gusset that is welded at the centerline of the beam shown in detail 3/S7 in Figure 8.30, we use the *Lb* dimension at the braced frame connection schedule 1/S8 in Figure 8.27. This dimension is indicated in Figure 8.27 as 22". Without formally calculating the exact size, we can choose to use a rectangle size that accommodates the purchase size of the plate. For example purposes, the plate is sized Pl 1" x 24 x 54".

To determine the length of the bracing TS 4 x 4 x 5/16 between the 2nd floor and the roof, calculate the diagonal length using the rise and run of that slope. The rise is the vertical dimension of the bracing start and stop. The run is the total horizontal dimension of the bracing.

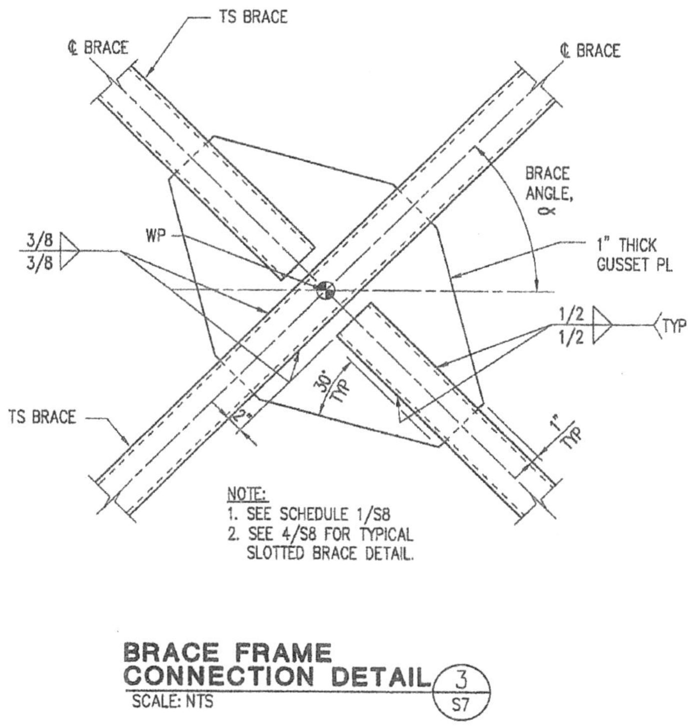

Figure 8.32 Brace frame connection detail

To calculate the run, start with the horizontal gridline span, which is the 20'-0" bay between grids 19 and 20; then divide by 2. Half of the bay dimension of 20'-0 is 10'-0". From that 10'-0", subtract 4 1/8", which is half of the W8 x 40 column width (the d dimension of this W8 x 40 beam is 8 1/4").

(20'-0" /2) = 10'-0"
(10'-0" – 4-1/8") = 9'-7 7/8"

Also subtract half of the *Lb* dimensions from the chart at the braced frame connection schedule at 1/S8 and at detail 3/S8 in Figure 8.28—the *Lb* dimensions vary at the different elevations. Figure 8.32 shows what detail 3/S8 looks like.

Another approach is to add the *Lb* dimensions shown in the chart in Figure 8.28 (17" at detail 1/S8 and 22" at detail 3/S8, totaling 39"), then divide by 2. The result is 19 1/2" (or 1'-7 1/2").

(9'-7-7/8" – 1'-7 1/2") = 8'-0 3/8"

The measure of 19 1/2" allows for both ends of the bracing, because the bracing only comes about half way in on the gusset.

10' – 4 1/8" – 19 1/2" = 8'-0 3/8"

The result is 8'-0 3/8", and this dimension is the run.

The rise is calculated by subtracting the 2nd floor elevation at 12'-4 1/2" from the top of steel elevation at 24'-8 1/2". The difference is 12'-4". Subtract the depth of the beam (14", or 1'-2") that is above this brace (the bracing frames from underneath). Also subtract half of the Lc dimension from the braced frame connection schedule in Figure 8.28, at both details 1/S8 and 3/S8. (These dimensions are 16 1/2" for 1/S8 and 12 1/2" for 3/S8; the sum of those two dimensions is 29". Divide by 2; the amount to subtract is 14 1/2", or 1'-2 1/2".) Thus,

12'-4" – 1'-2" – 1'2 1/2" = 9'-11 1/2"

The result is 9'-11 1/2" for the rise. Using the run of 8'-0 3/8", with the rise of 9'-11 1/2", the diagonal dimension then is calculated at 12'-9 1/2".

This amount was found using a construction calculator that has the capacity to figure chord lengths and rise/run of stairs. We are able to use these same features in calculating the length of the diagonal bracing.

Although this measure may not be the exact length of the bracing for when the steel is detailed, it will be close enough to a use-length for estimating purposes. The dimension of 12'-9 1/2" will be listed on the take off sheet for the bracing length.

The bracing from the ground level to the underside of the 2nd floor is calculated using the bottom of base plate as –1'-0". The top of steel elevation is 12'-4 1/2" for the total of 13'-4 1/2". From this, we subtract 1'-6" for the depth of the 18" beam. We then subtract half of the Lc dimensions at details 2/S8 and 4/S8 in Figure 8.28. (Here, 2/S8 is 14" and 4/S8 is 18", for a total of 32", which is then divided by 2 for a measure of 16", or 1'-4"). Thus, we have

13'-4 1/2" – 1'-6" – 1'4" = 10'-6 1/2"

The rise is 10'-6 1/2".

The run is calculated as 10'-0" (half the bay dimension) minus the Lb dimension. (Here, 2/S8 is 18" and 4/S8 is 17" for a total of 35". Dividing by 2 is 17 1/2", or 1'-5 1/2"). We also subtract half the W18 x 40 column width dimension of 8 1/4" being 4 1/8". Thus,

10'-0" – 1'-5 1/2" – 4 1/8" = 8'-2 3/8"

for a total of a 8'-2 3/8" run. Calculating an 8'-2 3/8" run with a 10'-6 1/2" rise leads to a resulting diagonal length of 13'-4 1/4".

In Figure 8.33's material take off listing, we have noted the braced frame as Braced Frame *A-7—Grid A* so we will know what detail the materials are listed for. We show the quantity of holes at the base and cap plates in the Detail column by indicating Base—4H and Cap—4H (meaning base plate, 4 holes, and cap plate, 4 holes). These notes indicate the labor time necessary for putting the holes in the plates. These holes indicate the quantity of the anchor bolts that will needed. Similarly, the bolts needed for the cap plate connection are counted by the holes in the cap plate.

Additional notes are included in the detail column showing *stiff* for stiffeners and *Conn* for connection plates. Notes providing reference to particular details indicate where this material is shown in the drawings.

We note the holes at the connection plates for labor purposes and for counting the bolts we will need. Small sketches may be included on the right

Labor	Materials	Detail	QTY	Mat. Thkns	Plate shown in inches Width	Length (shapes in ft/dec of the ft)	Extended Weight	Cost Per lb.	Extended Mat. Cos	Extended Hardware
	Braced Frame A-7 - Grid A									
Labor	**Between grids 19 and 20**	**Detail**	30.718	9	3.4	per ea.	Lbs			
1.6	W8 x 40	Columns	8			20	2499	0.48	$ 1,200	
1	Plate 1 x 15	Base - 4H	2	1	15.0	15	127	0.487	$ 62	
1	Plate 1/2 x 8 1/8	Cap - 4 H	2	0.5	8.1	8.2	19	0.487	$ 9	
1.422	WFL 5/16 x 2'-8	weld to cap/base	4			2.6667				
2	AB 7/8 x 2'-6 w/nut & washers	G105/ F1554	8					28		$ 224
0.2	A325 3/4 x 2-1/2" w/nut & washers	A325	4					1.99		$ 8
8	Plate 3/8"	stiffener 2/S14	16	0.375	3.8	7	45	0.487	$ 22	
5.33	WFL 1/4 x 1'-3	weld to col web	32			1.25				
0.8	Plate 1/4 x 8 5/16"	Conn. Pl/3H	4	0.25	8.3	12	28	0.487	$ 14	
0.4	Plate 5/16 x 4-1/2"	Conn. Pl/4H	2	0.3125	4.5	16	13	0.487	$ 6	
0.4	Plate 1/4 x 4-1/2"	Conn. Pl/3H	2	0.25	4.5	12	8	0.487	$ 4	
1.6	A325 3/4 x 2-1/2" w/nut & washers	A325	32					1.99		$ 64
2	Plate 1 x 20	1/S14 - gusset	4	1	20.0	24	543	0.487	$ 265	
1.333	WFL 1/4 x 1'-3	Weld to Col	8			1.25				
1.6	W14 x 30	Roof Beam - 6 H	1			19.2292	577	0.48	$ 277	
0.4	Plate 1/4 x 4-1/2"	Conn. Pl/3H	2	0.25	4.5	12	8	0.487	$ 4	
1.333	WFL 1/4 x 1'-0	Weld to Beam	4			1				
0.3	A325 3/4 x 2-1/2" w/nut & washers	A325	6					1.99		$ 12
31							3867		$ 1,862	$ 308

Figure 8.33 Braced frame columns material take off listing

Labor	Materials	Detail	QTY	Mat. Thkns	Plate shown in inches Width (shapes in ft/dec of the ft)	Length	Ext Wt	Cost Per lb.	Exten. Mat.cost	Exten Hdwre
	Braced Frame A-7 - Grid A - continued									
Labor	Between grids 19 and 20	Detail	34.7667	6	5.8	per ea.	Lbs			
0.8	W8 x 130	beam 2nd floor	1			19.25	2503	0.48	$ 1,201	
3	Plate 1 x 15	stiffeners	6	0.500	5.1875	16.75	74	0.487	$ 36	
3.6	WFL 3/8 x 2'-3	weld stiff to beam	12			2.25				
1	Plate 1/2 x 5	vert stiffener	2	0.500	5	12.5	18	0.487	$ 9	
0.8	WFL 3/8 x 1'-6	weld to cap/base	4			1.5				
1	Plate 1 x 24	Gusset plate	2	0.500	5	12.5	18	0.487	$ 9	
14.4	WFL 3/8 x 4'-6	weld to cap/base	2			54				
0.4	Plate 1/4 x 4 1/2"	Conn. Pl/3H	2	0.250	4.5	12	8	0.487	$ 4	
0.533	WFL 1/4 x 1'-0	weld to beam	4			1				
1.5	A325 7/8" x 2 1/2 w/nut & washer	A325	6					1.99		$ 12
1.6	TS 4 x 4 x 5/16 - Brace	1 1/8" - slot 11" BE	2			12.7917	379	0.48	$ 182	
1.6	TS 4 x 4 x 5/16 - Brace	1 1/8" - slot 11" BE	2			13.3542	395	0.48	$ 190	
	Gusset Embed BF - A									
0.2	Plate 1 1/4 x 13	gusset base pl - 8H	1	1.250	13	48	221	0.487	$ 108	
0.4	Plate 1/2 x 5	Center stiffeners	2	1.250	13	48	441	0.487	$ 215	
0.667	WFL 3/8 x 1'-11	weld to col web	4			1.25				
0.2	Plate 1 x 24	Vert Gusset	1	1.000	24	54	367	0.487	$ 179	
1.067	WFL 3/8 x 4'-0	weld to base plate	2			4				
2	AB 7/8" x 2'-6 w/nut & washers	G105 / F1554	8					28		$224
35							4422		$2,131	$236

Figure 8.34 Braced frame beam material take off listing

to show the reader the particular fabrication conditions for which we are listing materials.

The take off sheet in Figure 8.34 has the W18 x 130 beam for the 2nd floor and stiffener plates that go on both sides of the web of this beam. It also has the big gusset plate with the vertical center stiffeners for the bracing above. The connection plate for the W12 x 14 beam located near the center of this beam at the 2nd floor is included here as well.

The bracings are listed with the required slot size in the notes. The embed gusset for the braced frame type A shows the base plate with the 8 holes for the anchor bolts, and the gussets at Pl 1 x 24 x 4'-6" with the vertical stiffeners at the center at Pl 1/2 x 5 x 1'-6".

These holes for the anchor bolts at the base plate will be larger than the bolts in the oversize condition of 1/8", making the holes 1" diameter. These holes will have to be drilled, as is always the case when the holes are equal to or less than the thickness of the plate.

Floor and Roof Beams

Steel floor and roof beams are listed on the take off sheets using an east-west and north-south system. This system allows readers to easily match or locate beams from the take off in relationship to the structural drawings.

By grouping similar beams, we can shorten our take-off time and combine similar labor applications. Grouping makes our job as steel estimators a little less tedious when we go back and calculate our shop labor time.

We start by listing the beams, east to west on the page, column to column,

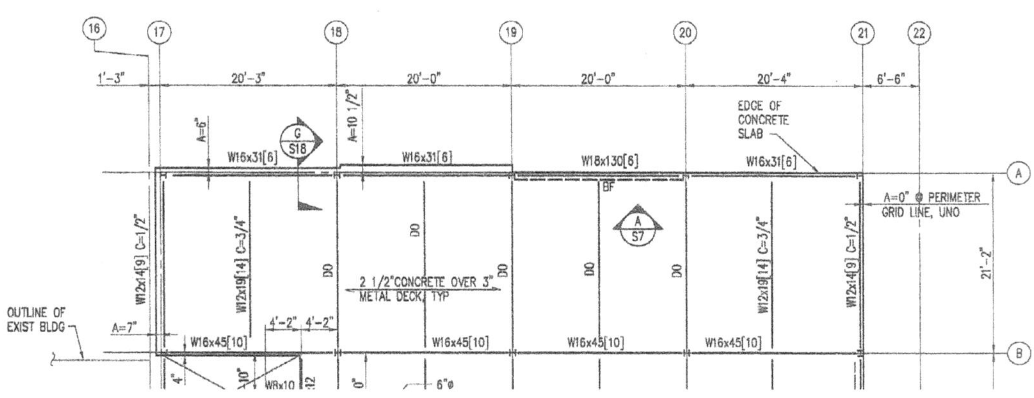

Figure 8.35 Close up of 2nd floor plan

by grid lines first. This will help us to list beams that are of the same length and generally with the same type of beam connections, making it easier to calculate labor for these beams.

Looking at this section of the 2nd floor plan in Figure 8.35, we see that the beams span between column flanges. To arrive at the actual length of the beam, we have to find the column size to deduct the width of that beam from the centerline of grid dimensions.

We found at the column take off that the typical sizes of these columns are W8 x 31 and W8 x 40. Depth of these beams varies by 1/8". The W8 x 30 is 8" deep from the top of the flange to the underside of the bottom flange, and the W8 x 40 is 8 1/8". This small amount of variance is not worth the time to calculate the exact use-length of these beams—our purpose is to create only an estimate. Exact lengths are left for the steel detailers to calculate.

Allowing for beam clearance of 1/2" each end, and using our nominal or typical column width of 8", we will subtract 9" from the bay dimensions gridline to gridline to arrive at an approximate fabricated length for these beams. Taking another look at this 2nd floor plan in Figure 8.36, we want to group similar beams with similar length and detail connections.

The shear plate connections for the beams that are connected to columns were listed in the column take off. These connections will not need to be listed with these beams.

The connections that are for beams that frame into other beams will be listed with the beams that they weld to. As a result, all the standard bolted connections will be with the parts to which they are welded, and the take off goes more quickly. The labor for those connections doesn't change even though they will be listed at a different place on our take off.

We list the W16 x 31 in our take off in Figure 8.36. We list only the beam because the connections for the ends of this beam to the columns are already

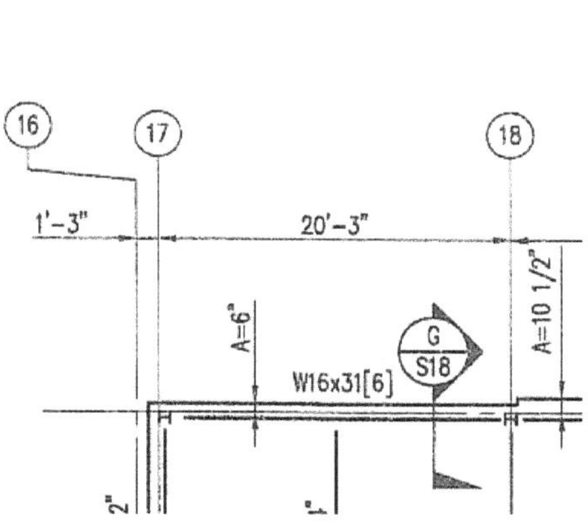

Figure 8.36
Close up of beam size spanning column to column

included in the column take off. We will include in the beam listing the connection for the W12 x 19 beam that frames into this one.

This method of take off allows the labor to be individually applied to each part. The beams are listed going east and west, column to column, and then east and west the ones that frame beam to beam. This format will help us to keep similar labor items together, thereby simplifying our labor calculations.

Figure 8.38 shows a sample take off with the beams column to column from Figure 8.37, then beam to beam all running east and west at the 2nd floor

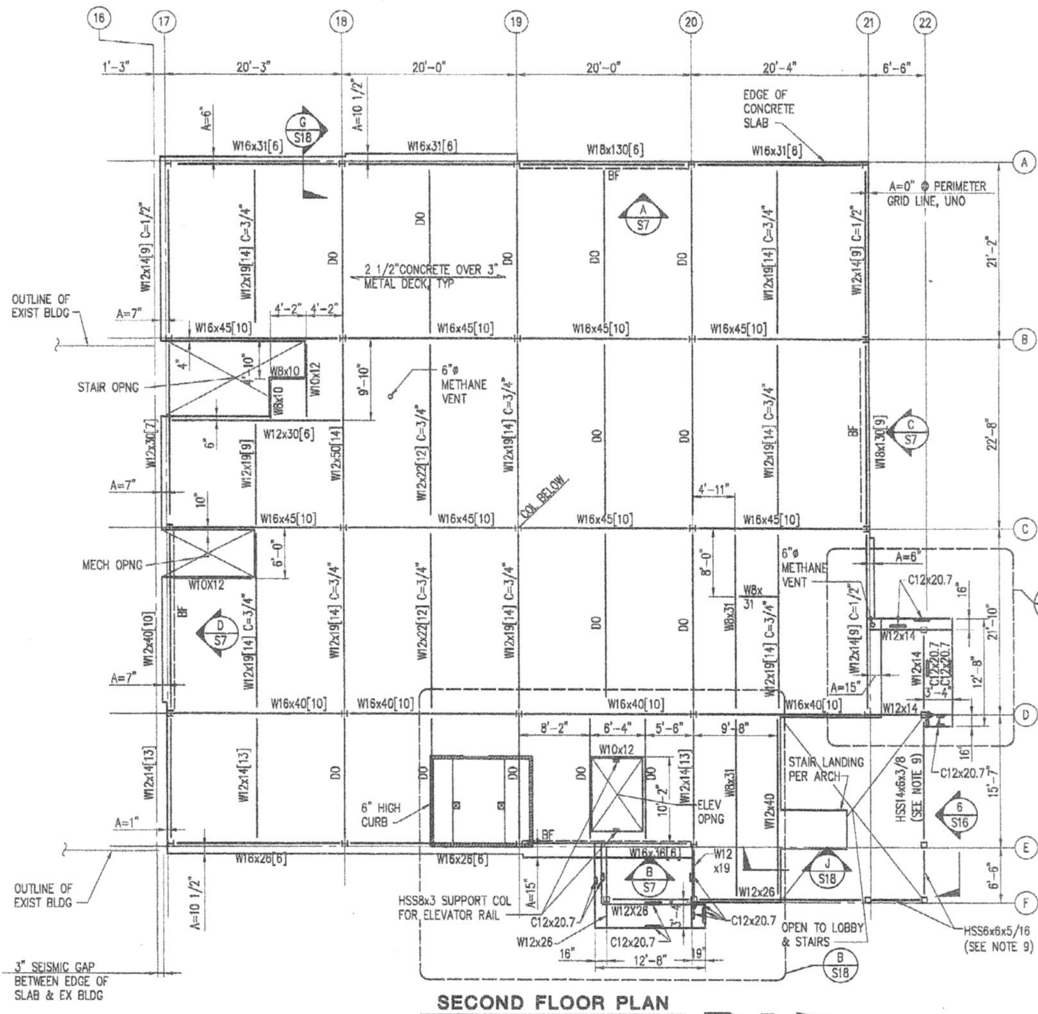

Figure 8.37 Second floor plan

Labor	Materials	Detail	QTY	Mat. Thkns	Plate shown in inches Width	Length (shapes in ft/dec of the ft)	Ext Wt	Cost Per lb.	Exten. Mat.cost	Exten Hdwre
	Beams East and West, grids 17 to 21, sheet S6									
Labor	Beams Column to column	Detail	67.04	6	11.2	per ea.	Lbs			
0.8	W16 x 31	8H	1			19.5	605	0.48	$ 290	
0.8	W16 x 31	8H	1			19.25	597	0.48	$ 286	
0.8	W16 x 31	8H	1			19.5833	607	0.48	$ 291	
1.6	W16 x 45	8H	2			19.5	1755	0.48	$ 842	
3.2	W16 x 45	8H	4			19.25	3465	0.48	$ 1,663	
1.6	W16 x 45	8H	2			19.5833	1762	0.48	$ 846	
0.8	W16 x 40	8H	1			19.5	780	0.48	$ 374	
1.6	W16 x 40	8H	2			19.25	1540	0.48	$ 739	
0.8	W16 x 40	8H	1			19.5833	783	0.48	$ 376	
0.8	W16 x 26	8H	1			19.5	507	0.48	$ 243	
1.6	W16 x 26	8H	2			19.25	1001	0.48	$ 480	
0.8	W12 x 26	6H	1			19.5833	509	0.48	$ 244	
6	Plate 1/4 x 4 1/2"	Conn. Pl/4H	30	0.250	4.5	16	153	0.487	$ 74	
10.64	WFL 1/4 x 1'-0	weld to beam	60			1.33				
	Beams - Beam to Beam, East and West									
0.8	W8 x 10	4H - C2E	1			4.375	44	0.48	$ 21	
0.8	W12 x 30	6H - C2E	1			20.125	604	0.48	$ 290	
0.8	W10 x 12	4H - C2E	1			9.875	119	0.48	$ 57	
0.8	W8 x 31	4H - C2E	1			4.9583	154	0.48	$ 74	
0.8	W10 x 12	4H - C2E	1			6.2083	74	0.48	$ 36	
1.8	Plate 1/4 x 4 1/2"	Conn. Pl/3H	9	0.250	4.5	12	34	0.487	$ 17	
2.4	WFL 1/4 x 1'-0	weld to beam	18			1				
27	A325 7/8" x 2 1/2 w/nut & washer	A325N	108					1.99		$215
67							15093		$7,246	$215

Figure 8.38 Sample of beam take-off listing

plan between grids A thru E, 17 thru 21.

The take off in Figure 8.38 lists the beams between grids 17 and 21, A thru E from Figure 8.37, going from east to west. Not listed are the beams that go with the braced frames—these were already listed in the take off with the fabrication for the braced frames. These beams have been listed as column to column first, then the intermediate beam to beams. Grouping the beams this way simplifies calculating the labor hours.

Look at the W16 x 31 listed first, and see the indicator 8 H in the Detail column. This designates that this beam will have 8 holes, with four holes at each end as allowed per the Standard Bolted Connection Schedule (Figure 8.23) we have reviewed previously. The number of shear studs or HAS shown on the beams, as indicated by the number inside the parenthesis, is not listed here. Generally the holes are not an item provided by the steel fabricator. However, they could be listed on the take off for the individual beam if the fabricator wanted to keep a count.

The listing *c2e* on the W8 x 10 in the Detail column is intended to mean cope 2 ends. This listing indicates that the beam has a notch cut out of it, cutting out part of the top flange and a section of web, and allowing for this beam to frame into other beams. The picture in Figure 8.39 shows a beam in elevation with a cope out of the top flange so that it will fit more closely to the connecting beam.

The length of this beam, 4'-2", is the dimension from the plan—we have deducted 1 1/2" for a set back on this beam to allow for clearance. This beam

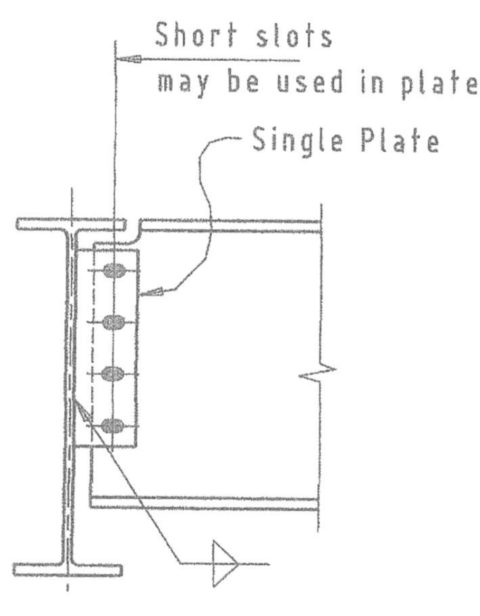

Figure 8.39
Wide flange to wide flange connection

will frame into other beams. Therefore, the deduction for the setback dimensions, the 1/2" clearance at each end, and the web thickness of the beam this member frames into, falls within an inch of the beam's actual use-length. For the ease of listing materials, we have not calculated the actual length. The shear plate connections are listed below the beams. We count all similar size beams, multiply that by 2, allowing for a connection at each end. This gives us a quick count of the connections for each size beam.

The connections that are welded to these beams are indicated as the A36 plate 1/4" x 4 1/2" x 1'-4" and the Plate 1/4" x 4 1/2" x 1'-0" for the three-hole and two-hole connections that will be required for the W12, the W10, and the W8 beams that are in this framing.

The wfl (weld fillet) is the note for the weld; 5/16" and 1/4" are the thickness of the weld called out in the Standard Bolted Connection schedule (Figure 8.23). This is followed by the length of the weld, which in this case matches the length of the part being welded. The length of the weld is not put in the Length column so that readers do not think of the weld as a material item.

The quantity of the welds is the sum of the connections multiplied x 2, one weld for each side of the shear plate being welded. This is a quick way to count the number of welds. Last, we multiply the number of shear plates by the number of holes for the part. This gives us an easy way to count the bolts. We total the hole counts to arrive at the quantity of bolts that need to be supplied for these connections. The bolts are listed last, along with the length, including nuts and washers, and the grade of the materials for the bolts.

The length of the bolts is determined by allowing for twice the bolt diameter plus the thickness of the materials that are being bolted together (also known as the grip). If the bolt is 7/8" diameter, twice that diameter is 1 3/4". Adding 3/8" for the thickness of the shear plate and 1/4" for the thickness of the beam web, the result is 2 3/8". Pick a size to the nearest 1/2", and that size bolt could be used for pricing. For this condition, 2 1/2" long bolts would be priced.

Now we have listed all of the beams going east and west in this plan. We have listed all of the standard bolted connections as welded to these beams in the shop. We have also listed all the welds associated with these parts and all the holes that need to be drilled or punched, as well as all the bolts needed for these connections. Having completed that task, we may now move on to the beams that run north and south. See Figure 8.40.

Labor	Materials	Detail	QTY	Mat. Thkns	Plate shown in inches Width	Length	Ext Wt	Cost Per lb.	Exten. Mat.cost	Exten Hdwre
					shapes in ft/dec of the ft					
	Beams North and South, grids A to E, sheet S6									
Labor	Beams Column to column	Detail	36.3778	6	6.1	per ea.	Lbs			
3.2	W12 x 14	6H	4			14.8333	831	0.48	$ 399	
2.4	W12 x 19	6H	3			21.0833	1202	0.48	$ 577	
0.8	W12 x 14	6H	1			21.0833	295	0.48	$ 142	
0.8	W12 x 30	6H	1			21.9167	658	0.48	$ 316	
0.8	W12 x 50	6H	1			21.9167	1096	0.48	$ 526	
1.6	W12 x 19	6H - Cam 3/4	2			21.9167	833	0.48	$ 400	
0.8	W12 x 14	6H - Cam 1/2	1			20.4167	286	0.48	$ 137	
3.2	W12 x 19	6H - Cam 3/4	4			20.4167	1552	0.48	$ 745	
	Beams - beam to beam grids A to E									
2.4	W12 x 19	6H - C2E	3			21.7083	1237	0.48	$ 594	
3.2	W12 x 14	6H - C2E	4			15.4583	866	0.48	$ 416	
0.8	W12 x 22	6H - C2E	1			21.7083	478	0.48	$ 229	
0.8	W8 x 31	4H - C2E	1			21.9583	681	0.48	$ 327	
0.8	W8 x 31	4H - C2E	1			21.7083	673	0.48	$ 323	
0.8	W12 x 19	6H - C2E	1			12.7083	241	0.48	$ 116	
0.8	W12 x 22	6H - Cam 3/4 - C2E	1			22.5417	496	0.48	$ 238	
1.6	W12 x 19	6H - Cam 3/4 - C2E	2			22.5417	857	0.48	$ 411	
3.2	W12 x 19	6H - Cam 3/4 - C2E	4			21.0417	1599	0.48	$ 768	
0.8	Plate 1/4 x 4 1/2"	Conn. Pl/3H	4	0.250	4.5	12	15	0.487	$ 7	
1.067	WFL 1/4 x 1'-0	weld to beam	8			1				
0.8	Plate 1/4 x 4 1/2"	Conn. Pl/2H	4	0.250	4.5	8	10	0.487	$ 5	
0.711	WFL 1/4 x 0'-8"	weld to beam	8			0.6667				
5	A325 7/8" x 2 1/2 w/nut & washer	A325N	20					1.99		$ 40
36							13904		$6,674	$ 40

Figure 8.40 Material listing for Beams Column to Column

Listing the steel beams continues in this manner for every floor and the roof. The list includes any canopy framing, elevator framing, and stair framing. Keeping all the large framing members listed together will make it easier to find parts and pieces later if the need arises.

Detail Material

After the beams are all listed, we move to the detail material. We use the Detail Material reference for everything steel that is not part of the main building framing. The needed information is found by reviewing the floor and roof details.

We'll start with the partial plan for the second floor. This section was reviewed in Chapter 7, so we are already familiar with many of the details.

In Figure 8.41, we review the close up of the 2nd floor plan. Review Section A/S17 that cuts through the elevator. We will use this detail to list the materials that need to be fabricated.

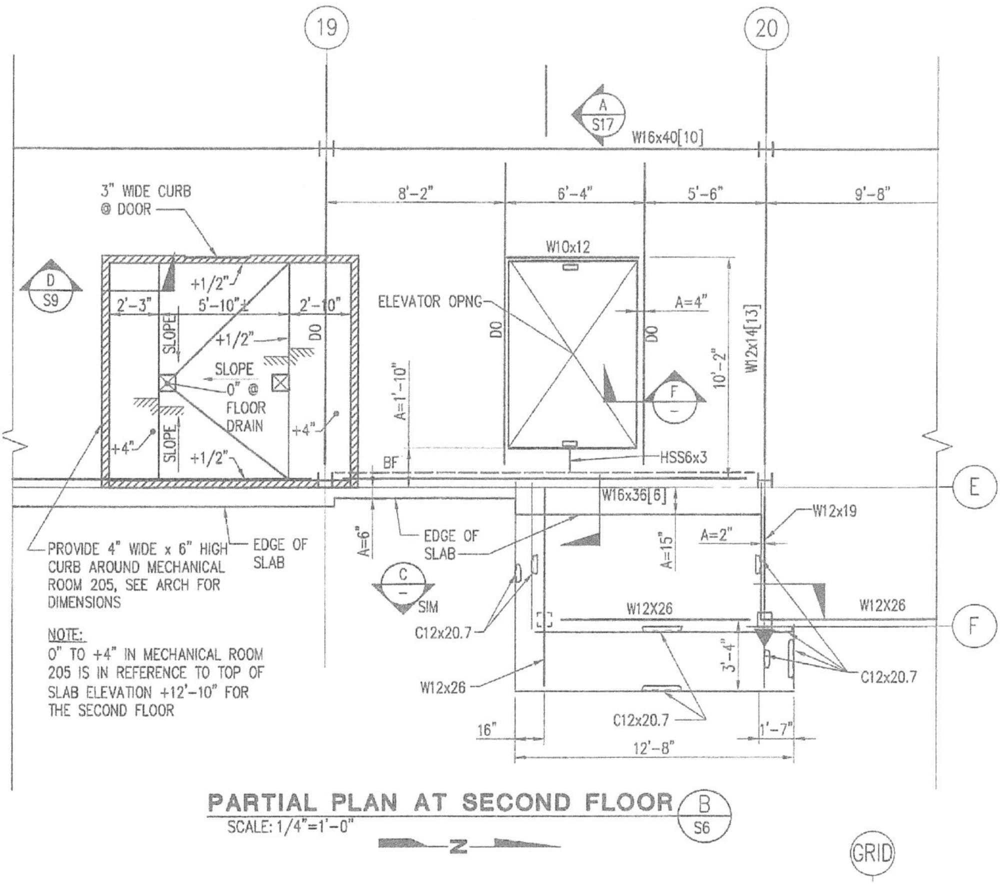

Figure 8.41 Partial plan at second floor

Starting with the columns (2) TS 8 x 3 x 1/4, determining the length will be somewhat of a challenge. Detail 5 in Figure 8.42 shows us the top of the columns, with detail 2 showing us the base.

Combine the dimensions of 12'-4 1/2" and 14'-0" from the top of the slab to the top of the roof beam for a total of 26'-4 1/2". We find from Detail 5 in Figure 8.42 that we need to subtract a dimension that will get us to the underside of the W8 elevator beam, and then add the 1'-6" from Detail 2 in Figure 8.42 or the additional length of the column where it connects to the sidewall of the elevator pit.

We can make this easy on ourselves by guessing that the W14 beam is 14" deep, and we will need to come up about 3" from the bottom of the W14 to get to the bottom of the W8: 14" minus 3" is 11". We already know we need to add 1'-6" from Detail 2 in Figure 8.42.

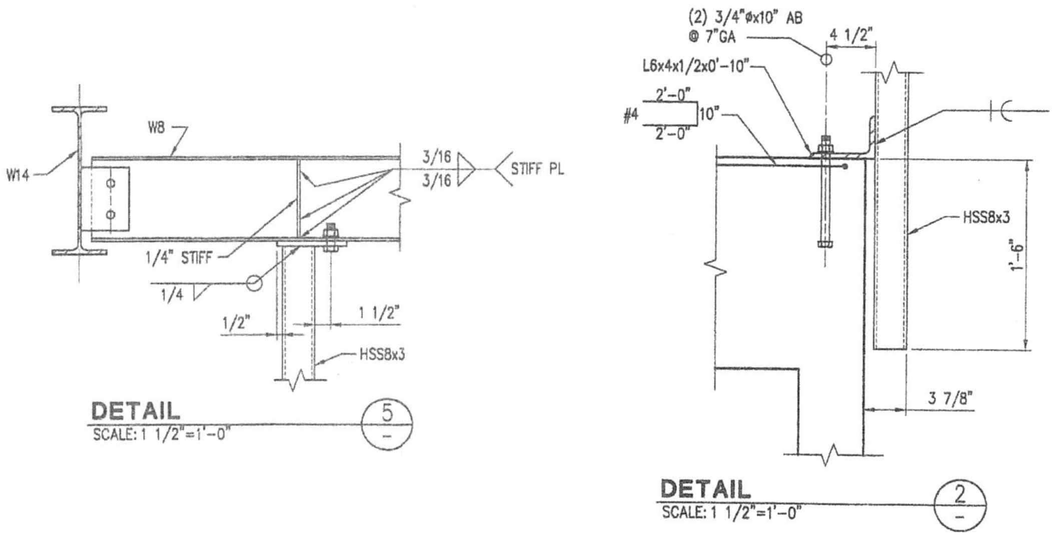

Figure 8.42 Cap and base connection for elevator column

The calculation for the column lengths then becomes 26'-4 1/2" minus 11" plus 1'-6" equals 26'-11 1/2". Add the cap plates shown in Details 4 and 5 in Figure 8.43. The material for the cap plate is called out as Bar 1/2 x 6 1/2 x 0'-9", but a piece of plate may be used. Four bolts, two for each cap plate, are 2" long to cover. Steel detailers will provide the actual length for fabrication.

To calculate the length of the W8 x 13 elevator beam, start with the dimension of 10'-2" shown in the roof plan in Figures 8.9 or 8.54; this measure is from the centerlines of the W14 x 26 beam at grid E between grids 19 and 20 and the W10 x 12 beam at grid E.75 (below grid D). Subtract 1" clearance for both ends of this beam. Detail 5 in Figure 8.44 shows the left end of the beam at grid line E whereas detail 4 shows the right hand end of the beam towards grid line D. Calculate the beam at 10'-0". This beam is also located at the top of Figure 8.43 between grids E and D.

The connections will be the same as what are used on the rest of the project—Plate 1/4 x 4 1/2 x 6" long, with two bolts at each end of the beam.

Add for the stiffeners shown in Details 4 and 5 in Figure 8.42 (both sides of the beam in Detail 5, and only the far side of the beam in detail 4), for (3) each Plate 1/4 x 1 13/16 x 7 1/2". Use the beam chart for the W8 x 13 to create a plate that fits between the top and bottom flanges, from the face of the web to the outside edge of the flange dimension in the same manner as was done for the braced frame calculations.

Finish with the rest of the details at this location by listing the materials that connect the column base with Detail 2 in Figure 8.45, and the intermediate supports at the 2nd floor, Figure 8.46.

The purpose of the TS 6 x 3 x 1/4 tubing is to keep the TS 8 x 3 x 1/4 vertical.

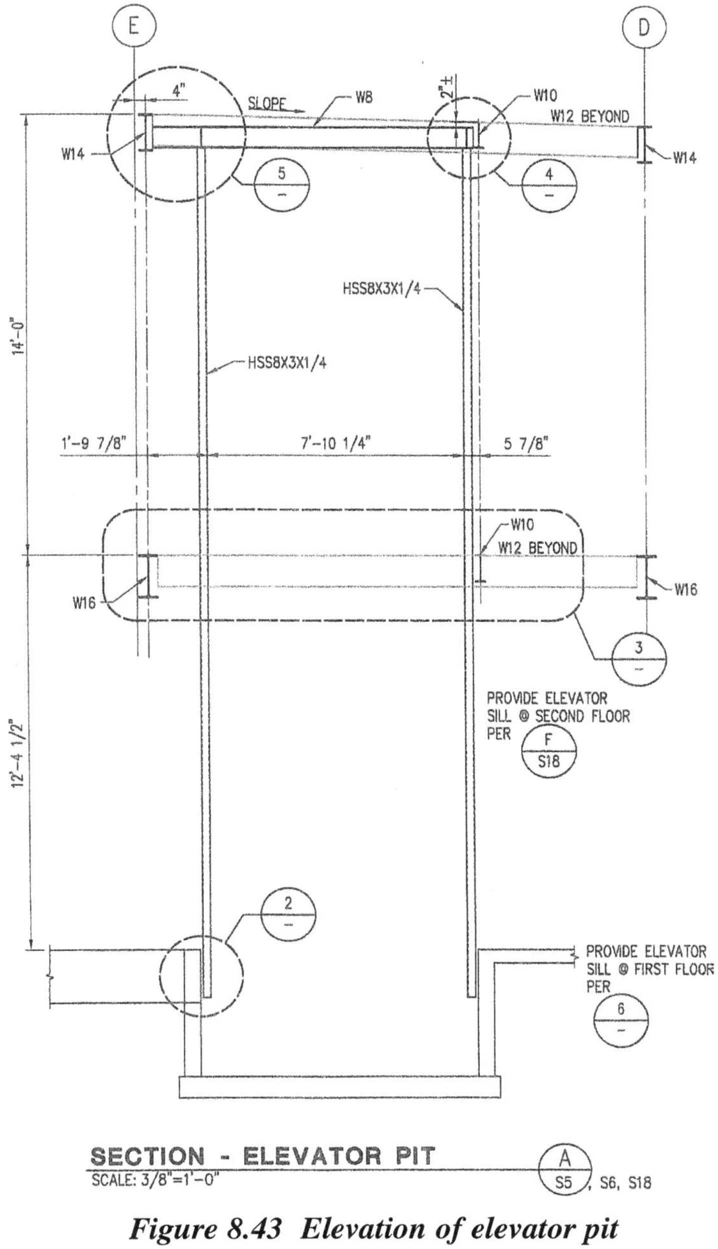

Figure 8.43 Elevation of elevator pit

The dimension given in Detail 3 in Figure 8.46 of 1'-9 7/8" is from the inside of the TS 8 x 3, to the centerline of the W14. The width of the column of 3" has to be subtracted from 1'-9 7/8", as well as half of the web thickness of the W14 x 26 beam (1/8"). Then subtract 1/2" for the thickness of the connection plate and finally subtract 1/4" clearance. The finished length result is 1'-6" for the length of the TS 6 x 3 x 1/4 at the left side of the column.

For the right side of the column, the starting dimension is 5 7/8". Subtract 3" for the width of the column, subtract 1/8" for half the thickness of the web of the W10 x 12 beam, subtract 1/2" for the thickness of the connection plate, and subtract the 1/4" clearance. That leaves a 2" length for the TS 6 x 3 x 1/4.

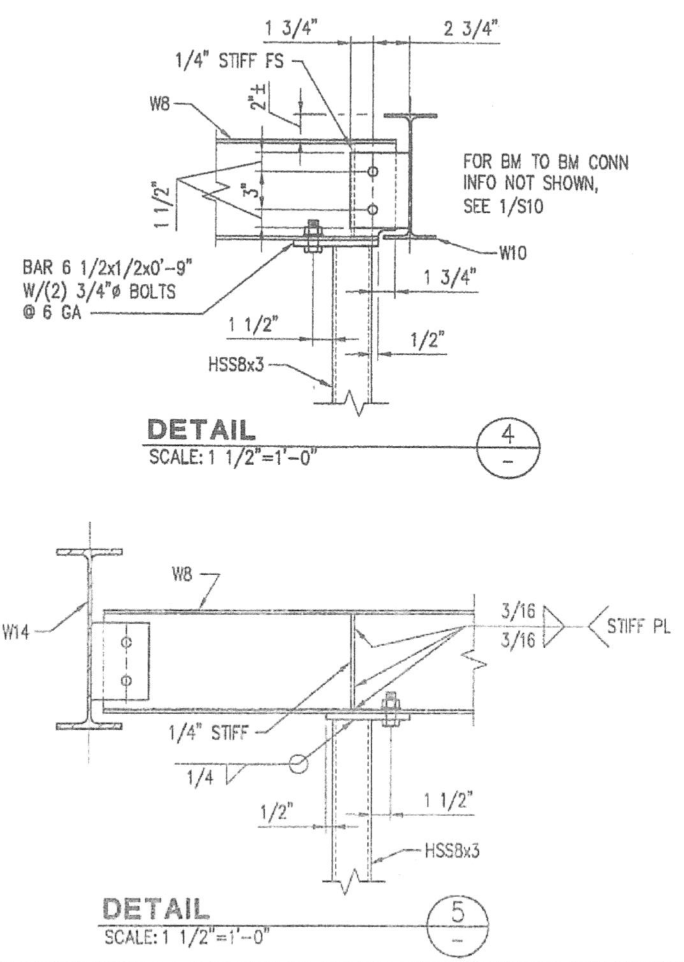

Figure 8.44 Cap plate connections for elevator support columns

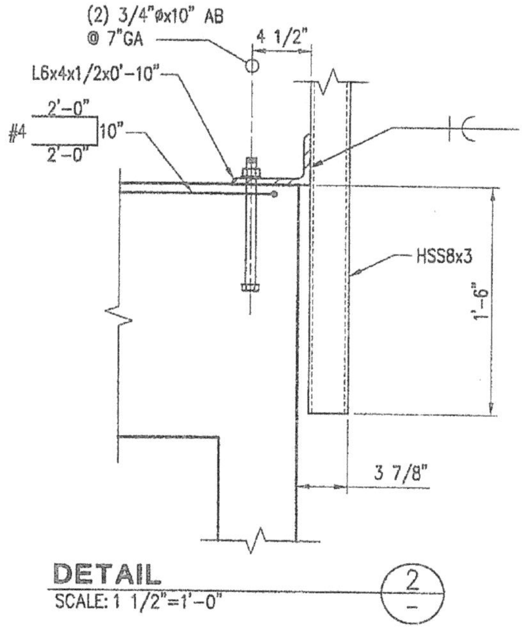

Figure 8.45 Elevator column base connection

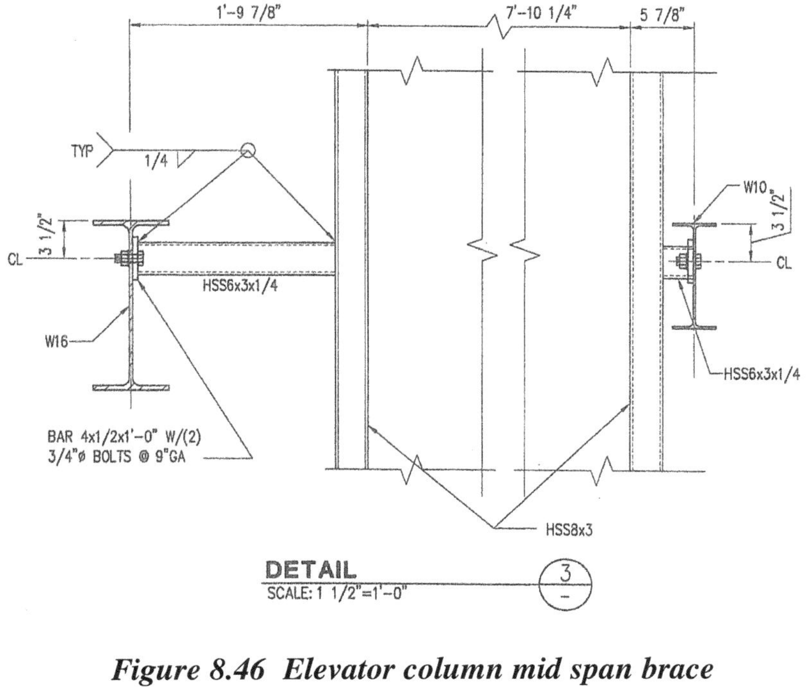

Figure 8.46 Elevator column mid span brace

Some estimators would thumbnail these lengths. However, by going through this exercise, you have learned how to more accurately measure the actual use lengths. In doing so, information can be discovered that may not normally be noticed.

The 2" dimension for the TS 6 x 3 x 1/4 at the right side connection of the column shown at Detail 3 is very small, creating a difficult situation for bolting in the field. The detail shows a 9" gauge between the bolts, with the column tube steel being 8" wide. It will be difficult to get the bolt in the hole and a tool in there with which to tighten it.

Although this may or may not end up as a problem, the reader is now aware of a potential problem that will probably result in the need for an RFI when the shop drawings are being created.

Sometimes it is only by performing calculations that these problems will become apparent. Otherwise, they would only be seen while processing the shop detail drawing under a time crunch. By being aware of these issues, steel estimators can save time by informing the concerned parties at the beginning of the job.

The material listing will then be for (2) TS 6 x 3 x 1/4, one at 1'-6", one at 2" with (2) plates 1/2 x 4" x 9" and (4) bolts 3/4 x 1 1/2" with nut and washers. The material listing of all this framing looks like the example in Figure 8.47.

	Materials		QTY	Mat. Thkns	Plate shown in inches Width	Length	Ext Wt	Cost Per lb.	Exten. Mat.cost	Exten Hdwre
					shapes in ft/dec of the ft					
	Elevator Pit at A/S5									
Labor	Column and beam	Detail	13.6445	3	4.5	per ea.	Lbs			
1.6	TS 8 x 3 x 1/4	A/S5	2			26.9583	934	0.62	$579	
0.489	WFL 1/4 x 1'-10"	weld to column	2			1.8333				
0.4	Plate 1/2 x 6 1/2	Cap plate detail 4 & 5	2	0.500	6.5	9	17	0.487	$ 8	
1	A325 3/4" x 2" w/nut & washer	A325N	4					1.99		$ 8
	at the base connection to the pit sidewalls 2/S17									
1.6	L 6 x 4 x 1/2	2/S17 - clips at floor	2			0.8333	27	0.62	$ 17	
1	AB 3/4 x 0'-10" w/nut & washer	A325N	4					2.99		$12
0.8	TS 6 x 3 x 1/4	3/S17 - RH side	1			1.75	24	0.62	$ 15	
0.8	TS 6 x 3 x 1/4	3/S17 - LH side	1			0.1667	2	0.62	$ 1	
0.8	WFL 1/4 x 1'-6"	weld to column and conn. Pl	4			1.5				
0.4	Plate 1/2 x 6 1/2	connection plate 3/S17	2	0.500	4	12	14	0.487	$ 7	
1	A325 3/4" x 2" w/nut & washer	A325N	4					1.99		$ 8
	Elevator Beam									
0.8	W8 x 13	4H - Roof plan 19-20, E to D	1			10	130	0.48	$ 62	
0.4	Plate 1/4 x 4 1/2"	Conn. Pl/2H	2	0.250	4.5	8	5	0.487	$ 2	
0.356	WFL 1/4 x 0'-8"	weld to W14 and W10 beam	4			0.6667				
1	A325 7/8" x 2 1/2 w/nut & washer	A325N	4					1.99		$ 8
0.6	Plate 1/4 x 1 13/16"	stiffeners 4, 5/S17	3	0.250	1.8125	7.5	3	0.487	$ 1	
0.6	WFL 3/16 x 11 1/4"	weld to beam	6			1				
14							1156		$693	$36

Figure 8.47 Material listing for elevator column and beam

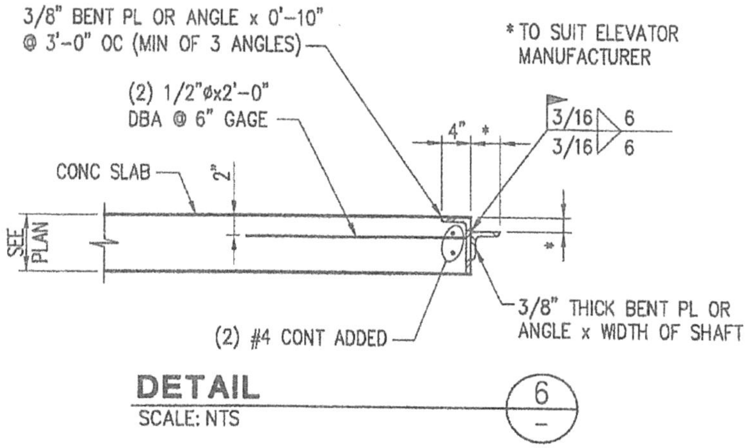

Figure 8.48 Elevator sill material

The Elevator Pit Section in Figure 8.43 has callouts for the elevator sill required at the first and second floors—Detail 6 in Figure 8.48 and F in Figure 8.49 show the bent steel plate.

Both of these details in Figures 8.48 and 8.49 call out a 3/8" thick plate with no other dimensions. Because they indicate that this is to be sized by the elevator manufacturer, we may either exclude this plate or give it a size. For

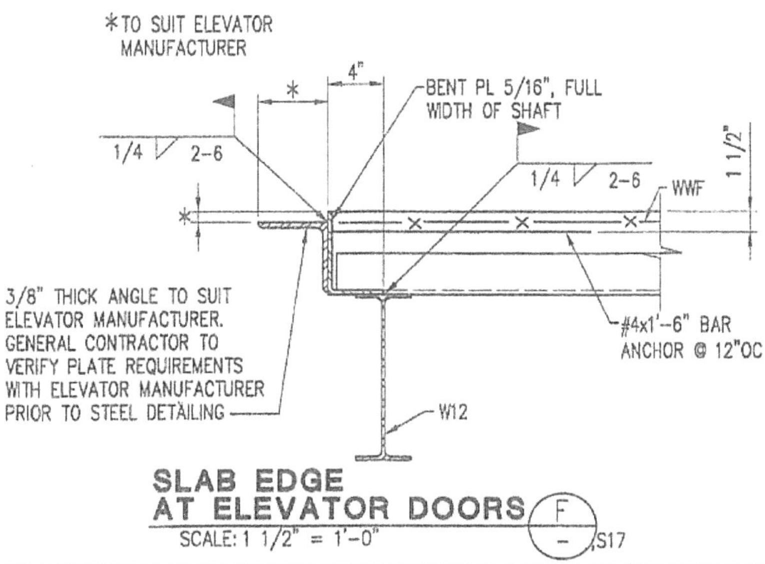

Figure 8.49 Slab edge at elevator doors

Labor	Materials	Detail	QTY		Mat. Thkns	Plate shown in inches Width	Length	Ext Wt	Cost Per lb.	Exten. Mat.cos	Exten Hdwre
						shapes in ft/dec of the ft					
	Elevator Sill angle and plate - see 6/S17, F/S18										
Labor	See detail A/S5	Detail	4.13333		3	1.4	per ea.	Lbs			
	Detial F/S18	ship loose									
0.4	Plate 5/16 x 10"	Bent 4 x 6 - full width shaft	2		0.313	10	120	212	0.487	$103	
1.333	Rebar #4 x 1'-6	at 1'-0 centers	10						1.99		$20
0.2	L 6 x 6 x 3/8	F/S18 - field weld	1				10	149	0.62	$ 92	
0.8	L 6 x 6 x 3/8	Detail 6/S17 - ship loose	1				10	149	0.62	$ 92	
0.6	Plate 3/8 x 10"	embed to concrete	3		0.375	10	10	32	0.487	$ 16	
0.8	Deformed Bar 1/2 dia x 2'-0		6						1.99		$12
4								542		$304	$32

Figure 8.50 Elevator sill material listing

this example, you might assume a 6” vertical and horizontal dimension to this plate or angle just for take-off purposes, and list that size in the bid letter.

To prepare a material listing for Detail F in Figure 8.49, we would use an L 6 x 6 x 3/8 x 10’-0”, and a bent plate 5/16 x 4 x 6 x 1’-0” with the rebar #4 x 1’-6” at 1’-0” centers. For Detail 6 in Figure 8.47, we would use the L 6 x 6 x 3/8 x 10’-0”, and the embed plate 3/8 x 10 (bent 6” x 4”) x 10” long with the 1/2 x 2’-0” deformed bar at the 6” gauge welded to the 10” piece of 3/8” plate. These will be embedded at 3 locations (3’-0” centers over the 10’-0” span) over the 10’-0” width of the elevator opening.

The take off for Details 6 and F are as shown in Figure 8.50.

Going to the roof plan in Figure .8.54, we find some exhaust fan openings between grids 17 and 18, B and C, which show no framing. We know from looking at the standard details there was some framing shown for such conditions. The typical slab opening details in Figure 8.52 are what is used for this condition.

Figure 8.52 has three openings. One is quite small, about 1’-0 or less, but the other two seem to be larger than 2’-0. Because we cannot scale the drawings, we determine the size of the openings by using the dimensions close by

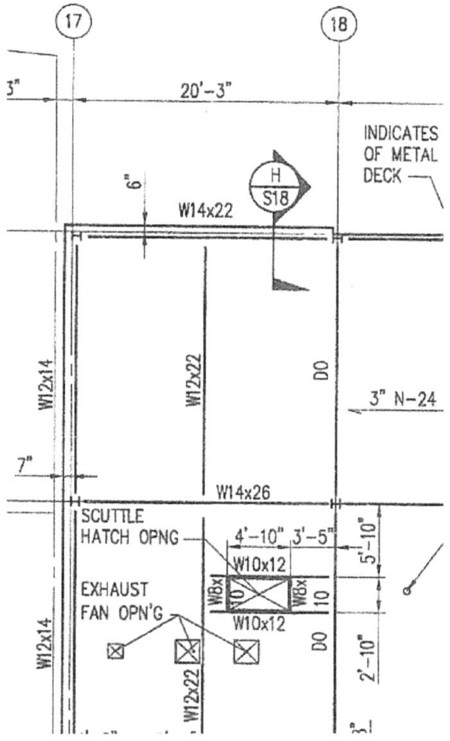

Figure 8.51
Small Roof Opening Frame

as a guide. With the notes and plan in Detail 6 in Figure 8.51, we have enough information to budget materials for the openings and write a qualification about them in the bid letter.

The small opening will use corner bar reinforcement per note 5 on the right side of Figure 8.52. As a result, this rebar mentioned in note 5 is excluded because the contractor will provide the rebar that is embedded in the concrete. The larger openings require the C4 x 5.4 and the C8 x 11.5 framing that installs between the roof framing beams. The materials are calculated by first establishing the dimensions between the beams.

In the roof plan in Figure 8.54, the dimension between bays 17 and 18 is 20'-3". The beams are assumed to be equally spaced; therefore, half of the 20'-3 is 10'-1 1/2". Subtract 1" clearance at each end that the designers have used in other details for this job. The space allowance between the beams is now 9'-11 1/2".

Because the openings are on opposite sides of the beam in the center of the bay, (4) C8 x 11.5 x 9'-11 1/2" will be needed for each side of the openings for the larger measurement. Going the short way for both sides of the opening, there will be (2) C4 x 5.4 required. Because we are not sure what the actual size of the opening will be, we can allow for something close, and qualify that length in the bid letter. We'll use 3'-0" just for take-off purposes.

Though the one opening looks like it may be close enough to the roof framing beam that only one C4 x 5.4 may be required, it is better to have it listed with our materials—the exact condition will not reveal itself until the steel detailers executes their work.

The two bolt connections are shown in Section 1 and Section 2 in Detail 6 in Figure 8.52. We can assume these connections are to be like the beam connections used in the framing details for the building.

Our take off then includes plate 1/4 x 4 1/2 x 6", with two holes each for the 7/8" A325N bolts for both ends of all the C8 x 11.5 framing pieces.

For the smalle r channel C4 x 5.4, the connection plates need to be adjusted for a single row of bolt connections, as noted in Section 2 from Detail 6. Using a 1/4 x 3" x 6" plate will allow enough room for the bolts. The material listing looks like the one shown in Figure 8.53.

In the bid letter, the inclusion for Roof Opening Frames could read as follows:

> 31) Steel framing for exhaust fan openings located on sheet S6 between grids B and C, 17 and 18, two locations using typical Slab Opening Detail 6/S15, (4) each C8 x 11.5 x 9'-11 1/2" and

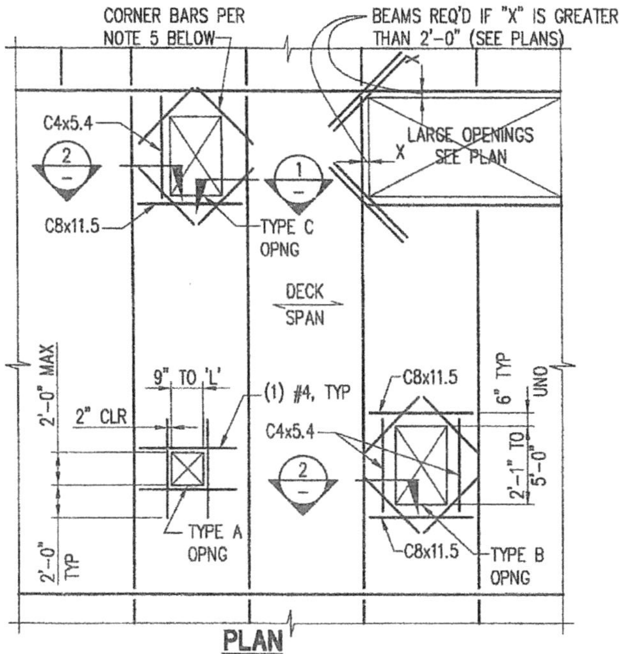

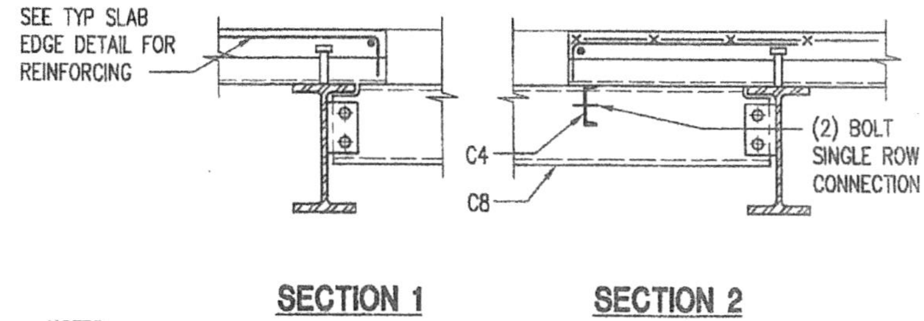

NOTES:

1. OPENINGS SMALLER THAN 9" DO NOT NEED REINFORCING.
2. OPENINGS BETWEEN 9" AND 2'-0", WITHOUT EDGE FRAMING, SHALL HAVE REINFORCING PER OPENING TYPE A.
3. BARS PARALLEL TO DECK FLUTES TO BE IN NEAREST LOWER FLUTE, 3/4" CLEAR FROM BOTTOM. BARS PERPENDICULAR TO FLUTES TO BE 3/4" CLEAR ABOVE DECKING.
4. OPENINGS LARGER THAN 2'-0" WILL REQUIRE EDGE FRAMING PER OPENING TYPES B OR C.
5. CORNER BARS TO BE PROVIDED AT ALL INSIDE CORNERS AS FOLLOWS:
   (1) #4x3'-0" FOR 6'-0" AND SMALLER OPENINGS
   (2) #4x4'-0" FOR OPENINGS LARGER THAN 6'-0".
6. BLOCK OUT UNSUPPORTED OPENINGS PRIOR TO PLACING CONCRETE. REMOVE BLOCKOUT AND CUT DECK AFTER CONCRETE HAS CURED.
7. CONTRACTOR TO COORDINATE OPENING SIZE AND LOCATION WITH ARCHITECTURAL DRAWINGS AND MECHANICAL AND ELECTRICAL CONTRACTORS.
8. EDGE FRAMING FOR OPENINGS LARGER THAN 5'-0" TO BE AS SHOWN ON FRAMING PLANS.

TYPICAL SLAB OPENING DETAILS 6/–
SCALE: NTS

Figure 8.52 Typical slab opening details

> (4) each C4 x 4.5 x 3'-0", with plate 1/4 x 4 1/2" x 6" for use with two bolt connections, using 7/8" A325N bolts x 2" with nut and washers, prime painted. Rebar for smaller openings or for opening reinforcement is not provided.

Be specific about the locations, material sizes, and quantities you have included in your proposal. If more or less materials are required during the execution of the construction, then the contractor will create a change order request to be executed accordingly.

Typical details sometimes force steel fabricators to guess. They should document their assumptions as clearly as possible so everyone knows what is included with their scope of work.

The building's roof plan in Figure 8.54 includes call outs for bent plate 3/16" material. This 3/16 bent plate, also called as edge form plate, is located between grids 17 and 18 along grid A. You can see the detail bubble H/S18, or Figure 8.55.

This plate is called out on the roof plan at only the one location indicated, but the information at the detail lets the reader know that the plate is also used at other locations. This tells us that the plate is used all along grids A, D, E, and F in this plan.

At grid A, this plate—spanning between grids 17 and 18 in Figure 8.54—uses the wider width of 6" in addition to the 4" from centerline of beam to the grid line. This makes this plate 10" at the horizontal with a 3" leg that bends up to meet the roof deck line. Because the dimension between grids is 20'-3",

	Materials		QTY	Mat. Thkns	Plate shown in inches Width	Length	Ext Wt	Cost Per lb.	Exten. Mat.cost	Exten Hdwre
					shapes in ft/dec of the ft					
	Opening frames per 6/S16 from S6									
Labor	Two each 2'-0 square	Detail	14.9333	2	7.5	per ea.	Lbs			
	Find at grids B-C, 17-18 on S6									
0.8	C8 x 11.5	6/S16	4			9.9167	456	0.62	$283	
1.6	Plate 1/4 x 4 1/2"	Conn. Pl/2H	8	0.250	4.5	6	15	0.487	$ 7	
1.067	WFL 1/4 x 1'-0	weld to beam	16			0.5				
4	A325 7/8" x 2" w/nut & washer	A325N	16					1.99		$32
0.8	C4 x 5.4		4			3	65	0.62	$ 40	
1.6	Plate 1/4 x 4 1/2"	Conn. Pl/2H	8	0.250	3	6	10	0.487	$ 5	
1.067	WFL 1/4 x 1'-0	weld to beam	16			0.5				
4	A325 7/8" x 2" w/nut & washer	A325N	16					1.99		$32
15							546		$335	$64

Figure 8.53 Material listing for roof opening frames

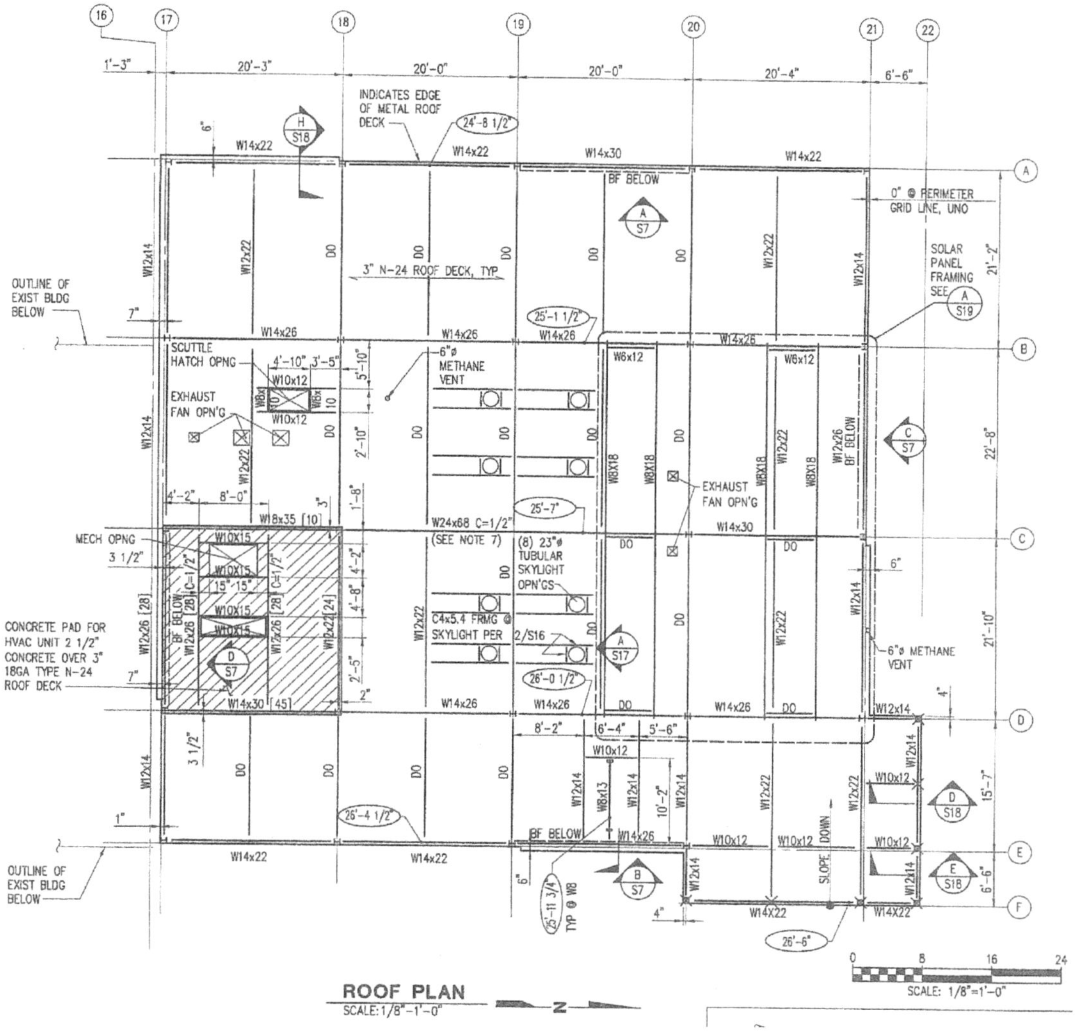

Figure 8.54 Roof plan

we know that the bent plate needs to be at least that long.

Notice that this plate may run beyond grid 17 to the left, although the designers don't show what exactly is happening there. We will make an allowance for the 7" dimension that is shown for the edge going the north/south direction between grids 16 and 17. Adding 7" to the 20'-3" between grids give us 20'-10" length to this plate.

In the detail in Figure 8.54, the plate is skip welded at both the top flange of the beam and the underside of the plate at the flange's edge. When calculat-

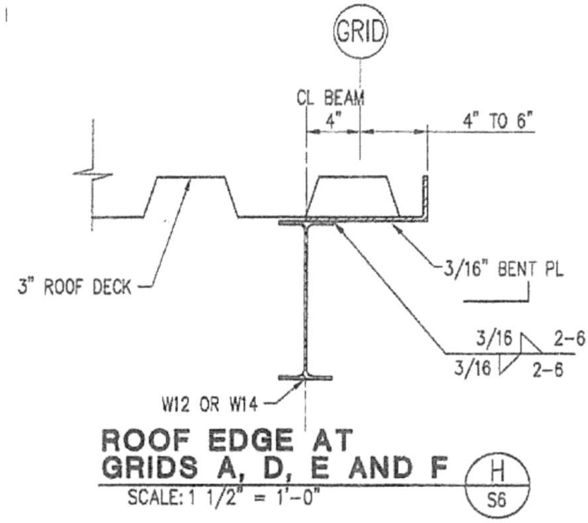

Figure 8.55 Roof edge closure plate at grids A, D, E, and F

ing skip welding, take the total length of the item to be welded and divide by 3. (Welding 2" every 6" reflects 1/3 of the total length to be welded.).

For the remainder of this grid line A, the plates stop at the grid line location that is 4" beyond the centerline of the beam. The 'stretch out' of these plates is 3/16" x 7" (4" plus 3") x 20'-0" between grids 18 and 19, and 19 and 20, then 20'-4" between grids 20 and 21.

Looking to grid E between grid 17 and 18 in Figure 8.54, we use the 3/16" x 7 plate x 20'-3", (no extension is shown past grid 17). Then between grids 18 and 19, we use Plate 3/16 x 7 x 20'-0". Between grids 19 and 20, there is the 6" extension shown, so this plate is 3/16 x 13 x 20'-0". At grid F between 20 and 21, this item could be combined with the same location at grid line A between grids 20 and 21. At grid lines D and F, we allow 6'-6" lengths of Plate 3/16 x 7 to cover both of these locations.

No detail is called out for what is used at grid line 17, 20, 21, and 22 in Figure 8.54. This lack of information leaves the reader wondering what is happening there. Steel estimators may choose to exclude all bent plates at these locations, or call the contractor or designers and find out. When in doubt, it is best to exclude it.

Although this take off is somewhat busy, it is as close to accurate as possible given the type of information we are shown to use when calculating the materials needed. The pricing or labor for bending the plates needs to be added in the material listing as well.

Labor	Materials	Detail	QTY	Mat. Thkns	Plate shown in inches Width (shapes in ft/dec of the ft)	Length	Ext Wt	Cost Per lb.	Exten. Mat.co	Exten Hdwre
	Bent plate at Roof detail H/S18									
Labor	Grids A, D, E, F only	Detail	17.6223	10	1.8	per ea.	Lbs			
	Along Grid A between 17 and 18									
0.2	Plate 3/16 x 13	Bent 3 x 10, long leg horiz.	1	0.188	13	20.8333	172	0.487	$ 84	
1.867	WFS 3/16 x 7'-0	weld to beam top and bottom	2			7.00				
	Along Grid A between 18 to 20									
0.4	Plate 3/16 x 7	Bent 4 x 3, long leg horiz.	2	0.188	7	20	178	0.487	$ 87	
3.556	WFS 3/16 x 6'-8	weld to beam top and bottom	4			6.67				
	Along Grid A and F between 20 to 21									
0.4	Plate 3/16 x 7	Bent 4 x 3, long leg horiz.	2	0.188	7	20.33	181	0.487	$ 88	
3.6	WFS 3/16 x 6'-9	weld to beam top and bottom	4			6.75				
	Along Grid line E between 17 and 18									
0.2	Plate 3/16 x 7	Bent 4 x 3, long leg horiz.	1	0.188	7	20.33	91	0.487	$ 44	
1.8	WFS 3/16 x 6'-9	weld to beam top and bottom	2			6.75				
	Along Grid E between 18 to 19									
0.2	Plate 3/16 x 7	Bent 4 x 3, long leg horiz.	1	0.188	7	20	89	0.487	$ 43	
1.778	WFS 3/16 x 6'-8	weld to beam top and bottom	2			6.67				
	Along Grid E between 19 and 20									
0.2	Plate 3/16 x 13	Bent 3 x 10, long leg horiz.	1	0.188	13	20.8333	172	0.487	$ 84	
1.867	WFS 3/16 x 7'-0	weld to beam top and bottom	2			7.00				
	Along Grid D and F between 21 and 22									
0.4	Plate 3/16 x 7	Bent 4 x 3, long leg horiz.	2	0.188	7	6.5	58	0.487	$ 28	
1.156	WFS 3/16 x 2'-2	weld to beam top and bottom	4			2.17				
18							942		$459	##

Figure 8.56 Material listing for closure plate at roof

The material listing would look like the one in Figure 8.56.

After going through several examples of reading the details and identifying the steel materials to be listed—along with the explanations of how to make the calculations—you should be increasingly familiar with these exercises. Every project will have different applications of these same conditions.

Summary

In this chapter, you have learned to read the structural steel drawings from the steel estimator's perspective. You have also discovered how to list the materials as they have been found in the drawings and subsequent details. By now, you should have genuine insight regarding the nuances of what information the drawings show and don't show. These same skills will apply when you read the architectural drawings, listing the materials found there.

9

Reading Architectural Drawings

The architectural metals or miscellaneous steel items for projects are all located in the architectural drawings. You will find the miscellaneous steel framing in the floor framing plan views and then go to the referred details.

For us to make certain that we have located all of the miscellaneous metals on a project, we have to review all of the details at the architectural set and refer back to the architectural plans to be sure that nothing is missed.

Architect designs reveal all the building dimensions and elevations. Engineers create their framework to fit those dimensions and elevations. Engineers also design all the other parts and pieces of steel that attach to the framework of the building to support other major building components as necessary.

If we find conflicts in dimensions and elevations between the architectural and structural drawings, we look to the architectural or A drawings for clarification. Material sizes and types for the structural framework are designated by engineers, who prepare their calculations to meet all the applied building codes. If we find conflicts in materials sizes and types for the steel framing, we look to the structural or S drawings as the governing authority for material sizes and types.

Typical Miscellaneous Steel Items Found in the Architectural Drawings

In the architectural drawings, we are looking for the following items made out of steel:

- Steel stairs, stair landings, stair railing
- Steel handrails
- Steel ramp framing
- Pipe bollards
- Embed angles at roll up doors
- Steel posts for short walls
- Steel ledger angles at the top of walls
- Toilet partition supports
- Steel supports for lavatory sinks and counters
- Steel supports for office desks, shelves, or counters
- Steel fabricated ladders, both interior and exterior
- Steel hanging supports for wall partitions
- Steel hanging supports for cameras, televisions, special lighting
- Steel framing for overhead equipment
- Steel framing for window or curtain walls
- Steel framing for large door openings
- Elevator sump pit frame and grating
- Elevator pit ladders
- Trench drain frames and grating
- Brick ledger angles
- Trash enclosure frames and gates
- Steel trellis framing
- Canopy framing that is not shown in the structural drawings
- Fall protection anchors
- Steel equipment supports not shown in structural drawings
- Roof top screen walls and parapet supports
- Generally anything built from steel that is 1/8" thick and heavier

You will notice some duplication between this list and the list in Chapter 7 (pages 67–182). We often find details shown in both places. Sometimes the intent is to better clarify the design conditions. This practice is confusing, however, because we are left wondering which conditions are actually going to be used.

As steel estimators, we usually consider the worst case scenario as the guide to real world conditions. Remember—ask the question, then document what is used in the material listing and the bid letter.

Always check the list of items in Section 055000: *Miscellaneous Steel* as the guide to providing the architectural metals. Quite often that section will list the specifications that reveal to the estimators bidding the project where to be on the alert.

Let's get started by looking at plans, details, and applicable material take off, working from the architectural drawings. The set of architectural drawings used in the following examples are a match to the structural drawings that were used as examples in Chapters 7 and 8.

Drawing Plan Review

Figure 9.1 provides the same first floor building plan that was used as the foundation in Chapter 7's structural steel example and Chapter 8's sample for the column and braced frame take off and material listing. We see gridline locations 16 through 22 at the top of the page and grids A through E at the right of the page.

In Figure 9.1 we see the building stairs located between grids B and C, 17 and 18 to the left, and again between grids 20 and 21, E and F. Detail *A* and *C/A*4.2 are called out in Figure 9.2.

Reviewing Stair Elevations and Details

These details in Figure 9.2 are plan views of the stairs from the ground floor. In both details, you can see the vertical lines representing the steps, called stair treads. The note at the dimension lines tell you how many treads are required to take a person from the ground floor up to the landing, and then from the landing up to the next floor.

Because these plan views are taken from the ground floor, the lines that make up the rest of the pictures above the ground floor are ghosted or dashed. These lines refer to elements that are above the area of this plan.

The section views *1/, 2/, 3/,* and *4/A4.2*—all called out in the Figure 9.2 plan views—show us the elevations of the stairs. From there we can see the height of this stair; detail cuts show us the materials that comprise these stairs.

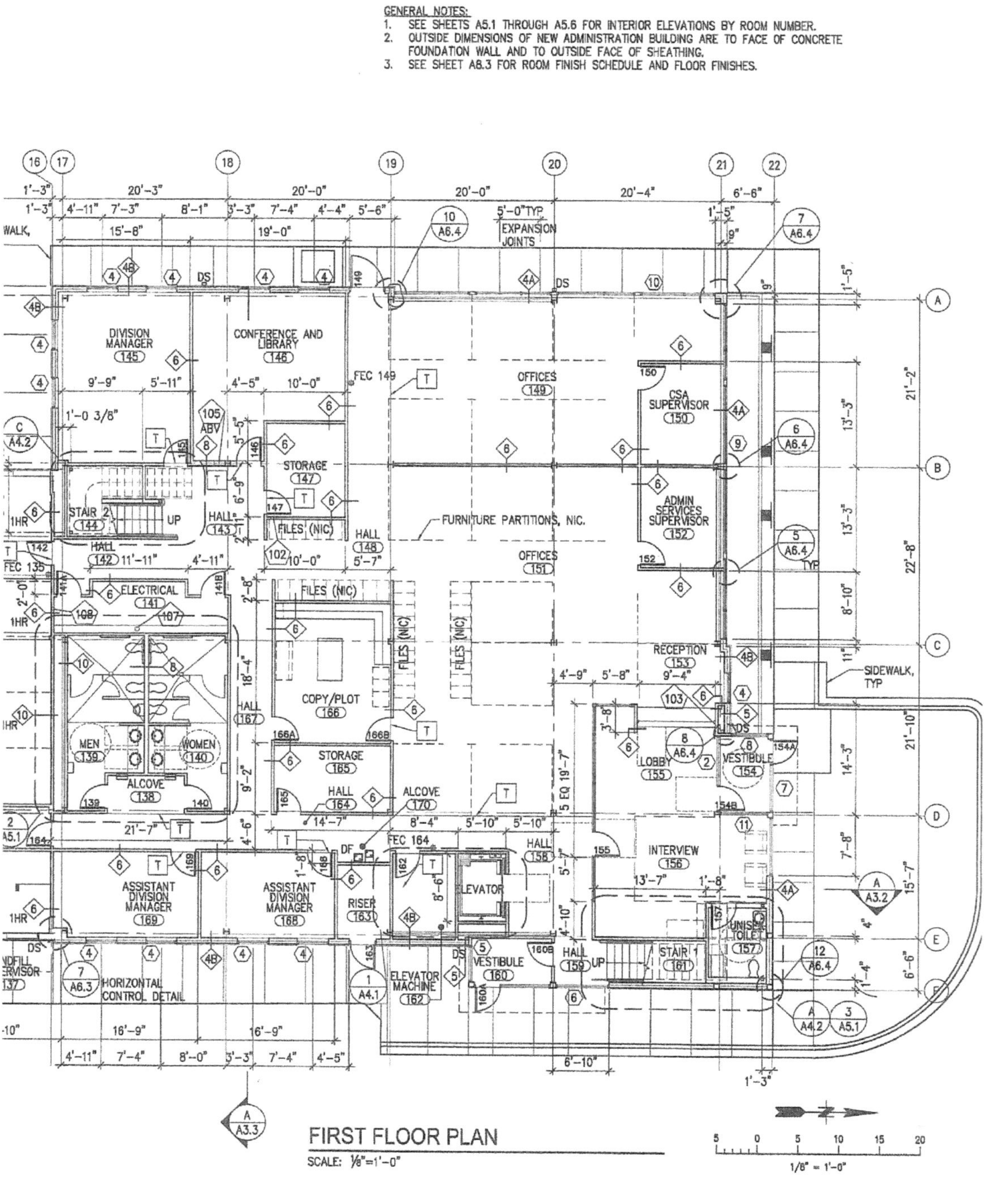

Figure 9.1 First floor plan

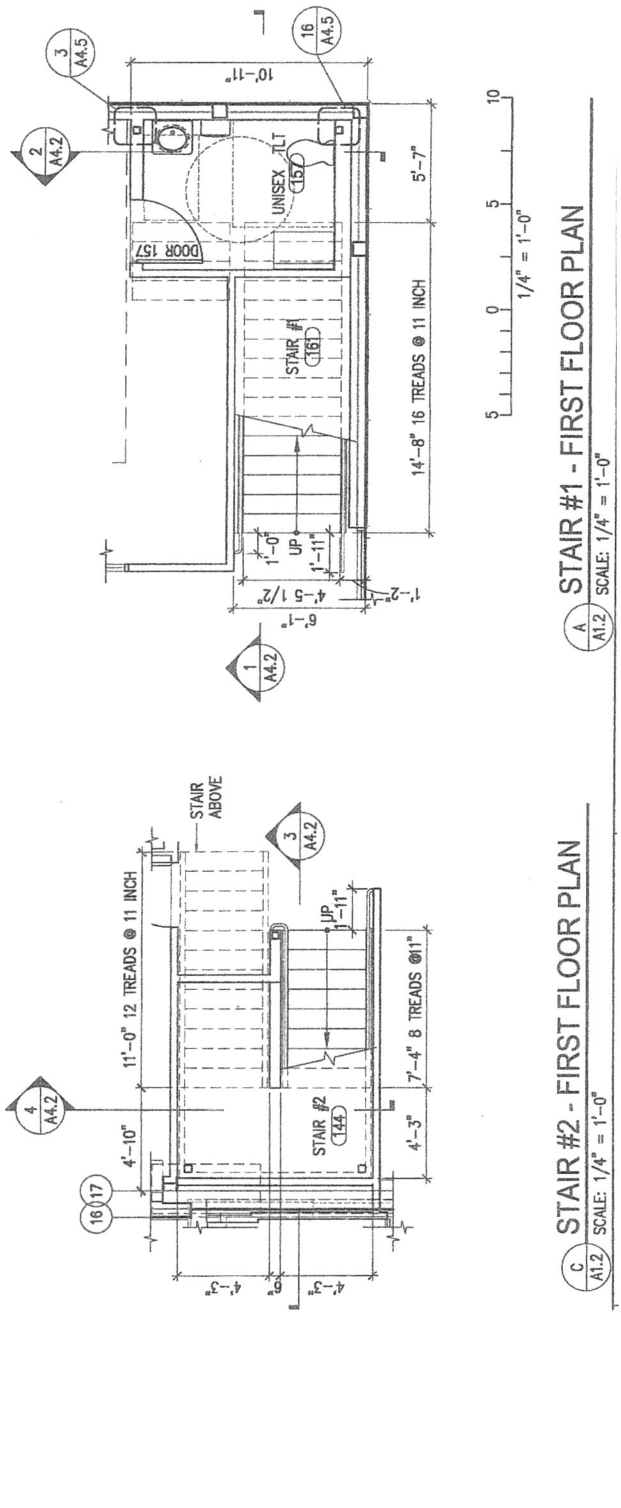

Figure 9.2 First floor plans for Stairs #1 and #2

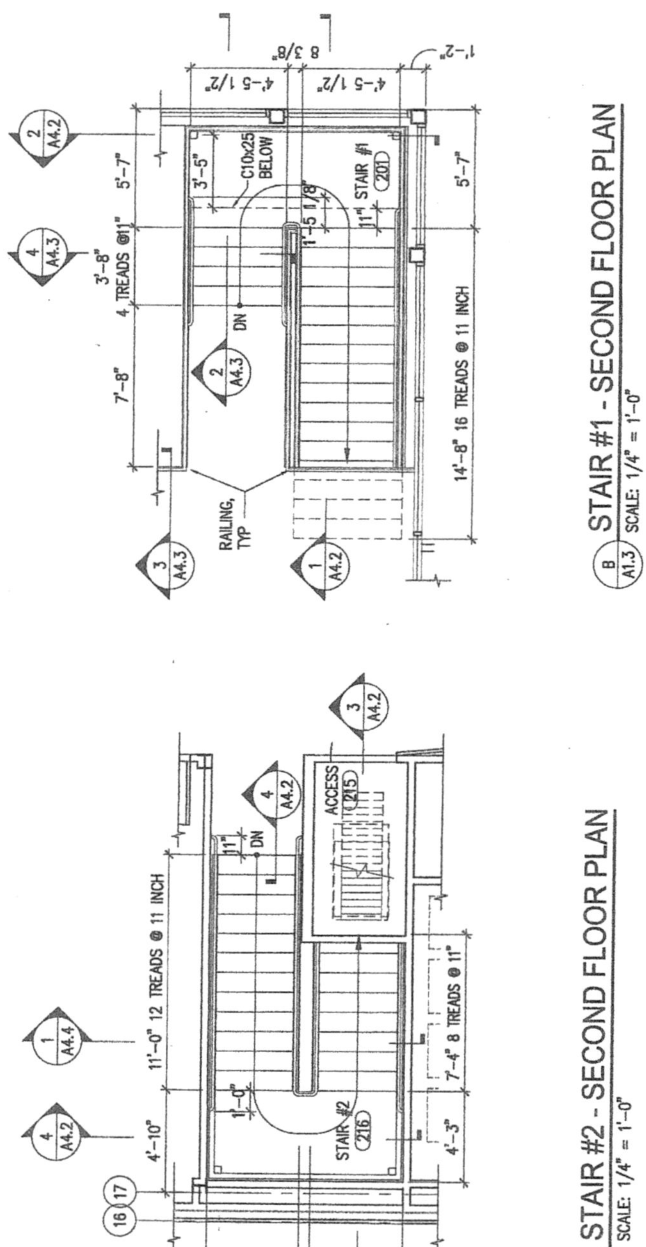

Figure 9.3 First and second floor stair plans

All of the plan views starting from the ground floor will typically call out these same elevation views. However, there may be other section cuts taken at the floor levels that might not be cut from anywhere else. All floor levels need to be checked for additional information.

The elevation cuts through the stairs in Figure 9.3 indicate how the stairs are to be constructed. The elevation cuts through Stair #1 are shown in *1/* and *2/A4.2* in Figure 9.4.

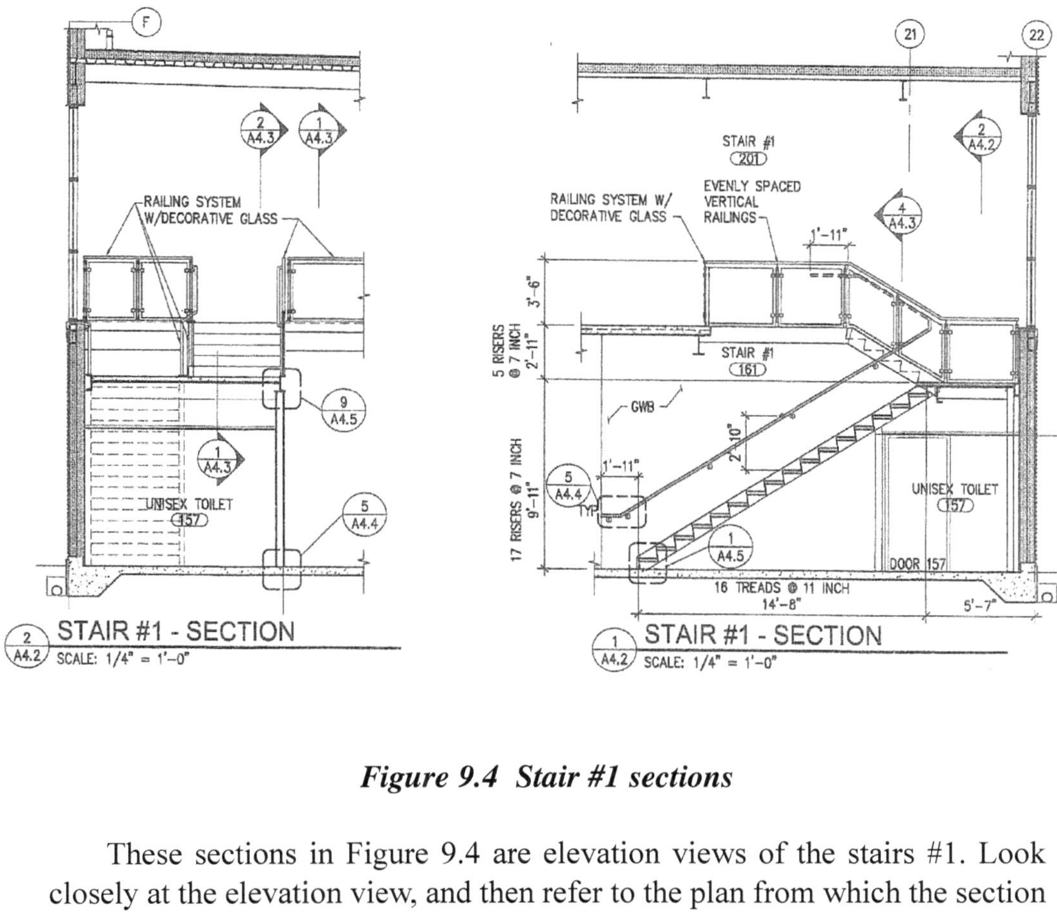

Figure 9.4 Stair #1 sections

These sections in Figure 9.4 are elevation views of the stairs #1. Look closely at the elevation view, and then refer to the plan from which the section is cut. In Figure 9.4, the section to the left shows the stairs when you are standing in front of them about to go up. The section on the right shows you the stairs as if you are looking from the side.

The details in Figure 9.5 are section cuts taken from the elevation views, giving a close up view of the stair construction.

These stair elevation views show you the steel materials being used to construct this stair. You can also see the hand railing in both the elevation and

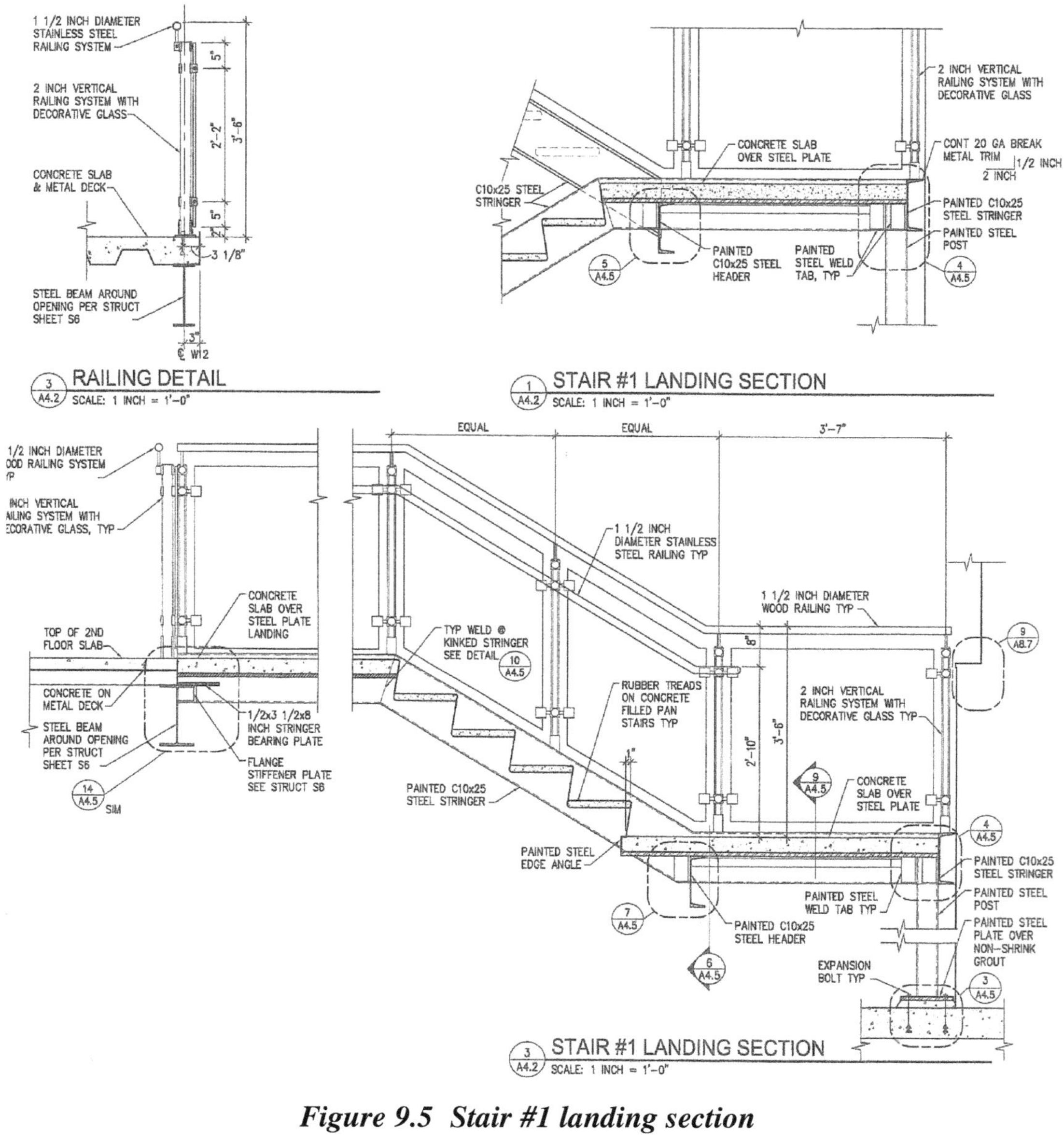

Figure 9.5 Stair #1 landing section

section views. The section cuts from these elevation views provide more detail regarding the steel fabrication.

We can see that the stair stringers are called out as C10 x 25. The same size channel used at the landings and tube steel is used as the landing posts. Look at the Stair #1 Landing Section *3/S4.2* in the upper right of Figure 9.5 to find the steel shown. The details in the lower right of Figure 9.5 show the stair framing connections and base plate areas. Figure 9.6 is the close up of the bottom of Stair #1 taken from the right side of Figure 9.4. Notice that the top detail in Figure 9.6 shows how the stair stringer is anchored to the ground. The bottom half of Figure 9.6 is cut from the picture above, and shows the location of the stair anchor bolts.

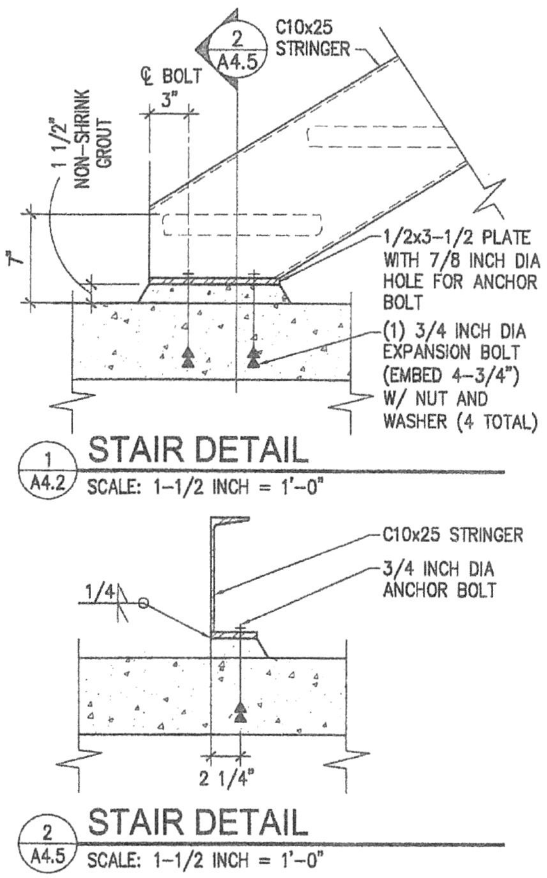

Figure 9.6 Stair detail

Detail *1/A4.2* in Figure 9.6 is the side view of where the stringer meets the ground, whereas *2/A4.5* is the end view of the stair stringer coming toward you. There is 1 1/2" grout under the base plate 1/2 x 3 1/2 that is welded to the stringer, and there are two expansion bolts 3/4" that embed 4 3/4" into the ground.

The length of the expansion bolt required in this condition is probably more in the line of 7 1/2" under the head of the bolt. (Because you have to allow for 1 1/2" grout, the thickness of the channel leg has to be considered.) The designers don't always give the length of the bolt—they allow the trades to make that determination based on the criteria shown. In this case, the architect indicates the requirement for 4 3/4" of the bolt to embed into concrete.

Also at the bottom of stair #1 is the information about the posts that support the stair landing as shown in Figure 9.7.

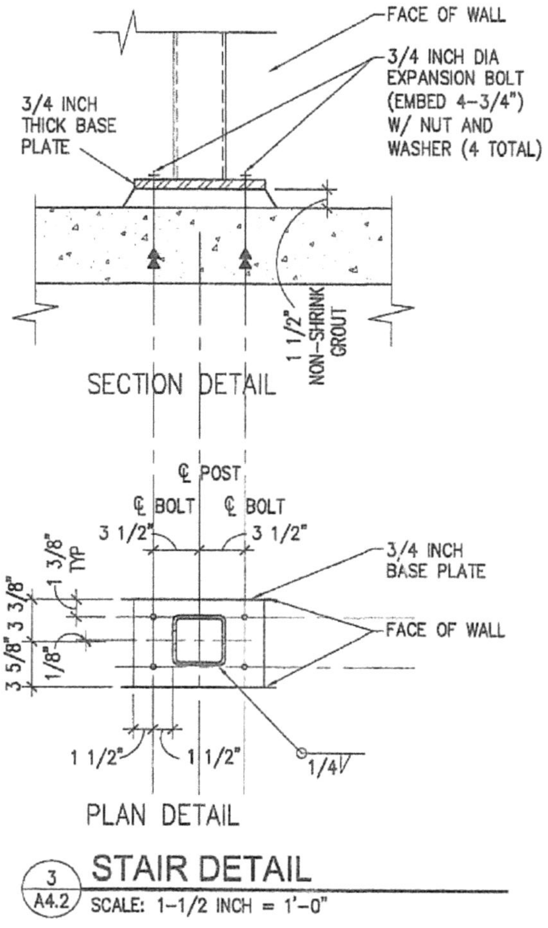

Figure 9.7
Stair support post detail

Figure 9.7 shows the elevation view and plan view of the tube steel support for the stair landing. The tube steel is not sized because it is probably intended as a typical detail. What is shown here is the thickness of the base plate, the dimensional locations of the holes for the anchor bolts, and the size and type of anchor bolts.

The materials not being sized may mean that they are sized in another detail or that the stairs may require engineering by the fabricator. Note, however, that the stair stringers and landings for this project have been designed and sized.

If the size of the support tubes are not found in other details, and there is no engineering requirement found in the specifications, then the size of tube steel supports for the landings must be assumed and identified in the bid letter.

If there is sufficient time during the bid process, you can question the designers and then address their answer by addendum. Otherwise, a steel size must be chosen and qualified in the bid letter.

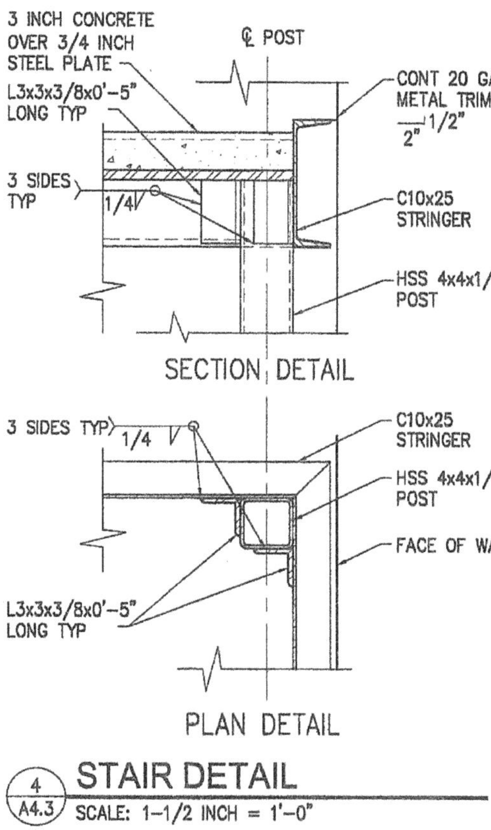

Figure 9.8
Stair support post to stair landing detail

The details in Figure 9.8 are at the top of the support posts for the stair landings. They show the size of the tube steel posts as TS 4 x 4 x 1/4. There are connection clips L 3 x 3 x 3/8 x 5" long that weld to the tube steel to the channel landing framing. In Figure 9.9, these details show us how the landing framing goes together. In Figure 9.10, this detail shows how the stair stringers are attached to the landing.

The connection in Figures 9.9 and 9.10 is the L 3 x 3 x 3/8 x 5" long angle, shop welded. The details in Figure 9.11 indicate that the stair landing is fabricated as a frame and is shop welded to the stair stringers, making the stair with the landing as a single frame.

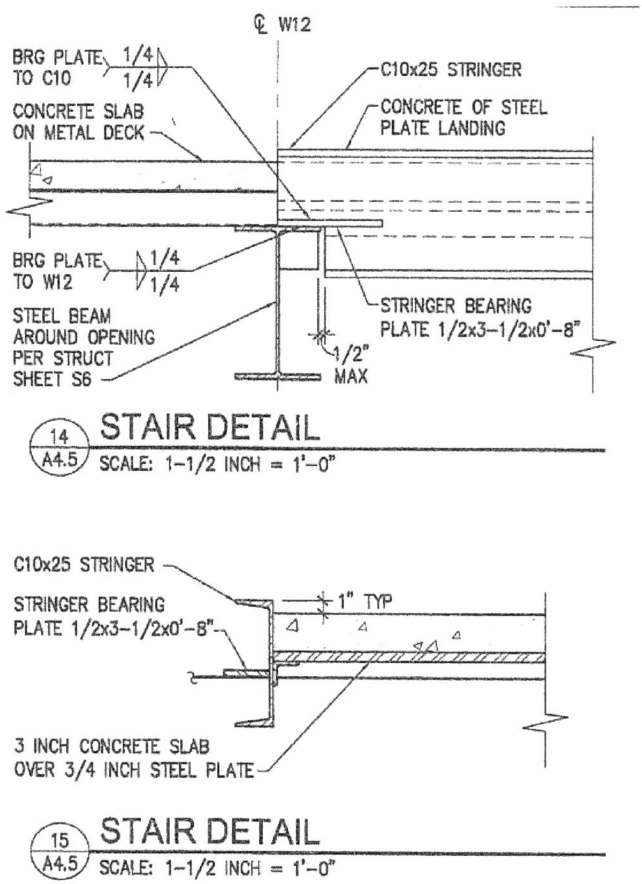

Figure 9.9* *End view of stair stringer to landing connections

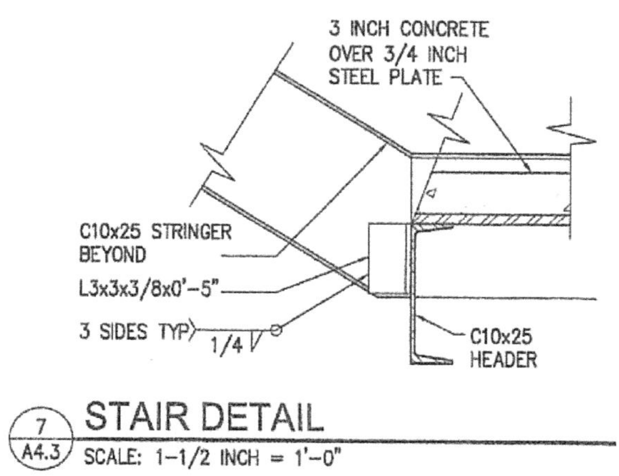

Figure 9.10* *Elevation view of stair stringers to landing detail

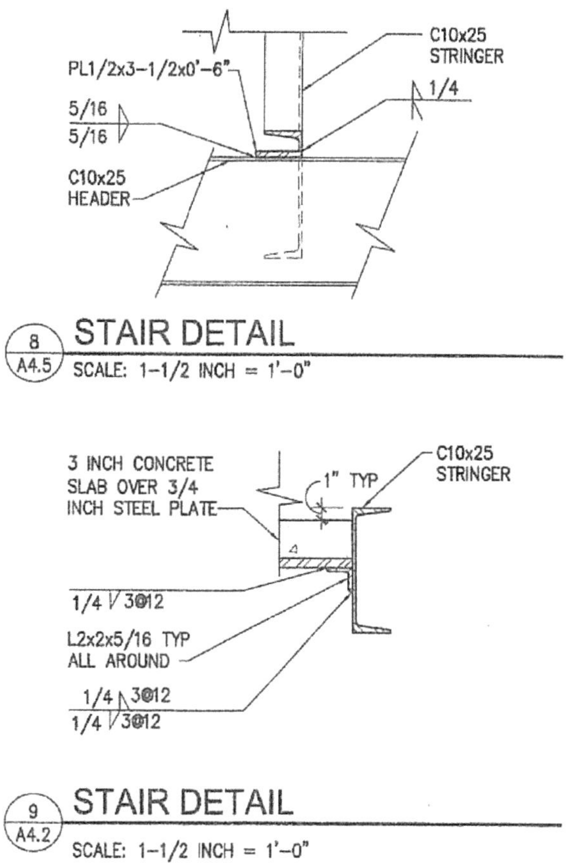

Figure 9.11 Stair landing framing detail

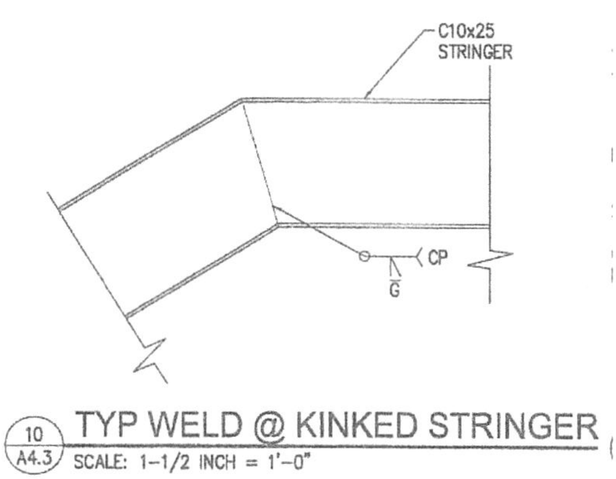

Figure 9.12 Typical complete penetration weld at kinked stair stringer

This construction of the stair is further confirmed by the detail in Figure 9.12.

This detail in Figure 9.12 shows the top of the stringer welded to the landing horizontal support framing member (which is the same size as the stringer itself). The weld is a single bevel weld, as indicated by the upside down V and the C underneath the V. The weld note, together with the CP note at the end of the tag line, is an indicator for a *complete penetration* weld. Because there is no flag attached to this weld directive, we know this weld is to be performed in the shop—there is no flag at the weld note indicating otherwise.

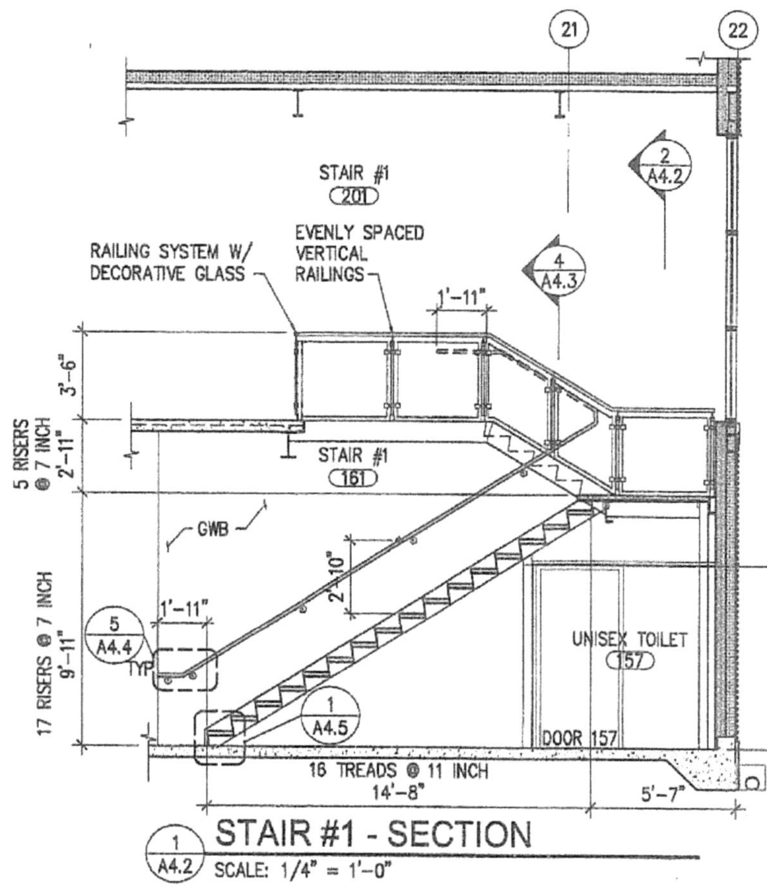

Figure 9.13 Elevation view of stair #1

Using this elevation in Figure 9.13 for Stair #1, we can begin listing the materials for this stair. The first run of stairs, going from the ground floor up to the first landing on the right, is listed first. We will use the information from the details we just reviewed to complete this task. The stringers are C10 x 25, and the length will be determined by the number of treads that complete this run of stairs.

We are on our own in determining the material for the stair pans because nothing is called out at the details for the thickness of this material. Stairs typically follow the *7-11 Rule:* Steps on a stair occur at height intervals about every 7", they are 11" deep, and they are usually 4'-0 wide. In real life, the elevation of the floors and intermediate landings will make the 7" height of the step vary, but not be by more than increments of an inch.

Steel plate used for forming stair tread pans is either 12-gauge or 14-gauge plate. It is bent in such a way that the pans will contain concrete and join together in their construction of the stair. An additional closure plate at the bottom called a kicker is used to complete the stair run.

The "stretch out" of the formed plate is the total width of the plate to be formed. The length is the width of the tread.

Stairs are typically 4'-0" wide to the inside measurement of the stair stringers and 4'-6" outside to outside legs of the channel stringers. The width of the stairs will be shown in the plan view of the stairs framing.

The stair pans and kicker made from the 12-gauge or 14-gauge plates have to be bent, either by the shop or an outside vendor. Therefore, the labor to form these plates will have to be added, either as shop hours or as a fixed cost provided by the vendor chosen to form these treads and kickers.

The treads are then supported by L 1 x 1 x 1/8 brackets welded vertically and horizontally. The supports are an inch shorter than the rise and run of the treads. These L 1 x 1 x 1/8 brackets will be either 6" or 10". They will support the tread on both sides when these angles are welded to the inside of the stair stringer.

Material Take Off Listing for Stairs

Figure 9.14 shows a sample of the take off material listing.

The (2) C10 x 25 x 18'-0" are the stair stringers that go from the ground floor to the intermediate landing. *P1E, or prepare one end,* is a note for the *CP* (complete penetration) weld for the top of the stringer to the landing. Pl 14ga x 24" x 4'-0" is for the treads. The picture of the bends in the detail column in the page's center allows the reader to see the intended shape. It shows how the 24" stretch out is to be utilized. There is also a picture for the kicker plate at the bottom of the stair.

Labor	Materials	Detail	QTY	Mat. Thkns	Plate shown in inches Width (shapes in ft/dec of the ft)	Length	Ext Wt	Cost Per lb.	Exten. Mat.cost	Exten Hdwre
	Steel Stair #1 - 1/A4.2									
Labor	Stair #1	Detail	34.3693	2	17.2	per ea.	Lbs			
0.4	C10 x 25	P1E	2			18	900	0.62	$558	
0.4	Plate 1/2 x 3 1/2 x 1'-0	2H	2	0.250	4.5	12	8	0.487	$ 4	
0.4	WFL 1/4 x 1'-6	weld to stringer	2			1.5				
1	Expansion anchors 3/4 x 0'-8"	A307 w/nut & washers	4					2.4		$10
6.4	L 1 x 1 x 1/8	tread supports	32			0.5	13	0.62	$ 8	
6.4	L 1 x 1 x 1/8	tread supports	32			0.8333	21	0.62	$ 13	
2.667	WFL 1/8 x 0'-6"	L to stringer	64			0.5				
4.444	WFL 1/8 x 0'-10"	L to stringer	64			0.8333				
6.4	Plate 14 ga x 24	Treads, bend 2 x 7 x 11x 2 x 2	16	0.075	24	54	438	0.487	$213	
0.4	Plate 14 ga x 24	kicker - Bent 2 x 7	1	0.075	9	24	5	0.487	$ 2	
5.333	WFL 1/8 x 2'-0	L to stringer	32			2				
0.125	WFL 1/8 x 0'-9	L to stringer	2			0.75				
34							1385		$799	$10

Figure 9.14
Sample of take off material listing for lower stair stringers

Labor	Materials	Detail	QTY	Mat. Thkns	Plate shown in inches Width (shapes in ft/dec of the ft)	Length	Ext Wt	Cost Per lb.	Exten. Mat.cost	Extn Hdwr
Labor	Landing for Stair #1	Detail	14.0768	2	1.0	per ea.	Lbs			
0.4	C10 x 25	3/A4.2, 4/A4.3	2			9.6667	483	0.62	$ 300	
0.4	C10 x 25	3/A4.2, 4/A4.3	2			5.5833	279	0.62	$ 173	
1.6	L 3 x 3 x 3/8	4/A4.3	8			0.4167	24	0.62	$ 15	
0.167	WFL 1/4 x 0'-10"	weld to C10	2			0.8333				
1.6	L 2 x 2 x 5/16	4/A4.3	8			9.5	296	0.62	$ 184	
1.6	L 2 x 2 x 5/16	4/A4.3	8			5.33	166	0.62	$ 103	
1.689	WFL 3/16 x 3'-2	Weld to frame	4			3.1667				
0.933	WFL 3/16 x 1'-9	Weld to frame	4			1.75				
0.4	TS 4 x 4 x 1/4	4/A4.3 - posts	2			9.9167	242	0.62	$ 150	
0.4	Plate 3/4 x 6 3/4	4H - base plate	2	0.250	6.75	10	10	0.487	$ 5	
0.355	WFL 1/4 x 1'-4"	weld to post	2			1.33				
2	Expansion anchors 3/4 x 0'-8"	A307 w/nut & washers	8					2.4		$19
0.8	L 3 x 3 x 3/8	4/A4.3	4			0.4167	12	0.62	$ 7	
0.667	WFL 1/4 x 0'-10"	weld to col	8			0.8333				
0.4	Plate 3/4 x 5'-7 x 9'-9	Deck plate per 5, 6/A4.3	2	0.750	67	116	3299	0.487	$1,607	
0.667	WFS 1/4 3-12 x 10'-2	Weld to channel landing frame	8			0.8333				
14							4812		$2,543	$19

Figure 9.15
Sample of take-off material listing for stair landing

	Materials		QTY	Mat. Thkns	Plate shown in inches Width	Length	Ext Wt	Cost Per lb.	Exten. Mat.cost	Exten Hdwre
					shapes in ft/dec of the ft					
	Steel Stair #1 - 1/A4.2 - continued									
Labor	Stair #1	Detail	23.76	2	11.9	per ea.	Lbs			
0.4	C10 x 25	P1E	2			6	300	0.62	$ 186	
0.8	Plate 1/2 x 3 1/2 x 1'-0	2H	4	0.250	4.5	12	15	0.487	$ 7	
0.8	WFL 1/4 x 1'-6	weld to stringer	4			1.5				
2	A325 7/8" x 2 1/2 w/nut & wash	A325N	8				1.99			$ 15.92
2	L 1 x 1 x 1/8	tread supports	10			0.5	4	0.62	$ 2	
2	L 1 x 1 x 1/8	tread supports	10			0.8333	7	0.62	$ 4	
0.833	WFL 1/8 x 0'-6"	L to stringer	20			0.5				
1.389	WFL 1/8 x 0'-10"	L to stringer	20			0.8333				
1.6	Plate 14 ga x 24	Treads, bend 2 x 7 x 11x 2 x 2	4	0.075	24	54	110	0.487	$ 53	
0.8	Plate 14 ga x 24	kicker - Bent 2 x 7	2	0.075	9	24	9	0.487	$ 4	
1.333	WFL 1/8 x 2'-0	L to stringer	8			2				
0.25	WFL 1/8 x 0'-9	L to stringer	4			0.75				
0.4	C10 x 25	3/A4.2, 4/A4.3	2			7.6667	383	0.62	$ 238	
0.4	C10 x 25	3/A4.2, 4/A4.3	2			4.5	225	0.62	$ 140	
1.6	L 3 x 3 x 3/8	4/A4.3	8			0.4167	24	0.62	$ 15	
1.333	WFL 1/4 x 0'-10"	weld to C10	16			0.8333				
0.4	L 2 x 2 x 5/16	4/A4.3	2			9.5	74	0.62	$ 46	
0.8	L 2 x 2 x 5/16	4/A4.3	4			5.33	83	0.62	$ 52	
1.689	WFL 3/16 x 3'-2	Weld to frame	4			3.1667				
1.867	WFL 3/16 x 1'-9	Weld to frame	8			1.75				
0.4	Plate 3/4 x 4'- 6 x 7'-8	Deck plate per 5, 6/A4.3	2	0.750	54	92	2109	0.487	$ 1,027	
0.667	WFS 1/4 3-12 x 10'-2	Weld to channel landing frame	8			0.8333				
24							3345		$ 1,774	$ 15.92

Figure 9.16 Sample of material listing for upper stair stringers, landing to the 2nd floor

The length of the stringers is based on using 1'-0" for each tread plus 1'-0" for each end, allowing for the special shapes that have to be cut to complete the stair stringer itself. If the stringer is to support 16 treads, the stringer then would be about 18'-0" long to accommodate the demands. Whatever length this is, the actual use length will have to be taken out of a 20'-0" stick; being off a bit on our length doesn't matter that much for estimating purposes.

Stair landings are usually 5'-0" x 10'-0". On the 2nd floor of this stair #1 plan, the dimensions appear to be 5'-7" x 9'-8" long. As we saw in the details, the landing is built from C10 x 25 as well, with an L 2 x 2 x 5/16 at the inside perimeter to support the decking that will be attached to them to hold in the concrete floor at the landing.

We will add L 2 x 2 x 5/16 cross member supports at mid-span for added support. It is a good idea to have mid-span supports about every 4'-0" when you have decking and concrete on top. The elevation in Figure 9.13 shows two of the TS 4 x 4 x 1/4 supports up against the wall to support the side of the landing there. You have to list them, as in Figure 9.15.

Continuing with the stair take off, we next list the stair that goes from the intermediate landing up to the 2nd floor (see Figure 9.16).

The railing for stair #1 in Figure 9.17 is a glass railing with stainless steel grab bars. This railing could be a buy out item, not made by the steel fabricator. The vendor that does this type of work will most likely be bidding directly to the end user or general contractor. The estimator will need to check on that provision. If the steel fabricator is not providing the railing, it should be listed in the bid letter with the items to be excluded.

Reviewing Drawings for Steel Handrail Details

The railing at Stair #2 in Figure 9.18 is indicated as being a steel railing.

Figure 9.18 shows the elevations of Stair #2. See the close up of the railing at Stair #2 in Figure 9.19. This section is called out from the left hand side of the stair at Figure 9.18

The note on the railing in Figure 9.19 is 1 1/2" *dia painted steel hand rail typ.* What is usually used at this condition is 1 1/2" standard pipe, which is almost a 2" outside diameter. 1 1/2" diameter is close to the outside diameter of 1 1/4" standard pipe. There may be an indication at the specification section for the miscellaneous metals or for the steel railing that the designers specifically want to be used. For this example, we will use 1 1/2" standard pipe for the purposes of preparing the take off for this railing.

To list the materials for the railing at this stair, we will need to refer to the elevation 3/A4.2, per Figure 9.21, to determine the lengths of the railing.

The number of treads from the ground floor to the landing is 8. For the railing, we will allow 1'-0" for each tread and 1'-0" for each end; this leads to a total of 10'-0". The number of treads from the landing up to the 2nd floor is 12; adding 1'-0" for each end makes the length 14'-0". The railing will be on both sides of the stairs. We will also need to make bends in this rail, usually by welding in elbows that can be purchased in the same diameters as the pipe. The railing will be fastened to the wall with brackets. These brackets are either welded or mechanically fastened to the railing (mechanically fastened is with self tapping screws) at a maximum of 6'-0" centers to give adequate support for the railing. (Self tapping screws are used for mechanically fastened brackets.) Figure 9.22 shows a sample take off listing.

The listing for pipe railing in Figure 9.22 shows the continuous length of rail at 8'-0" with the returns or the horizontal pieces at 1'-0". The change from horizontal to skewed in the rail requires the use of a 45-degree elbow. Some architects don't mind if the fabricator foregoes the use of these 45-degree ells and allows for the pipe simply to be cut at a 45 degree angle; this approach saves labor and makes the railing less expensive.

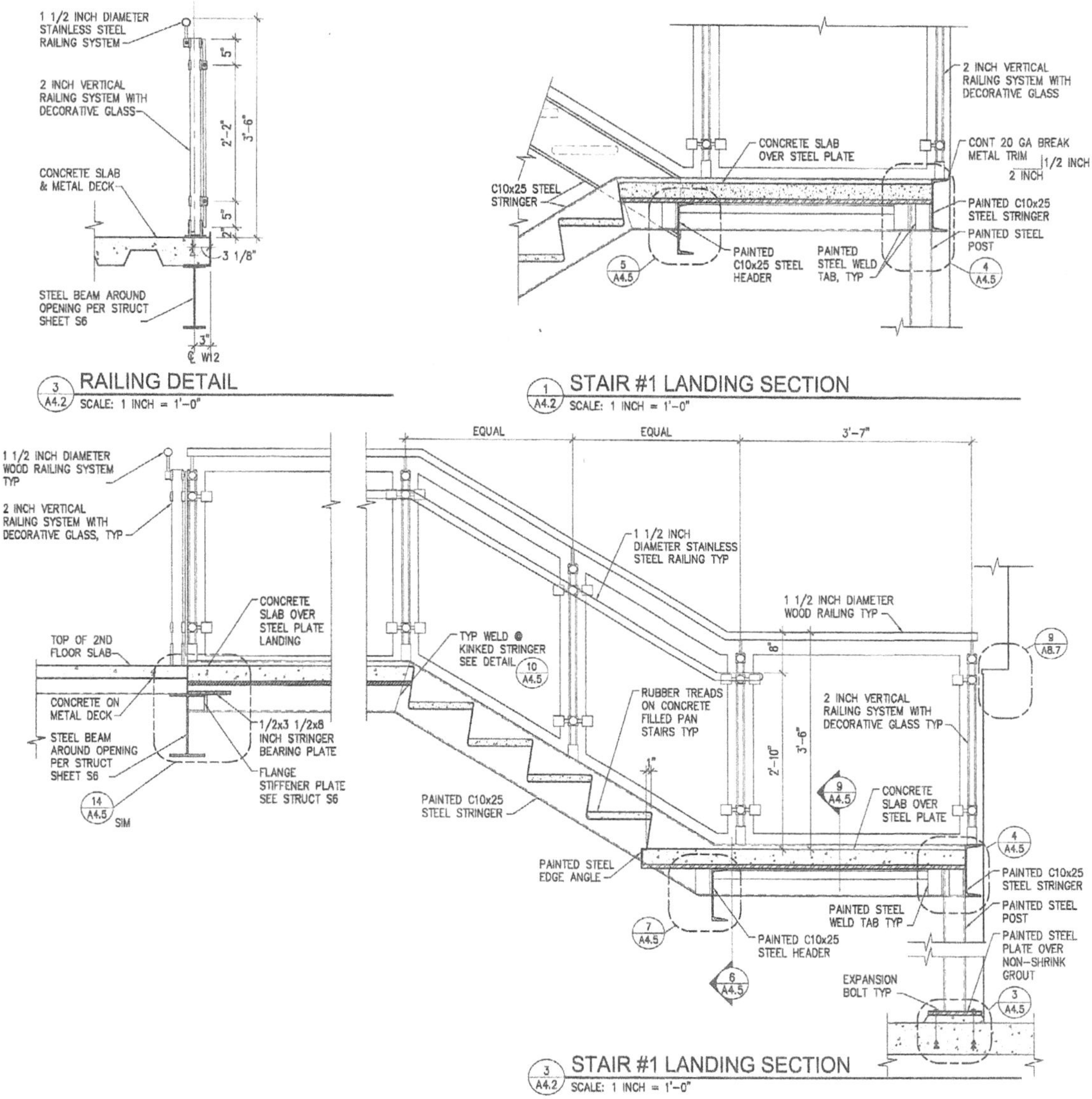

Figure 9.17 Stair #1 landing section

The 90-degree ells are for the return to the wall instead of a blunt end. Regardless of how the railing ends are completed, this take off suffices for pricing.

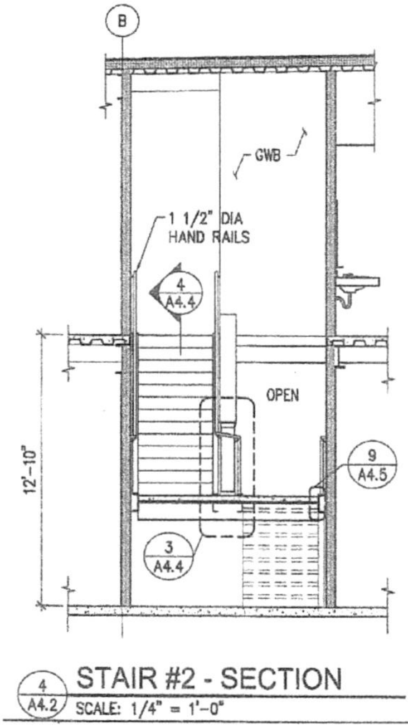

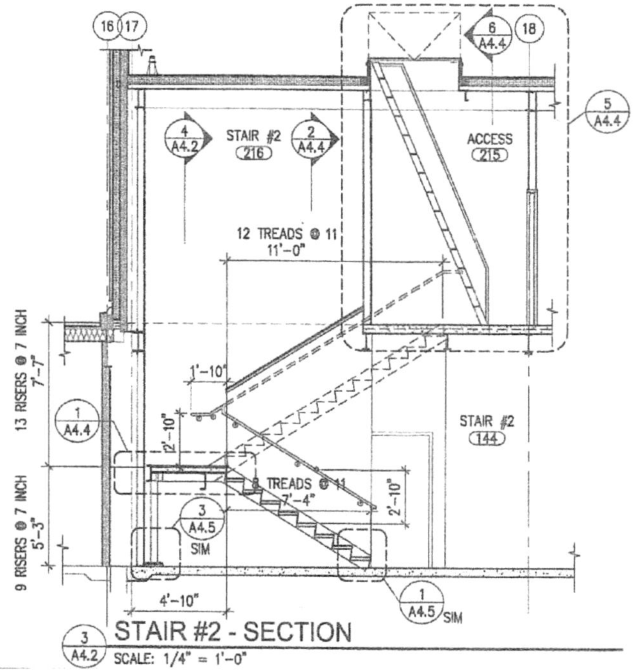

Figure 9.18 Stair #2 sections

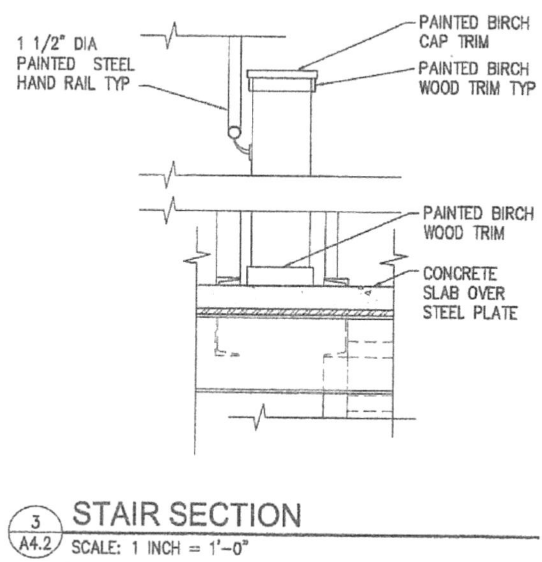

Figure 9.19 Stair section

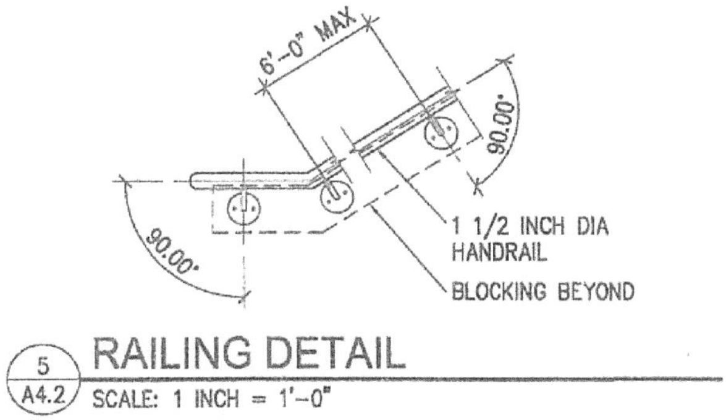

Figure 9.20 shows another detail of this railing.

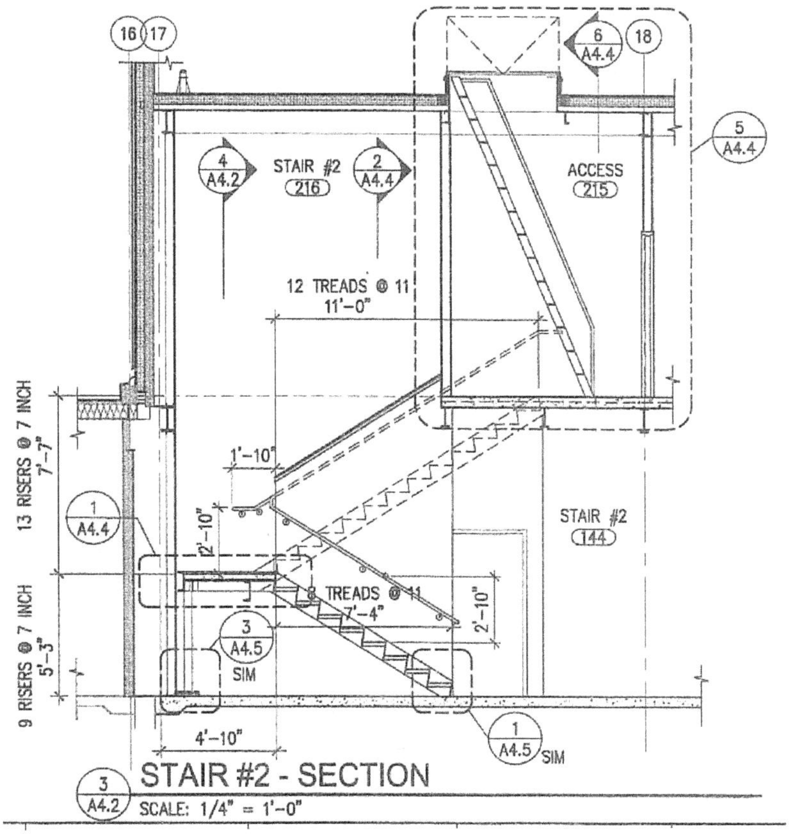

Figure 9.21 Stair #2 elevation

	Materials		QTY		Mat. Thkns	Plate shown in inches Width	Length	Ext Wt	Cost Per lb.	Exten. Mat.cos	Ext. Hdwr
						shapes in ft/dec of the ft					
Labor	Rail at Stair #2 - prime paint	Detail	12.5334		2	6.3	per ea.	Lbs			
0.4	1 1/2" STP	A53 - pipe - skewed rail	2				8	44	0.62	$ 27	
0.8	1 1/2" STP	A53 - pipe - return	4				1	11	0.62	$ 7	
0.8	1 1/2" elbows 45 deg		4								
0.8	1 1/2" elbows 90 deg		4								
0.667	WFL 1/4 x 4 1/2"	weld straight to return	16				0.4167				
2	1 1/2" wall brackets		10								
0.8	WFL 1/4 x 0'-6 "	weld brackets	16				0.5				
0.4	1 1/2" STP	A53 - pipe - skewed rail	2				12	65	0.62	$ 40	
0.8	1 1/2" STP	A53 - pipe - return	4				1	11	0.62	$ 7	
0.8	1 1/2" elbows 45 deg		4								
0.8	1 1/2" elbows 90 deg		4								
0.667	WFL 1/4 x 4 1/2"	weld straight to return	16				0.4167				
2	1 1/2" wall brackets		10								
0.8	WFL 1/4 x 0'-6 "	weld brackets	16				0.5				
13								131		$ 81	$-

Figure 9.22 Sample material take off listing for stair railing

Reviewing Drawings for Ladder Details

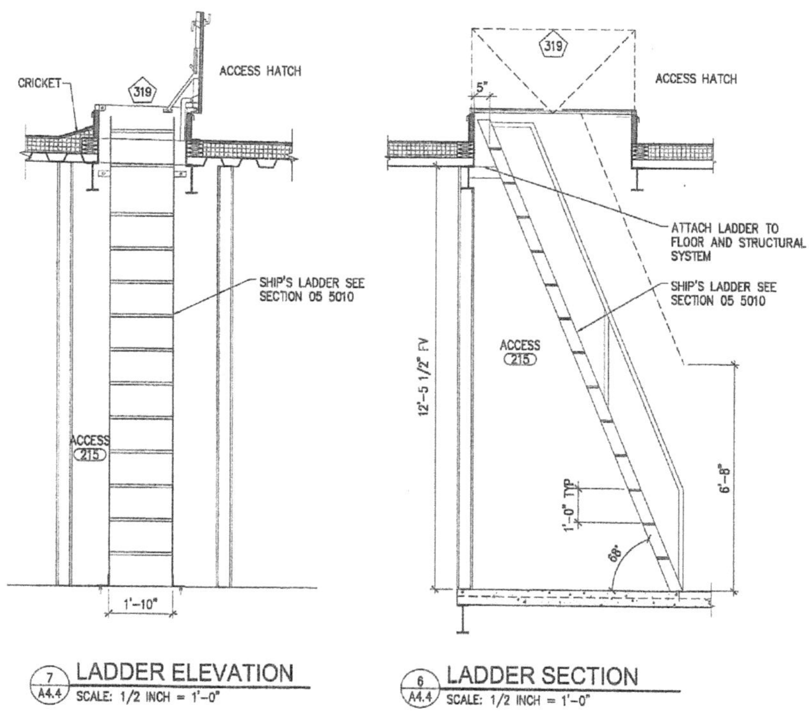

Figure 9.23 Ladder elevation and section

The ladder in Figure 9.23 is found at the top of Stair #2; it goes from the 2nd floor to the roof of the building. It is called a ship's ladder, and is often required to be constructed out of aluminum. The ladder may be a buyout item, either by the steel fabricator or by the general contractor. If it is a buyout item, don't forget to add the cost of packaging and shipping it; those costs will be extra.

If this ladder is to be a steel-fabricated item and there are no details, the steel fabricators may have a style of ladder that they propose for such items. Otherwise, the materials for this ladder could be listed as: (2) MC 8 x 18.7, purchased ship's ladder treads from a grating supplier that are 3 diamond tread (8 1/8" wide) x 1'-10", and 1 1/4" diameter single-line pipe rail with posts at 6'-0 centers, with appropriate anchor plates and fasteners to both the floor and roof attachments.

The take off in Figure 9.24 represents a suggestion of material sizes and types. These materials need to be listed in the bid letter with the item to the included.

	Materials		QTY		Mat. Thkns	Plate shown in inches Width	Length	Ext Wt	Cost Per lb.	Exten. Mat.co	Ext. Hdwr
						shapes in ft/dec of the ft					
Labor	Ship's ladder 3/A4	Detail	12.396		1	12.4	per ea.	Lbs			
0.4	MC8 x 18.7	5/A4.4	2				14.5	542	0.62	$336	
0.8	L 2 x 2 x 1/4	clips - 2H	4				1	13	0.62	$ 8	
0.333	WFL 3/16 x 0'-6	weld clips to stringer	8				0.4167				
0.533	A307 3/4 x 2 1/2 w/ nut & Washer	A307 w/nut & washers	4					1.99			$7.96
0.533	Expansion anchors 3/4 x 0'-8"	A307 w/nut & washers	4					2.4			$ 10
2.8	Grating treads x 8 1/4"		14								
1.4	WFL 1/4 x 0'-6 "	weld brackets	28				0.5				
	single line grab rail										
0.4	1 1/4" STP	A53 - pipe - skewed rail	2				14	64	0.62	$ 39	
1.2	1 1/4" STP	A53 - pipe - posts	6				3	41	0.62	$ 25	
0.4	1 1/2" elbows 45 deg		2								
0.4	1 1/2" elbows 90 deg		2								
0.396	WFL 1/4 x 0'-4"	weld straight to return	12				0.33				
2	1 1/2" wall brackets		10								
0.8	WFL 1/4 x 0'-6 "	weld brackets	16				0.5				
12								664		$409	$ 18

Figure 9.24 Suggested material listing for ladder

Figure 9.24 lists the following: the MC8 x 18.7 x 14'-6"; the clip angles L 2 x 2 x 1/4 as connections that weld to the channels both top and bottom, with holes for the 3/4" expansion anchors at the floor; the 3/4" A307 bolts at the top; and the grating treads that will be welded to the MC8 x 18.7. The grab rail at both sides of the ladder is listed as 1 1/4" diameter standard pipe extending the length of the ladder with three posts each side. The materials are not exact, but close enough for pricing.

Reviewing Drawings for Pipe Bollard Details

The plan view in Figure 9.25 shows how to find pipe bollards in the plan view of either the civil C drawings or at the architectural A drawings:

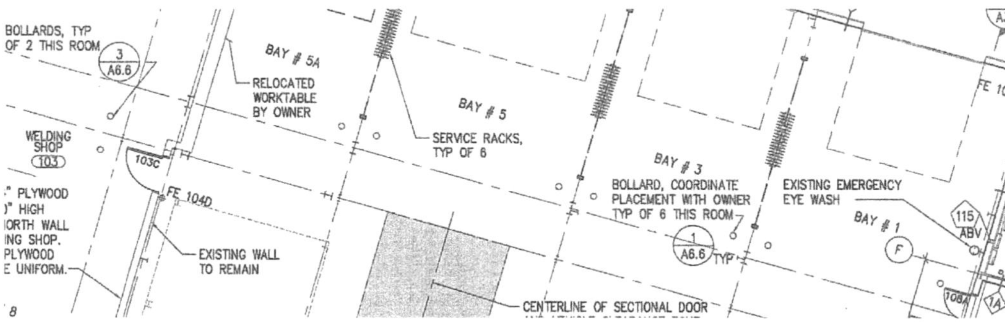

Figure 9.25 Pipe bollards

The details in Figure 9.26 show callouts for two different types of pipe bollards. Both Details *1* and *3/A6.6* are called out in Figure 9.25's plan. Note that the designers provide you with a quantity for each type in Figure 9.25. Always double check designe-supplied quantities for accuracy.

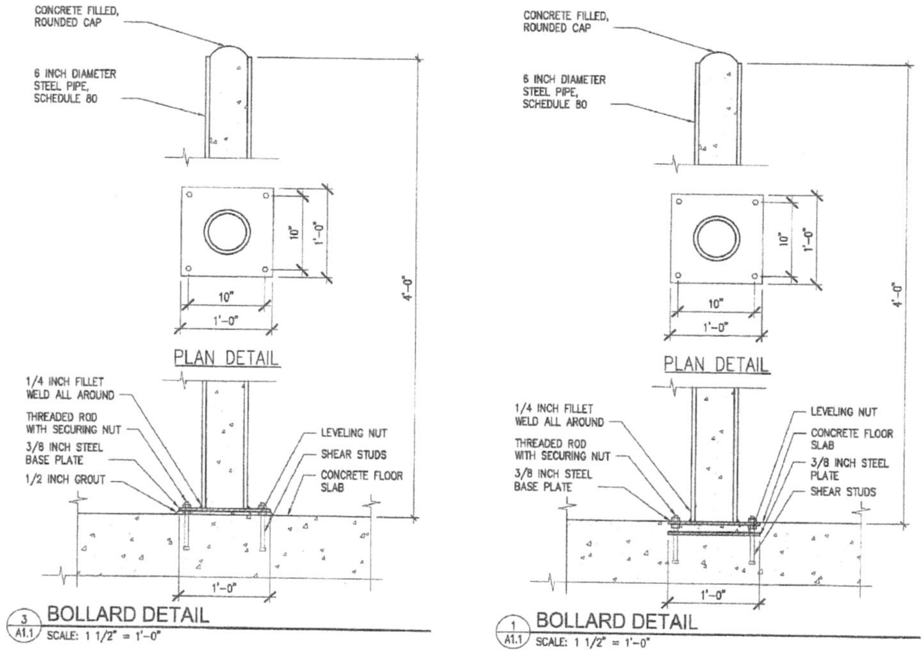

Figure 9.26 Bollard details

Figure 9.26 shows the details of the pipe bollards. The difference between the two bollards is that the picture on the right of Figure 9.26 shows an embedded plate whereas the picture to the left does not.

The take off for these bollards looks like the sample in Figure 9.27:

					Mat.	Plate shown in inches		Ext	Cost	Exten.	Ext.
	Materials		QTY		Thkns	Width	Length	Wt	Per lb.	Mat.cos	Hdwr
						shapes in ft/dec of the ft					
	both Detials 1, 3/A1.1										
Labor	Pipe Bollards - from A1.1	Detail	11.675		1	11.7	per ea.	Lbs			
1.4	6" diameter STP	1/A1.1 - pipe bollards	7				4	531	0.62	$ 329	
1.4	Plate 3/8 x 12 x 12	4H - base plate	7		0.375	12	12	107	0.487	$ 52	
1.458	WFL 1/4 x 2'-1 1/8	weld base to bollard	7				2.0833				
1.4	Plate 3/8 x 12 x 12	4H - embed plate	7		0.375	12	12	107	0.487	$ 52	
3.733	3/4" dia. Threaded rod anchors	w/DBL nut & Washers	28					0.9			$ 25
0.4	6" diameter STP	3/A1.1 - pipe bollards	2				4	152	0.62	$ 94	
0.4	Plate 3/8 x 12 x 12	4H - base plate	2		0.375	12	12	31	0.487	$ 15	
0.417	WFL 1/4 x 2'-1 1/8	weld straight to return	2				2.0833				
1.067	3/4" dia. Threaded rod anchors	w/DBL nut & Washers	8					0.9			$ 7
12								929		$ 542	$ 32

Figure 9.27 Sample take off for pipe bollards

The 6" diameter standard pipe is listed with the base plate Pl 3/8 x 12 x 1'-0" welded to the bottom, having four holes for the anchor bolts. For the right hand picture at Figure 9.26, there is an additional anchor plate of Pl 3/8 x 12 x 1'-0" with four holes. The anchor bolts are threaded studs with an approximate length of 6", including double nuts and washers to accommodate this detail. The material listing for the left hand picture in Figure 9.26 does not have this additional plate.

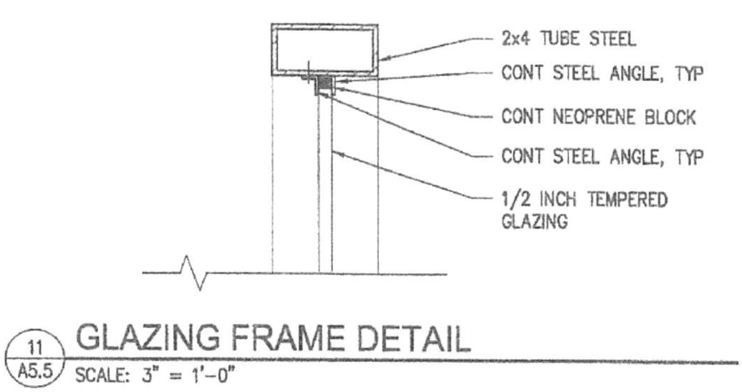

Figure 9.28 Glazing frame top detail

Reviewing Drawings for Unidentified Steel Framing Items

While reviewing the architectural details for miscellaneous steel items, details containing steel may appear that will require some research to locate. An example of this is seen in Figure 9.28.

The Glazing Frame Detail *11/A6.5* in Figure 9.28 is for framing windows. When steel estimators see this information, they are on alert for more details referencing the same. This glazing or glass work is a partition, probably inside the building.

The first point of research should be with the internal elevations of the building. This detail is a final close up from an elevation view—it is that elevation view that needs to be located.

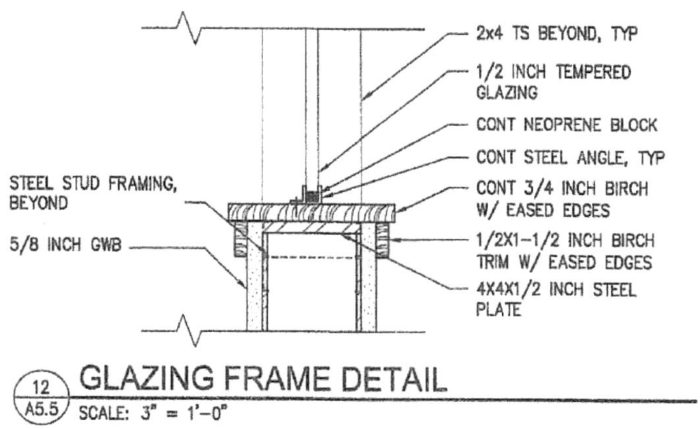

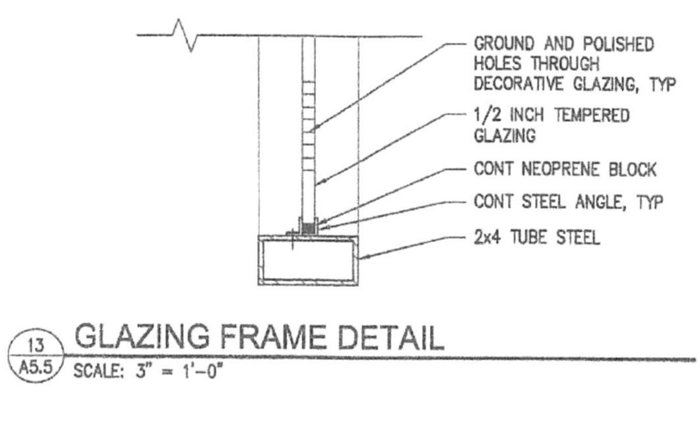

Figure 9.29 Glazing frame base at counter details

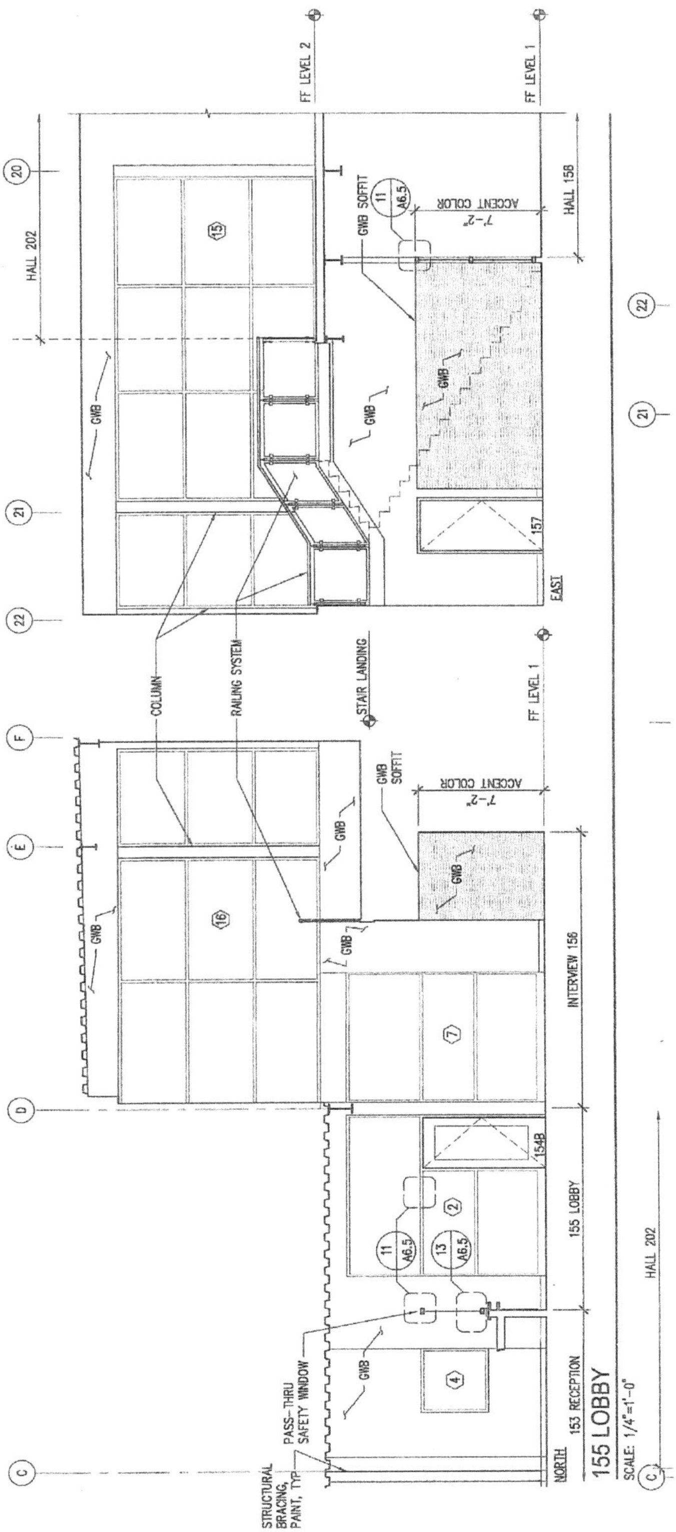

Figure 9.30 Lobby Area 155, office area elevation

Fortunately this detail bubble has a reference back to the drawing where the detail is called out. These details are shown as *11/A6.5, 12/A6.5,* and *13/A6.5*—see Figure 9.29, referencing back to sheet A5.5 for the origin.

Figure 9.30 shows the office area; see the close up of the plan view in Figure 9.32. See the details *11/A6.5* and *13/A6.5* in the left hand side of this Figure 9.30 between grid lines C and D on the first floor. This framing is for window (glazing) supports that are shown in Figures 9.28 and 9.29.

Notice how the left side of this picture shows the elevation between grid lines C and F, then look to the tiny note that says *NORTH* at the bottom left. Looking at the right side of Figure 9.30, we see grids 20 through 22 at the top. The tiny note underneath this picture says *EAST.*

We know from the *NORTH* and *EAST* indicators that we are looking at the elevation of the Lobby 155 from two different directions within this detail. The grid lines at the top of these pictures show us how to match these elevations in Figure 9.30 with the floor plan view in Figure 9.32.

The north elevation shows you the glazing frame detail reference in *11/A5.5* (our Figure 9.28) is TS 4 x 2 as it supports the windows above the

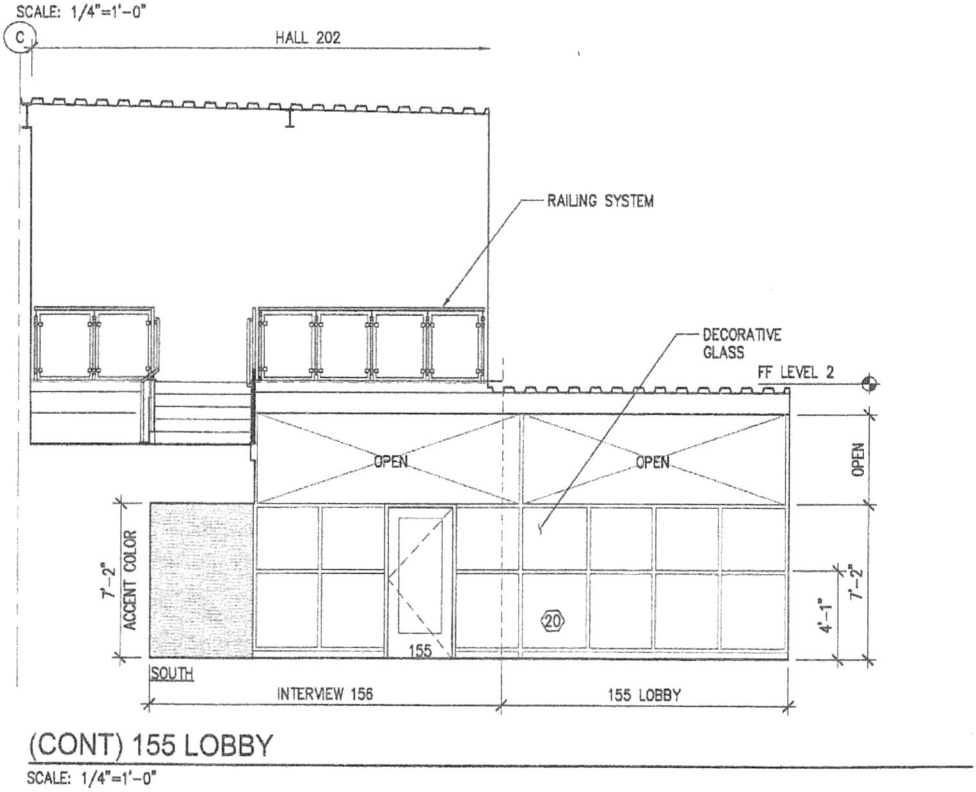

Figure 9.31 Lobby Area 155, elevation view

reception counter (details *11/A6.5* and *13/A6.5*). The east elevation shows you the TS 4 x 2 that supports the window above the door (detail 11/A6.5). Detail *11/A6.5* is shown at the west elevation in Figure 9.31.

Now we have two elevations of the building where this steel occurs. We have already covered what Figure 9.30 shows. Figure 9.31 also identifies the elevation location as 155 Lobby. Using the grid lines at the top of Figure 9.31

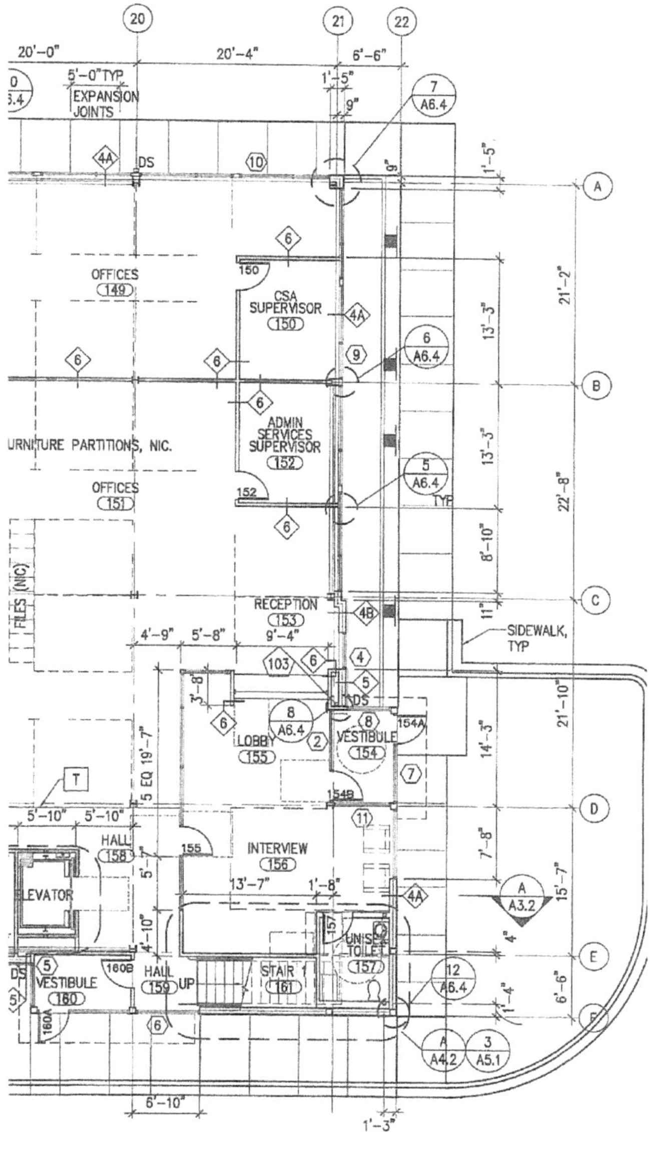

Figure 9.32 First floor plan close up

as locators, we can go to the floor plan in Figure 9.32 to see where the steel is located.

In Figure 9.32, the plan view of the first floor shows the 155 Lobby between grids C and D, 20 to 22.

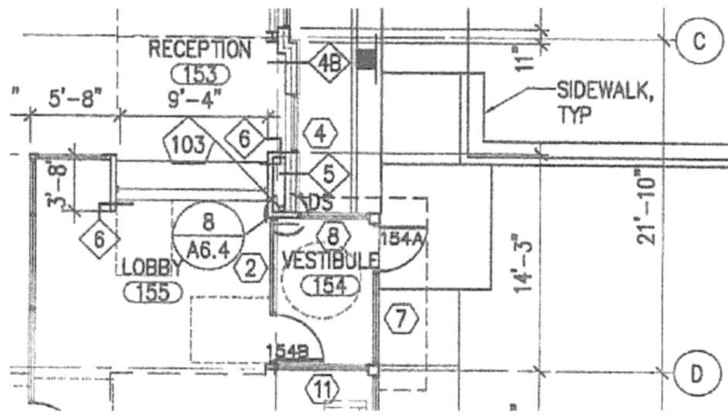

Figure 9.33 Lobby 155 close up

The plan view in Figure 9.32 has no elevation cuts or detail references, although we know from the grid line locations and the room location of the elevation identified as 155 Lobby that this is the location of the steel. Figure 9.33 shows a close up of the Lobby 155 area. The details given for the steel are very obscure and create lots of questions. We see the horizontal tubing, but what we need to know is how the horizontal tubes are supported and some details that show the connection material size and type.

Looking back to the structural drawings for any possible reference or provision of steel supports reveals nothing in the way of detail for this framing. There are a couple of tube steel posts shown at grid C.5 and D at grid 22 in the structural foundation and first floor plans. However, there is no reference or indication that the steel shown here at the architectural drawings exist.

Questions regarding the steel required for this item can be asked of the designers, if time allows prior to bid. If there is not enough time and no one has located this steel per addendum prior to bid time, it can be excluded.

If you decide to make an allowance for this steel, some assumptions could be made and qualified as an inclusion item in the bid letter. A price separate from our bid total would be supplied; the customer could choose to add it to the base price or not.

Offering an alternate price eliminates the risk of driving our base price in a zone that would make it appear less competitive. In this way you have covered all that may be required in this job without seeming less competitive by providing items your competition may not.

Going back to details *11/A6.5, 12/A6.5,* and *13/A6.5* in Figures 9.28 and 9.29, we can see that we need two horizontal TS 4 x 2 referenced from Details 11 and 13. Meanwhile, Detail 12 shows that a vertical steel TS post is there as well. We could make a frame of TS 4 x 3 x 3/16 (assuming the thickness of this tube steel, because nothing is shown).

The typical height of a window wall indicated at these elevations is 7'-2". Using this height, we can make a frame. We need the two posts of TS 4 x 2 x 3/16 x 7'-2", with base plates of 1/4 x 8 x 10. (The dimensions of base plates are calculated by using the outside dimension of the tubing and adding three inches each way to allow for 3/4" diameter fasteners.) We also need two pieces of TS 4 x 2 x 3/16 x 9'-4". (In Figure 9.32, the first floor plan at A1.2 has the dimension of 9'-4" for the width of the reception desk.) A base plate with anchor bolts needs to be included. The tube steel frame for the glazing we have been talking about here may be built and delivered as a welded frame.

Referring back to Figure 9.29, the cont. steel angle called out in the top half of this picture that frames the glazing itself could be supplied by the window installer. This angle should be excluded because it is not sized. If it needs to be added, an L 1 x 1 x 1/8" on both sides might be used.

In Figure 9.28, the TS for Detail *11/A6.5* at the north elevation can be field attached at the top of the door level—the tube steel posts at the structural drawings would support that tubing. Still, we will list connections for it.

The length of the TS 4 x 2 spanning those columns is 10'-7", as calculated from the dimensions from the lobby area. The take off for the tube steel framing supporting the glazing or windows in Lobby 155 is seen in Figure 9.34.

	Materials		QTY		Mat. Thkns	Plate shown in inches Width	Length	Ext Wt	Cost Per lb.	Exten. Mat.cos	Ext. Hdwr
						shapes in ft/dec of the ft					
	11, 12, 13/A6.5 Lobby window elevations A5.5										
Labor	see room 153 Reception	Detail	6.4		1	6.4	per ea.	Lbs			
0.4	TS 4 x 2 x 3/16	Posts 11, 12, 13/A6.5	2				7.1667	98	0.62	$ 61	
0.4	TS 4 x 2 x 3/16	Horiz	2				9.33	128	0.62	$ 79	
0.4	TS 4 x 2 x 3/16	intermediate	2				3.0833	42	0.62	$ 26	
0.4	Plate 1/4 x 8 x 0'-10	4H - base plate	2		0.250	8	10	11	0.487	$ 6	
0.8	WFL 3/16 x 1'-0	weld frame together	8				1				
2	Expansion anchors 3/4 x 0'-8"	A307 w/nut & washers	8						2.4		$ 19
	Grid C.5 and D										
0.4	TS 4 x 2 x 3/16	Posts - 11/A6.5	2				10.5833	145	0.62	$ 90	
0.8	L 2 x 2 x 3/16	clips - 2H	4				0.33	3	0.62	$ 2	
0.8	WFL 3/16 x 0'-4"	weld clips to post	8				1				
6								429		$264	$ 19

Figure 9.34 Sample take off for windows frames in Lobby 155

The exercise we have just gone through is exactly the type of situation that, if not addressed by addendum at the time of bid, will create RFIs at the time of construction. Steel estimating may be done based on assumptions, but steel detailers cannot proceed on assumptions. Clarifications will need to be made prior to the work being fabricated.

Figure 9.35 shows another elevation view inside the building. This reception area is located at the grid line intersection of C and 21 in Figure 9.32. Notice the counter supports highlighted at this elevation. These supports may or may not be steel fabricated items. The fact that they *could* be a steel fabricated item puts us on alert, as counter brackets often are. The lack of section cuts in this picture will make us want to add these counter brackets to our list of exclusions in our bid letter.

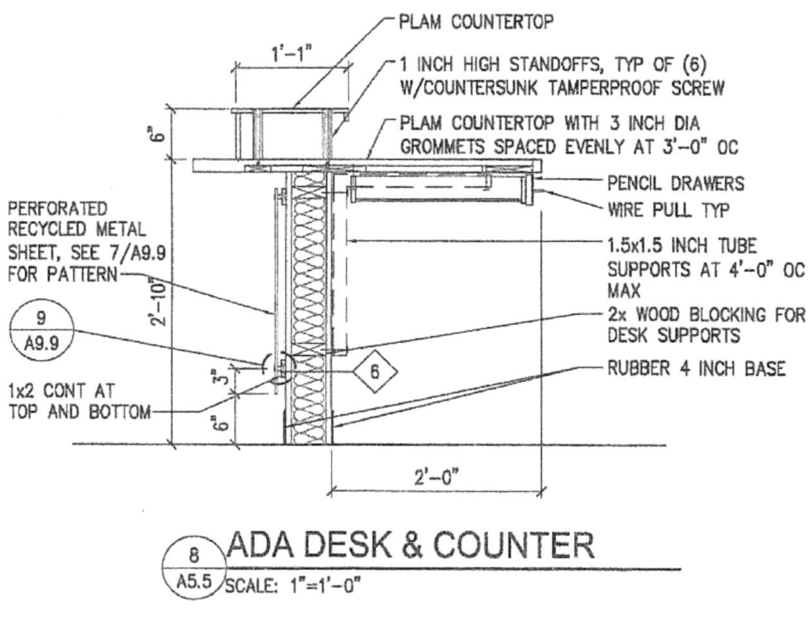

Figure 9.35 Reception area

Reviewing Drawings for Counter Bracket Details

As we are reviewing the drawings, we find the counter brackets in other details in the drawings. These counter bracket details are shown in Figure 9.36. We know that they are the bracket details that we are looking for due to the *ADA Desk* and *Counter* reference noted in both Figures 9.35 and 9.36.

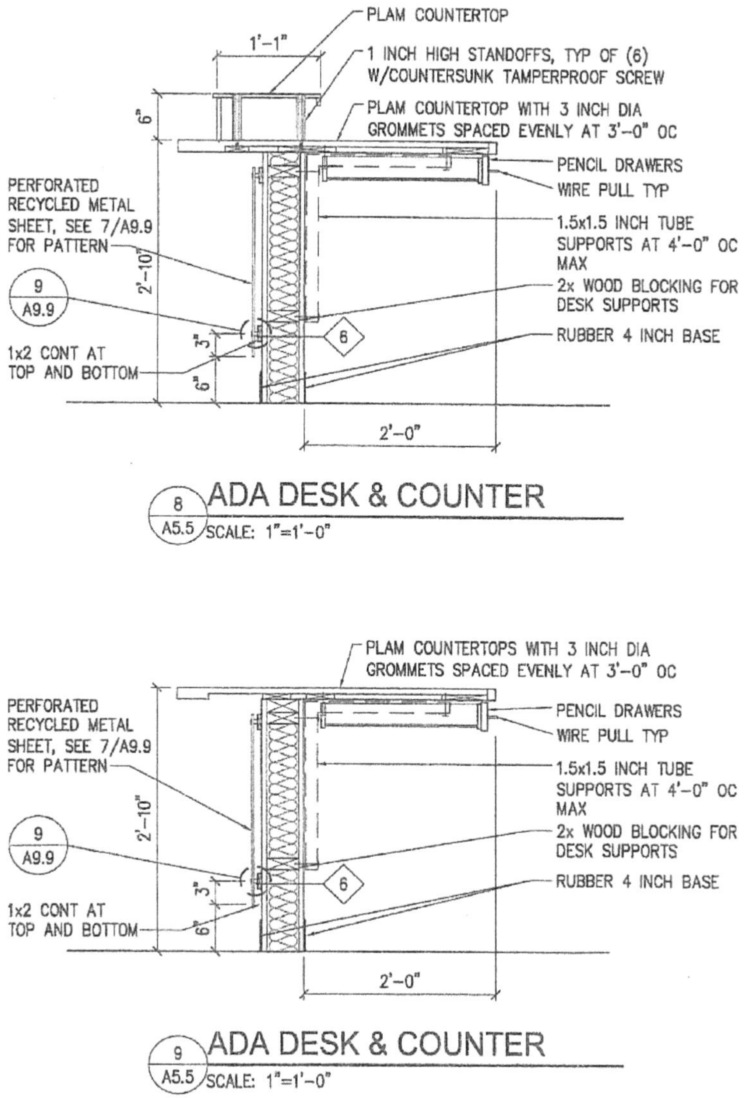

Figure 9.36 ADA desk and counter

While proceeding with the review of the architectural drawings, more bracket details are located shown in Figure 9.37. Now we need to find out where the designers are using these brackets by looking for these detail call outs from the interior elevations of the building. The clue in locating brackets for the ADA Desk and Counter was with the titles used in the details. Now we are looking for these brackets using either the detail section cut call out or the title of the call out, because the designers has now shown us they are utilizing both methods.

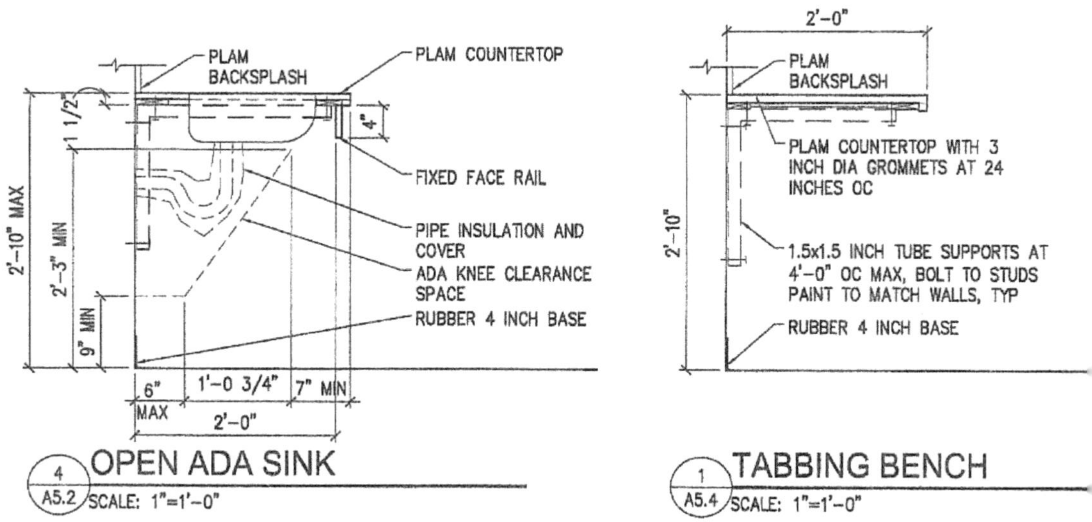

Figure 9.37 Open ADA sink and tabbing bench

These supports are used at the sinks in the bathrooms. Therefore, we will go to the interior elevations to look for any section cuts showing the sink area. We have already found the counter supports for the reception area 153—three of them are shown at the elevation.

All of these details in Figures 9.36 and 9.37 show the TS 1 1/2 x 1 1/2 supports at 4'-0" centers.

The elevations in Figures 9.39 and 9.40 show the lavatories that need supports at 4/A5.8 in Figure 9.29, taken from the plan view called out at 111 Women's and 118 Men's in Figure 9.38.

The pictures in Figures 9.42 and 9.43 are used to show the room elevations from the plan in Figure 9.41. The counters will also need the supports shown in Figure 9.37.

In Figure 9.44, the elevation view of what is called 130 Tabbing has the section cut we were looking for (referenced in Figure 9.37). See the section cut here noted as 1/A5.8.

The elevation in Figure 9.45 comes from the plan view in Figure 9.44.

You can see the little squares in the wall running left to right in the center of this picture—those are the tube steel brackets that are called out in Figure 9.44. These brackets are the ones shown in detail in Figure 9.37.

Figure 9.45 is found within Figure 9.46, located between grids 13 and 14 below grid line F. The designers did not provide a section cut at the Tabbing 130. We can tell by the room title in Figures 9.44 and 9.45 that these details are to be used for the Tabbing 130 area.

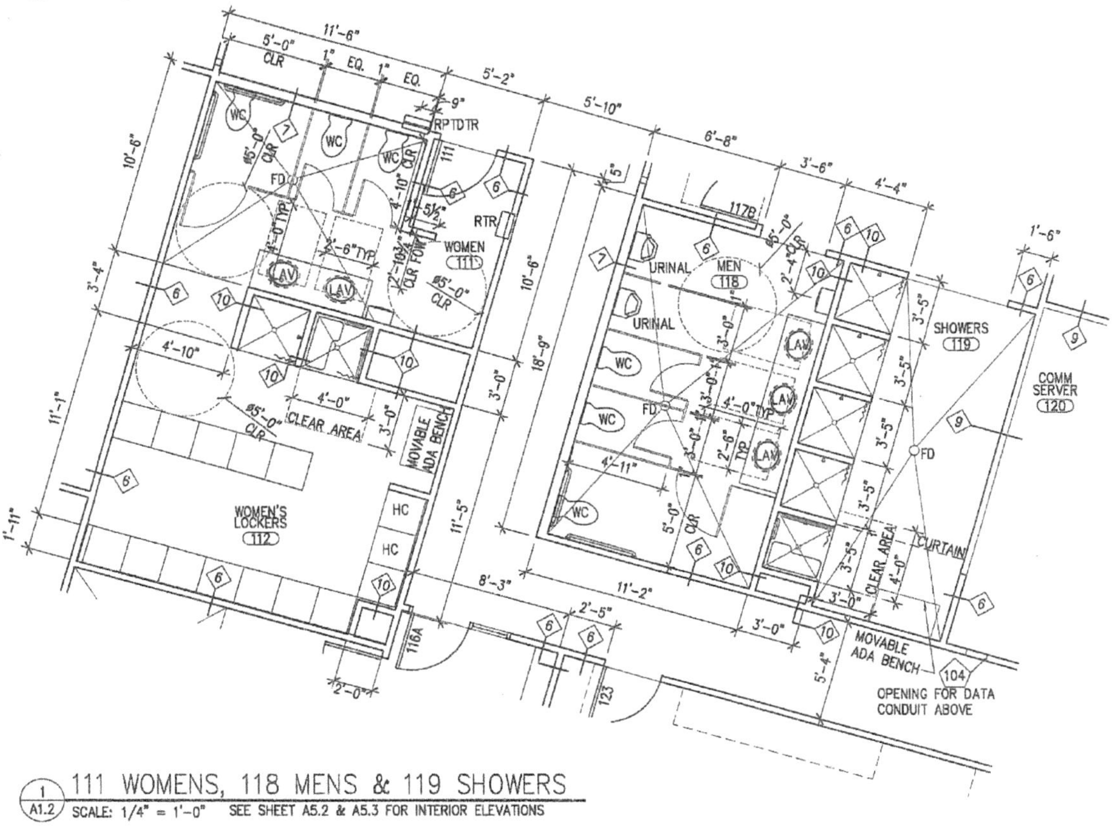

Figure 9.38 Plan view of the men's and women's lavatories

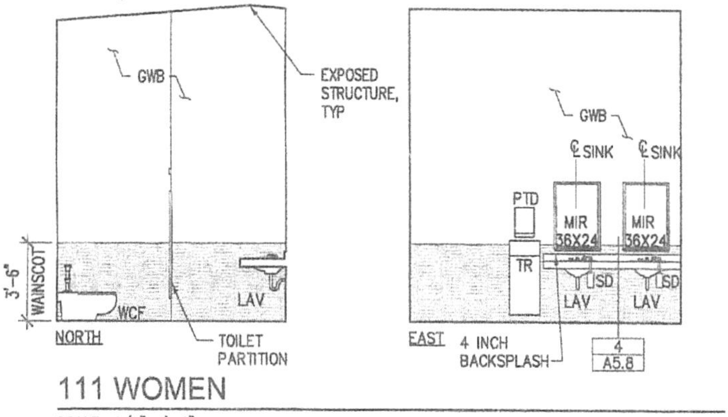

Figure 9.39 Interior elevation women's lavatory #111

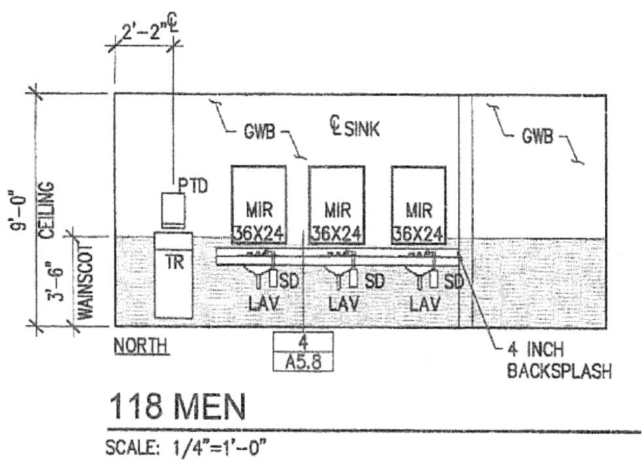

Figure 9.40 Interior elevation men's lavatory #118

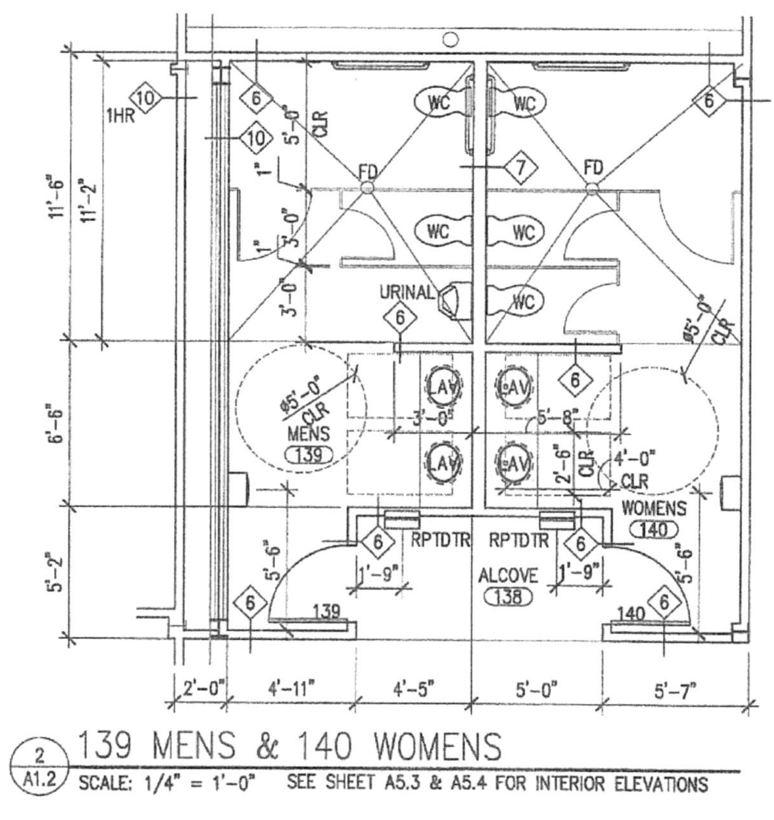

Figure 9.41 Lavatories #139 and #140 plan view

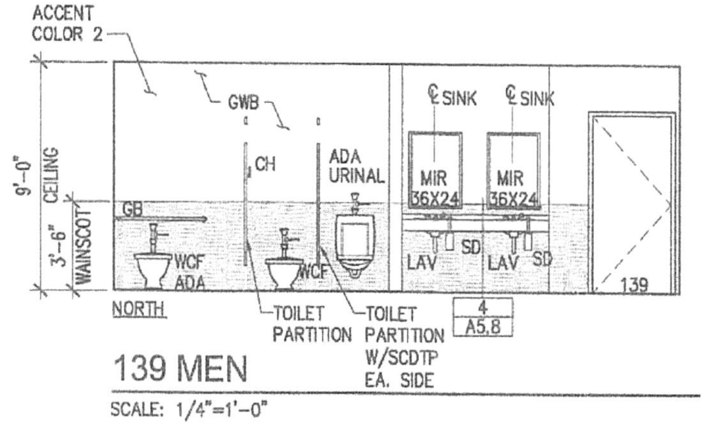

Figure 9.42 Men's lavatory #139 Elevation View

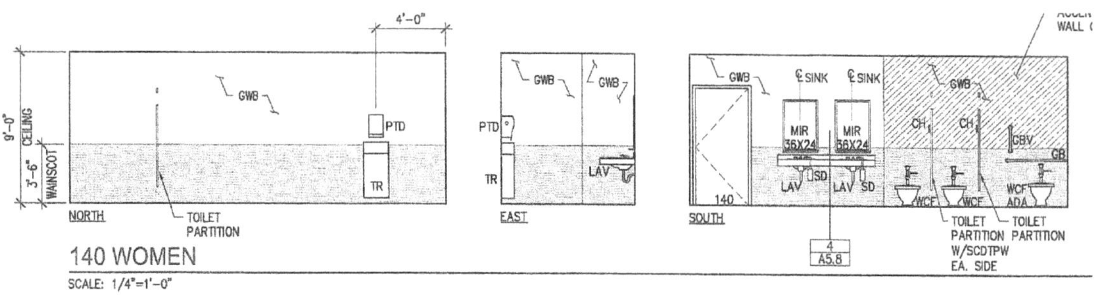

Figure 9.43 Women's lavatory #140 Elevation View

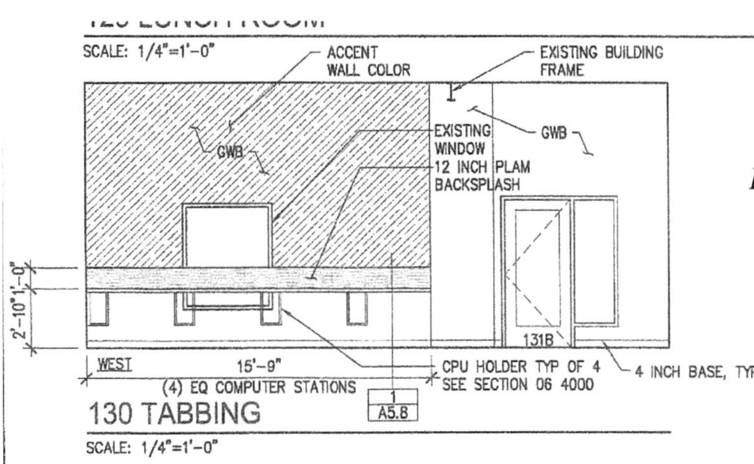

Figure 9.44 Elevation view of tabbing area

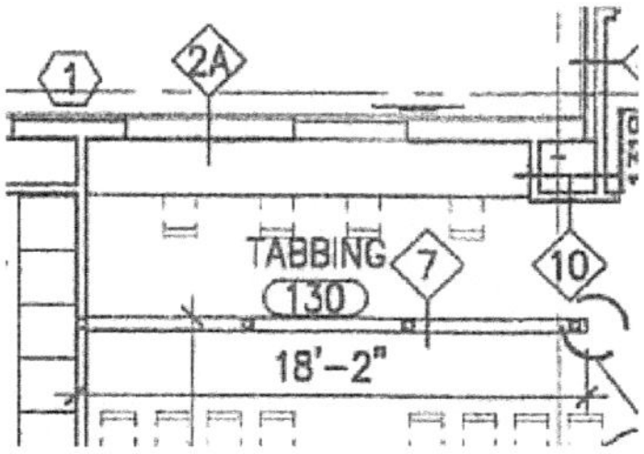

Figure 9.45
Plan view of tabbing counter

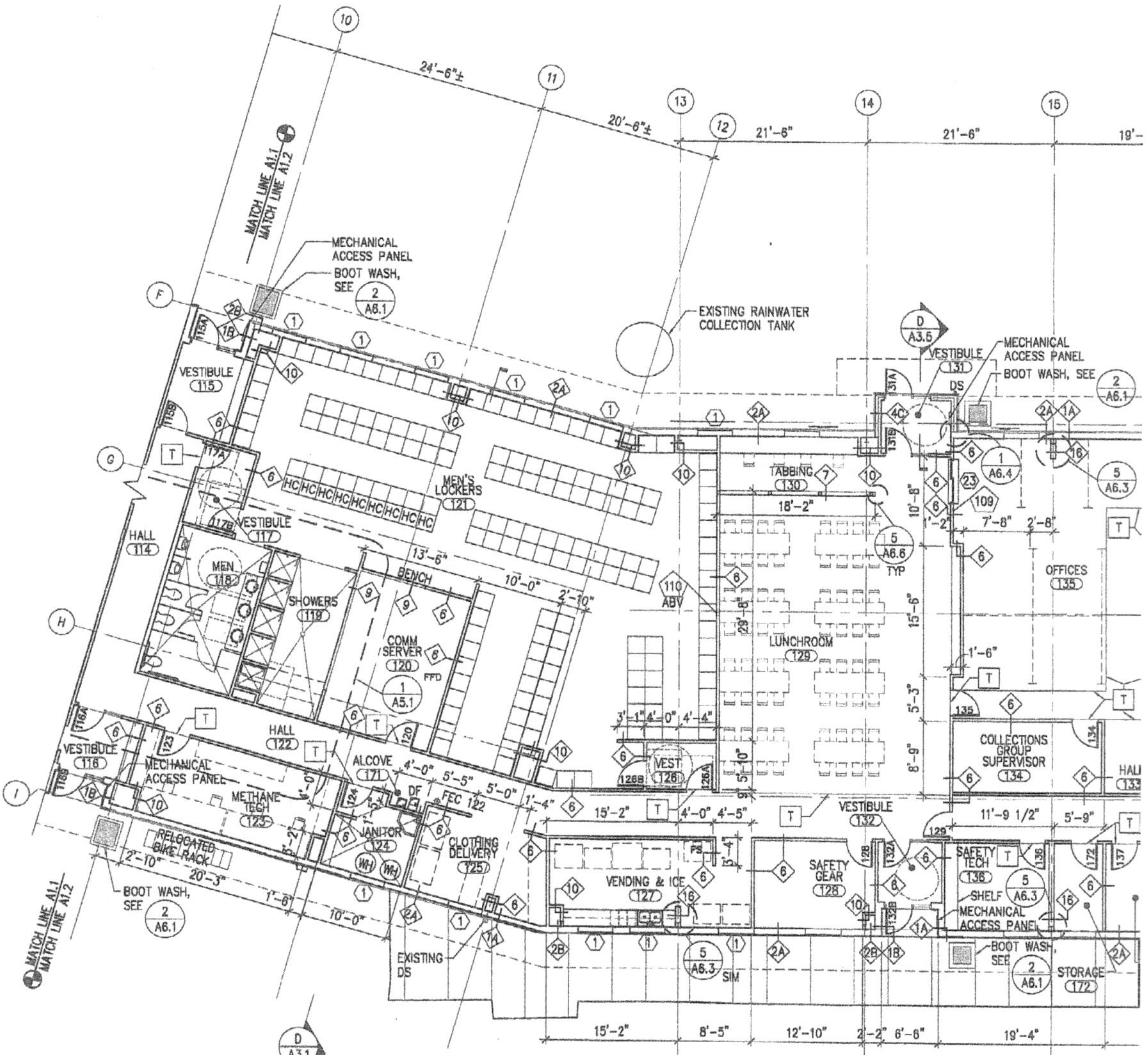

Figure 9.46 Left side of first floor plan

Labor	Materials	Detail	QTY		Mat. Thkns	Plate shown in inches Width (shapes in ft/dec of the ft)	Length	Ext Wt	Cost Per lb.	Exten. Mat.cos	Ext. Hdwr
	ADA Desk and Counter Brackets 8, 9/A5.8										
Labor		Detail	30		1	30.0	per ea.	Lbs			
	(3) Brackets at 153 Reception from A5.5 - 1 location										
1.2	TS 1 1/2 x 1 1/2 x .120	"L" Bracket	6				7.1667	95	0.62	$ 59	
0.3	WFL 1/8 x 0'-6	Weld to "L" shape	6				0.5				
3	A307 3/4 x 3" w/ nut & washer	A307 w/nut & washers	12						0.9		$ 11
	(4) Brackets at 130 Tabbing A5.4 - 1 location										
1.6	TS 1 1/2 x 1 1/2 x .120	"L" Bracket	8				7.1667	127	0.62	$ 79	
0.4	WFL 1/8 x 0'-6	Weld to "L" shape	8				0.5				
4	A307 3/4 x 3" w/ nut & washer	A307 w/nut & washers	16						0.9		$ 14
	(13 total) Brackets at 111 - Womens, 118 Mens, 139 Mens, 140 Womens, 210 Mens, 211 Womens										
5.2	TS 1 1/2 x 1 1/2 x .120	"L" Bracket	26				7.1667	412	0.62	$255	
1.3	WFL 1/8 x 0'-6	Weld to "L" shape	26				0.5				
13	A307 3/4 x 3" w/ nut & washer	A307 w/nut & washers	52						0.9		$ 47
30								634		$393	$ 72

Figure 9.47 Sample Material take off listing for brackets

Figure 9.47 is a sample of what the take off looks like for the brackets shown at Figure 9.36 and 9.37, found in the details in Figures 9.35 through 9.45.

In the bid letter, we will need to qualify the material thickness of the TS 1 1/2 x 1 1/2 x .120 wall because none is specified. A provision of the fasteners must be noted because they too are not called out. The material size was chosen as a common size from a steel supplier's material book. The brackets will be listed in the exclusions and offered as an additional alternate price.

Figure 9.48 shows a sample weight chart for tube steel, in this case the TS 1 1/2 x 1 1/2 x 11 ga or .120 wall tubing. The top line shows the TS 1-1/2 x 1-1/2 representing the outside dimensions of the tube steel. The numbers in the first column represent the different wall thicknesses for this size of tube steel that are available. The numbers in the second column represent the decimal equivalent of the tube thicknesses. The third column shows how much each tube steel item listed weighs per foot.

Figure 9.48 shows the material size chart used to pick the size of tube steel that we want for counter brackets.

1½" x 1½"		
18	.049	.9669
16	.065	1.269
15	.072	1.380
14	.083	1.600
13	.095	1.815
12	.109	1.992
11	.120	2.210
.145	.145	2.672
3/16	.188	3.040
1/4	.250	4.067

Figure 9.48
Weight chart for tube steel

The quantity was calculated as 20 each total, based on (3) for the ADA desk, (4) used at the Tabbing Desk, and (13) each for the lavatories, with the quantity based on the brackets being used at 4'-0" centers.

Sometimes the information on fabricated items is limited. Sometimes we are unable to verify prior to bid time what the designer is really looking for. In these cases, making an assumption—as long as it is specifically identified at the bid letter—will generally work out.

Reviewing Drawings for Special Fabrication Items

The following example shows how the counter brackets might be described at the bid letter inclusions:

- TS 1 1/2 x 1 1/2 x .120 brackets with 1'-6 legs, provided with (4) 3/4" A307 fasteners, prime painted, (3) each at room 153–Reception per details 8, 9/A5.8; (4) each at Tabbing area room 130 on A5.4; (13) each at lavatories per details 1, 2, 4/A5.1–(4 total) at 111 Men and 118 Women, (4 total) 139 Men and 140 Women, and (5 total) at 210 Men and 211 Women shown at elevations on A5.2, A5.3, and A5.4. Offered as an additonal alternate price each of $00.00.

This description includes what is being provided, the referenced details, the quantity, and the offer of an additive alternate price each. It makes a statement about what assumptions are being made for the price that is provided.

The details of the bootwash in Figures 9.49, 9.50, and 9.51 may or may not be an item supplied by the steel fabricator. Because it is made from steel, however, and has steel grating, we will assume it is. There is an embedded frame with a grate and a piece of railing there to hold on to while the workers are scraping their boots.

We would exclude the actual boot scraper, but we can make some simple assumptions about the grating and the embedded angle when materials are not sized. These grates are exposed to the weather and will probably need to be galvanized as well. All these assumptions would be qualified in the bid letter.

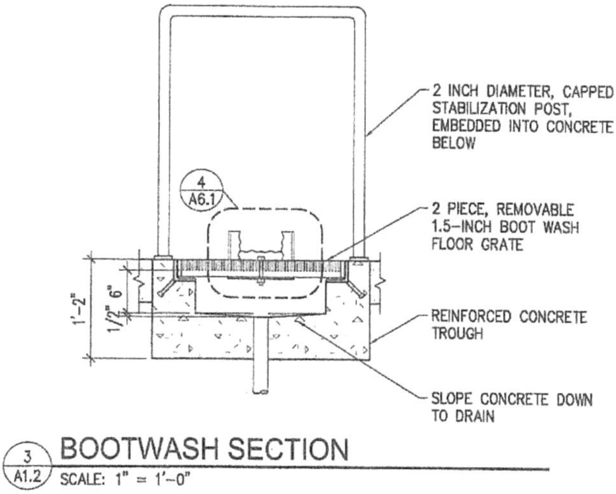

Figure 9.49 Bootwash section

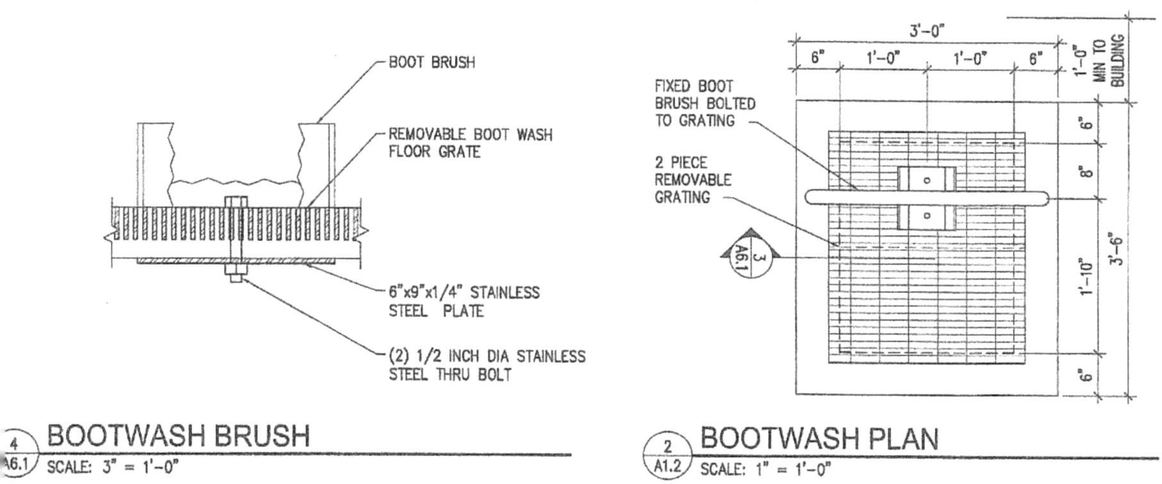

Figure 9.50 Bootwash brush and bootwash plan

- Bootwash embed frame and grate details 2, 3, 4/A6.1 from plan at A1.2, two locations, materials not sized or typed but assumed as A36 steel, galvanized, L 1 1/2 x 1 1/2 x 1/4 frame, 2'-3" x 2'-9" with HAS 1/2 x 3 at 9" centers, Grating provided 2 piece removable, 19-W-4 style 1 1/4 x 3/16 bearing bars, and a 1 1/2" standard pipe grab bar, excluding the boot brush. Add alternate price $0,000

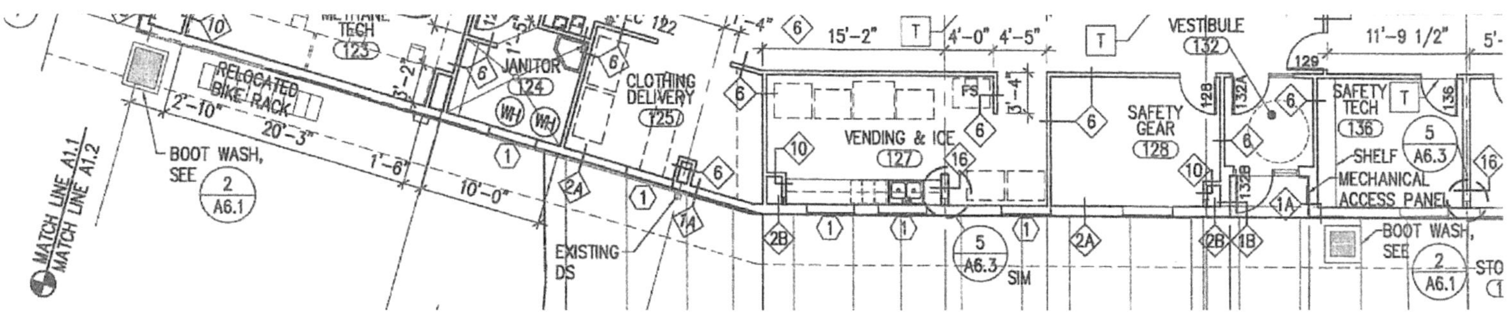

Figure 9.51 Details of the bootwash area

Labor	Materials	Detail	QTY	Mat. Thkns	Plate shown in inches Width	Length	Ext Wt	Cost Per lb.	Exten. Mat.cos	Ext. Hdwr
					shapes in ft/dec of the ft					
	A1.2 - Boot Wash Frame and Grate - GALVANIZED									
Labor	2, 3, 4/A6.1	Detail	5.93332	1	5.9	per ea.	Lbs			
	(2) two each from plan view on A1.2 - GALVANIZED									
0.4	L 1 1/2 x 1 1/2 x 1/4	Embed Angle	2			2.2917	11	0.62	$ 7	
0.4	L 1 1/2 x 1 1/2 x 1/4	Embed Angle	2			2.7917	13	0.62	$ 8	
3	HAS 1/2 x 3	A307 w/nut & washers	12					0.9		$ 11
0.4	WFL 3/16 x 0'-3	weld frame	16			0.25				
0.3	Grating 19-W-4 1'-4 1/2 x 2'- 3 1/2	1 1/4 x 3/16 bb @ 4" cntr	2		1.3	2.25				
	Grab Bar									
0.4	1 1/2" STP	horizontal	2			2.75	15	0.62	$ 9	
0.8	1 1/2" STP	Posts	4			4	44	0.62	$ 27	
0.233	WFL 3/16 x 0'-4 1/2	weld posts to horiz	4			0.5833				
	EXCLUDE BOOT BRUSH									
6							82		$ 51	$ 11

Figure 9.52 Sample take off for bootwash steel

The items listed cover all our assumptions. Offering the bootwash as an add alternate in the bid letter will keep our base bid price looking competitive.

A phone call to the designers asking for information prior to bid time is necessary. If this call is not possible, then estimators will need to make assumptions, qualify them, and offer a price to add.

A sample of the take off for the bootwash steel is in Figure 9.52. For the grate to sit inside the frame that spans the opening, the frame has to be made a bit larger. This frame has mitered corners, making it look like a picture frame. To frame the bootwash opening pit, the outermost edge of angle goes past the physical opening area that will be equal to twice the angle size. We are using L 1-1/2 angle. For this example, a total of 3" extra needs to be added to the size of the opening to get to the actual use length of the steel used to make the bootwash embed frame.

The grating has to fit inside the frame. Therefore, it will be smaller than the inside of the frame by 1/2" at both width and length, allowing for clearance. The grating will then slip in easily. We used 1 1/2" standard pipe railing for the grab bar, and made the posts 4'-0" so they could go in the ground about 6".

The take off listing is for two items: the embed frame with grating and the grab bar. If we find more, then we only have to change the quantity indication rather than relist all the materials with those extended quantities.

Summary

In this chapter, we reviewed specific fabricated items. We demonstrated how to research the drawings to find the information required for the fabrication of the steel items. We also discussed writing bid letter inclusions when dealing with the need for detailed explanations. Examples showed how to list these fabricated items on the take off sheets for pricing.

Most of the ambiguities on projects will be found in the architectural drawings; the need to be creative will arise. Ask questions and get clarification when you can. Cover yourself by being diligent about the drawing and specification reviews. Document all that is shown (and not shown) using the bid letter inclusions and exclusions. Be specific with the quantities and lengths, material types and sizes, and product finish on the fabrication. All of these steps go a long way toward creating a complete and comprehensive quote for architectural metals.

10

Fabrication Labor

Labor Hours

Shop fabrication labor hours are a key element to every job. The items purchased to fabricate the job have absolute costs, but shop fabrication labor is the largest variable cost. In theory, all of the steel fabricators bidding on a project will have the same materials. However, every steel fabricator will have its own interpretation of how much labor is needed to fabricate the steel.

In this chapter, we review both the labor times used by estimators to calculate shop labor hours and the theory by which they apply labor times to the parts to be fabricated. The listing of materials—or take off—is the easy part. Calculating the time your shop needs to fabricate the steel is the hard part.

There are specific elements to consider while estimating your shop labor time. Consider the sample material take off sheet in Chapter 8, Figure 8.1, in which each labor function is itemized. Let's review each function with regard to the actual application of labor.

Labor times are broken down to a number of minutes per function. All the necessary functions are timed. Then these times are totaled, giving you your labor hours. Functions per part will vary. However, they are easily identified by the details and information you get by using the take off sheet where you have listed all your parts to be fabricated.

To explain how to estimate shop labor times for individual applications, we will walk though the basic process, starting with the steel already at the fabrication shop, offloaded, and staged. By "staged" we mean the steel is located at a work station ready for shop hands to begin their work.

Labor Application

Estimating a Beam

The fabricator will have a shop drawing for working on a particular beam. First, the beam is laid out—in other words, the piece of material is marked with the required fabrication elements shown on the shop drawing. The designated layout worker performs this function.

The beam is already marked by the person who cut the beam to length. Some shops purchase beams cut to length and piece marked by the steel supplier. Others cut their own steel, depending on equipment capacity. Outsourcing the process of cutting materials to length may save time to fit fabrication schedules or money on shop labor. Shop fabricators should insure that the part they are about to lay out is the same part as indicated on the drawing they are using.

Working from left to right, the fabricators lay out any locations for holes, copes, stiffeners, and any other steel items that need to be welded to the beam. Next, they drill or beam punch the holes as marked, then cut copes if they are required. In turn, the welders tack into place and weld the stiffeners and any other needed items to the beam. Once all the necessary functions of fabrication have been completed on the beam, the supervisor examines the beam to verify all is correct, after which the beam moves on to be either staged for painting, if required, or loaded onto trailers for trucking to the jobsite.

We can estimate the layout time for each function from the beam itself simply by counting them and listing them next to the beam in our take off sheet. These functions include drilling holes, adding copes or special cuts, welding stiffeners, and welding any attached items separately.

The labor applications described in the following section, *Labor Operations,* should be used as general guidelines. The labor time allowed for each function is based on theoretical applications which have been used by shops as standards for estimating purposes. These times are applied for typical fabrication functions. They enable you to calculate a reasonable total labor time to be applied per part.

The total labor time may have to be tempered for the capacity of the shop that will be doing the work. For example, handling could be difficult with no overhead equipment and limited forklift capacity. In that case, more time has to be added for moving and staging materials. Maybe the shop doesn't have a beam punch and all holes will have to be drilled. Or maybe the shop doesn't have an ironworker and all parts have to be saw cut. All these shop elements affect labor time; they may add to the total time estimated by the use of the theoretical times.

Labor Operations

This section lists labor functions as shown on the take off sheet presented in Chapter 8, Figure 8.1, for listing steel materials. The references to detail material is a category for small parts that do not fit in the wide flange, channel, angle, and plate categories, such as flat bar, round bar, and MC channel.

Theoretical Labor Times by Functions

LO—Layout. When the shop worker has a drawing for a steel beam or part, that part needs to be marked for location of holes, special cuts, and welded parts. The time needed to lay out the piece to be fabricated for beams is estimated at minutes per foot.

The formula is:

Length of beam x 1/2 minute per foot per face (per side of beam)

Add 3 minutes for each additional face (this means flipping the beam to the other side). Also, add an additional 5–8 minutes for the labor to turn the beam over.

Applying this formula to a 20'-0" beam, the estimated layout time is 10 minutes for the first side, 13 minutes for the other side, plus 5–8 minutes to flip the beam over. The reason the second side takes longer is that fabricators spend time setting up for the layout, making sure they are working from the correct end and re-checking their work.

CUT—Cut to Length. Large beams are usually cut one at a time. Gang cutting may be executed with small angles, channels, round, and flat bar with the saw (if the shop doesn't have an ironworker). Small parts can be cut with the use of an ironworker, which makes the cutting time go much more quickly than being completed by saw. Some ironworkers have a bending capacity for small parts as well. The ironworker is limited by "throat capacity," the same as any saw.

Labor for Cutting

Table 10.1 is helpful when estimating labor/time for cutting.

Table 10.1 Labor/Time for Cutting

Labor	Estimated Time for Cutting
Shear up to 1/2" maximum	1 minute per foot
Clips, small	120 clips per hour or 1/2 min each
Flat or round bar	180 bars per hour
Burn by hand	3 minutes per foot
Burn by machine	2 minutes per foot
Plates over 1"	3 minutes per foot
Plates over 2 1/2"	4 minutes per foot
Wide flange, channel, angle	3 minutes per foot

COPE. A cope or a notch may be put in the ends of the beam to allow it to fit or frame into another beam or column. Allow 5–8 minutes for each cope (the time needed varies according to the size of the cope and the size of the beam, channel, or angle to be coped.) Coping is usually done by hand torch.

PUN—Punching of holes by machine. Table 10.2 summarizes the estimated time for punching, whether done with a beam punch or by an ironworker.

Table 10.2 Labor Time for Punching

Labor	Estimated Time for Punching
Single holes or hole in beams	3 minutes each
Detail material	1/2 minute per hole (large quantity)
Wide flange, channels, angles	Layout time plus 10%
Drilling	4 minutes per hole
Tapping	5 minutes per hole
Plates	1/2 minute per hole

HDL—Handle. This step is required if the beam or part either has to be turned or rolled over during the fabrication process, or must be moved to a special section of the shop. Table 10.3 summarizes the time estimates.

Table 10.3 Labor Time for Handling

Labor	Estimated Time for Handling
Wide flange and channel	Length x 1 minute per foot
Angles up to 4”	1/4 minute per foot
Channel and angles up t 8”	1/2 minute per foot
Detail material	1 minute each
Plates	1/2 minute per square foot
Turning beams	Add time required and per face

CAM—Camber. To camber a beam is to put an arch in it so it has a rise with the high point in the center. Add 1/2 hour per camber.

F/ASS—Fit and/or Assemble. Table 10.4 summarizes the estimated time for fitting and assembly.

Table 10.4 Labor Time for F/ASS

Labor	Estimated Time for F/ASS
Basic work	Length x 1/2 minute per foot
Move part to area of fit up	Estimate time case by case
Fit up of clips and stiffeners	Estimate time case by case
Clips and stiffeners, base plates	5–10 minutes each
Gussets, bevels	10–15 minutes each
Large gussets	20–30 minutes each
Stair stringers	30 minutes each
Full-size stick for wide flange	3 minutes per foot
Layout for detail materials	1–4 minutes each
Angles to 4” leg	1/2 minute per foot

WELD—Time needed to weld. Weld times are based on the thickness of the weld (3/16, 1/4, 5/16, 3/8, 1/2, and so on) and the length of the weld. The number of passes needed to accumulate the particular thickness of weld is important here. Figure 10.1 shows a welding symbol.

Manual Welding.

Time is per pass and includes start and stop, when welding in the flat position (see Figure 10.2). Table 10.5 summarizes the time estimates for manual welding in the flat position.

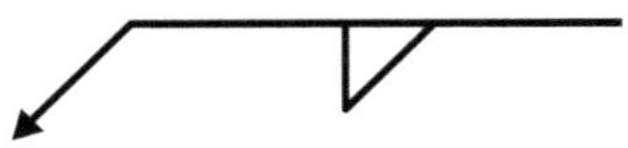

Figure 10.1
Fillet weld welding symbol

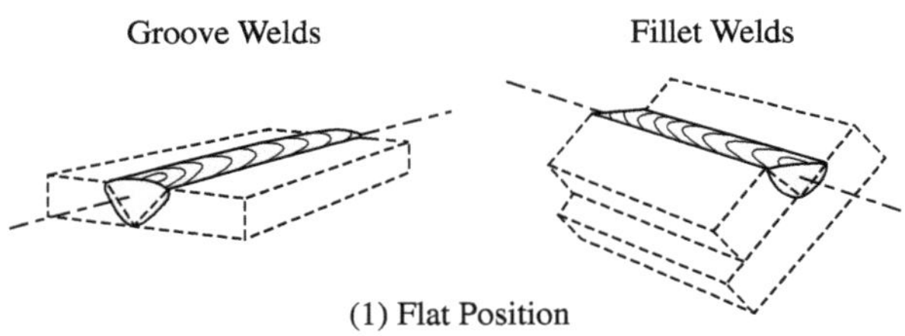

Figure 10.2 Welding in the flat position

Table 10.5 Labor Time for Flat Position Welding

Labor	**Estimated Time for Welding Flat Position**
1/4" fillet	8 minutes per foot
3/16" fillet	6.5 minutes per foot
5/16" fillet	10 minutes per foot
1/8" fillet	5 minutes per foot
11/16 single bevel weld	50 minutes per foot
1/4" single bevel weld	15 minutes per foot
3/8" single bevel weld	20 minutes per foot
Average hand welding	6 minutes per foot
Average machine welding	3 minutes per foot
Tack in place	1 minute per inch

Weld positions other than flat will require additional labor time. Review the following examples of how to calculate these special conditions.

Horizontal position.

The formula for welds in the horizontal position multiplies the total time of the flat position by 1.15% (see Table 10.6). Figure 10.3 shows the horizontal position welds.

Table 10.6 Labor Times for Horizontal Position Welding

Labor	Estimated Time for Welding Horizontal Position
1/4" fillet	8 minutes per foot x 1.15 = 9.2 minutes per foot
3/16" fillet	6.5 minutes per foot x 1.15 = 7.47 minutes per foot

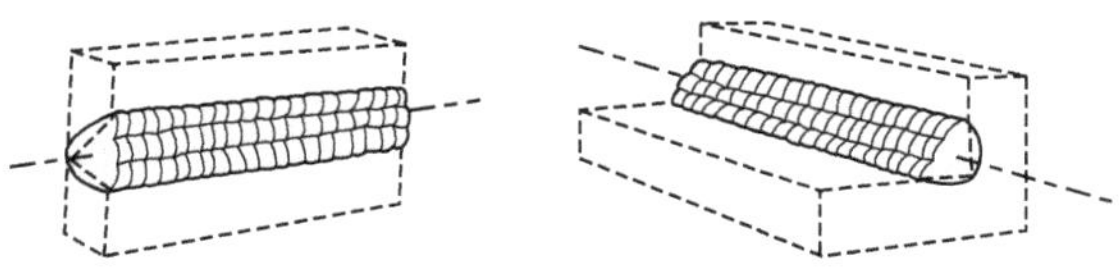

(2) Horizontal Position

Figure 10.3 Welding in the horizontal position

Vertical position.

The formula for welds in the vertical position multiplies the total time of the flat position by 1.75% (see Table 10.7). Figure 10.4 shows the vertical position welds.

Table 10.7 Labor Times for Vertical Position Welding

Labor	Estimated Time for Welding Vertical Position
1/4" fillet	8 minutes per foot x 1.75 = 14 minutes per foot
3/16" fillet	6.5 minutes per foot x 1.75= 11.37 minutes per foot

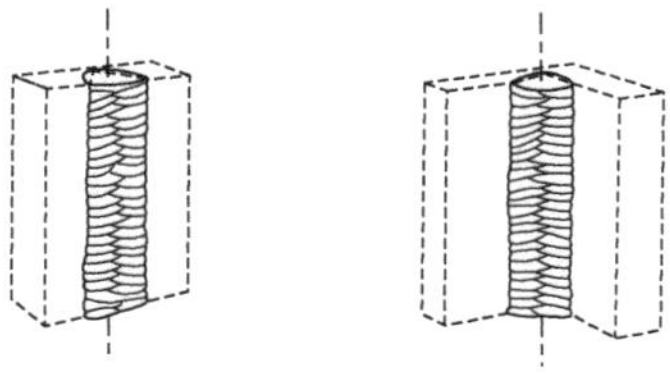

Figure 10.4
Welding in the vertical position

Out of position.

The formula for welds out of position multiplies the total time of the flat position weld by 2.55% (see Table 10.8).

Table 10.8 Labor Times for Out-of-Position Welding

Labor	Estimated Time for Welding Out of Position
1/4" fillet	8 minutes per foot x 2.55 = 20.4 minutes per foot
3/16" fillet	6.5 minutes per foot x 2.55 = 16.57 minutes per foot

This out-of-position condition applies to situations where the welder has difficult access.

Overhead position.

The formula for welds in the overhead position multiplies the total time of the flat position weld by 2.25% (see Table 10.9). Figure 10.5 shows the overhead position welds.

Table 10.9 Labor Times for Overhead Position Welding

Labor	Estimated Time for Welding Overhead Position
1/4" fillet	8 minutes per foot x 2.25 = 18 minutes per foot
3/16" fillet	6.5 minutes per foot x 2.25= 14.62 minutes per foot

Back gouge and last pass welding are included in weld rates.

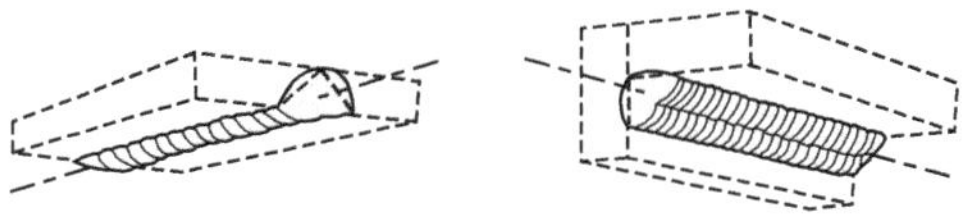

(4) Overhead Position

Figure 10.5
Welding in the overhead position

Heavy fillet welds

It will also be necessary to consider the number of passes for accumulative weld thickness. Table 10.10 summarizes a variety of thicknesses, using 3/16" fillet weld as a base that is then adjusted for larger welds.

Table 10.10 Labor Times for Heavy Fillet Welds

Labor	Estimated Time for Welding To Accumulated Thickness
3/16" fillet	6.5 minutes per foot
1/2" fillet	6.5 minutes per foot x 4 passes
	plus 3 minutes stop and start
	= 26 minute per foot + 3 minutes = 29 min/ft total
5/8" fillet	6.5 minutes per foot x 5 passes = 32.5 min/ft total
3/4" fillet	6.5 minutes per foot x 5 passes
	plus 3 minutes stop and start
	= 32.5 minutes per foot + 3 minutes = 35.5 min/ft total

Figure 10.6 shows samples of heavy fillet welds. The heat caused by heavy welds may distort the base material. Thus, you should be aware that the additional care to prevent issues with the base materials slows weld time down considerably. Welders will stagger their weld locations while performing heavy welds to even out the heat being applied to the steel.

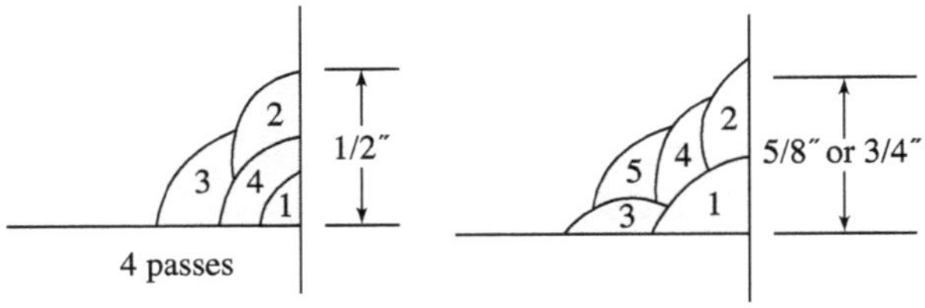

Figure 10.6
Heavy fillet welds

Additional Welds

Table 10.11 summarizes single bevel, Vee groove, and double bevel welds.

Table 10.11 Labor Times for Various Welds

Labor	**Estimated Time for Welding**
Single bevel weld: 1/2"	15 minutes per foot x 4 passes = 60 min/ft total
Single bevel weld: 3/4"	
back gouge and last pass	15 minutes per foot x 8 passes = 120 min/ft total
Vee groove weld: 3/8"	
back gouge and last pass	1/4" single bevel weld at
Vee groove weld: 3/4"	15 minutes per foot x 4 = 60 min/ft
at 60 degrees	
back gouge and last pass	3/16" single bevel weld at Vee groove weld: 3/4"
at 90 degrees	6.5 minutes per foot x 13 = 84.5 min/ft
back gouge and last pass	3/16" fillet 6.5 minutes per foot x 12 = 78 min/ft
Vee groove weld: 1 1/2"	
at 60 degrees back gouge	
and last pass	3/16" fillet 6.5 minutes per foot x 27 = 175.5 min/ft
Double bevel weld: 1/2"	1/4" fillet 8 minutes per foot x 4 = 64 min/ft
Double bevel weld: 1"	1/4" fillet 8 minutes per foot x 8 = 64 min/ft
Double bevel weld: 2"	1/4" fillet8 minutes per foot x 22 = 176 min/ft

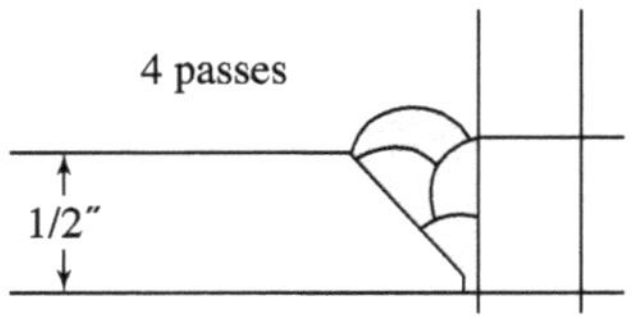

Figure 10.7
Single bevel weld for 1/2"

Figure 10.7 shows the build up of weld required to achieve a 1/2" single bevel weld whereas Figure 10.8 shows a 3/4" single bevel weld. The little plate under the weld is called a back up bar.

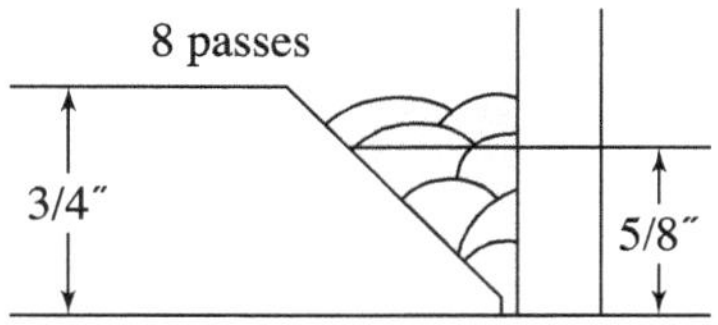

Figure 10.8
Single bevel weld for 3/4"

Figure 10.9 shows the build up of weld required to achieve the 3/8" vee groove weld. In turn, Figure 10.10 shows the build up of weld required to achieve the 3/4" Vee groove weld at 60 degrees. Figure 10.11 shows the build up of weld required to achieve the 3/4" Vee groove weld at 90 degrees. Figure 10.12 shows the build up of weld required to achieve the 1 1/2" Vee groove weld at 60 degrees.

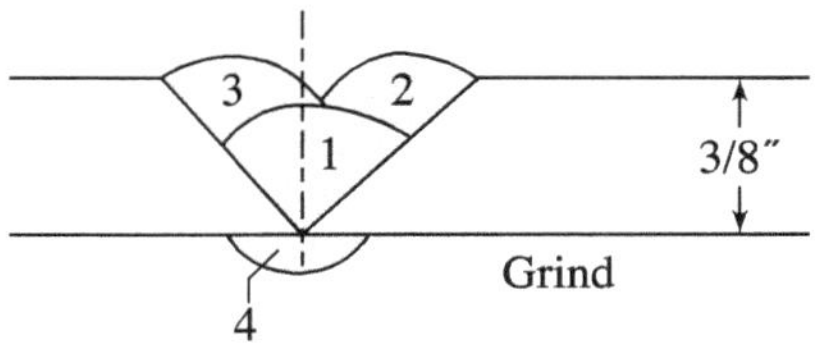

Figure 10.9
Vee groove weld for 3/8"

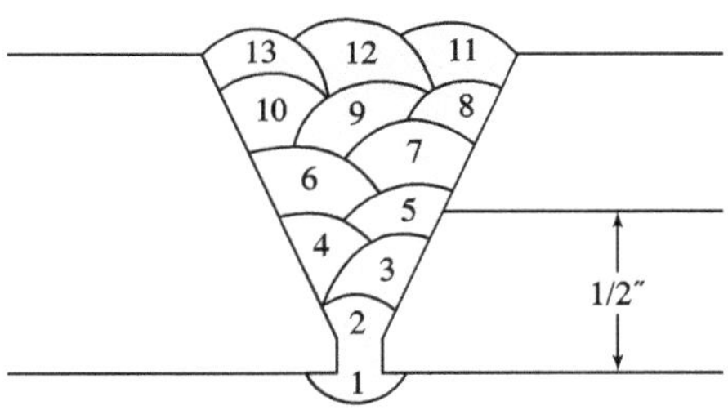

Figure 10.10
Vee groove weld for 3/4" at 60 degrees

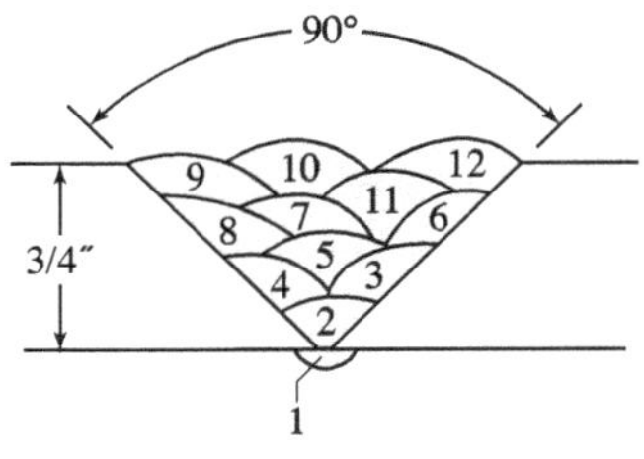

Figure 10.11
Vee groove weld for 3/4" at 90 degrees

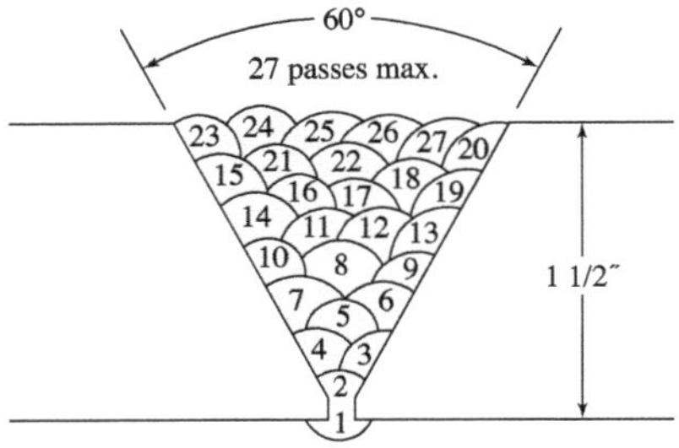

Figure 10.12
Vee groove weld for 1 1/2" at 60 degrees

When estimating double bevel welds, the welds need to be counted for both sides of the material. Figure 10.13 shows the double bevel weld and the welding symbol. Figure 10.14 shows the weld build up required for the double beveled 1" thick plate. Figure 10.15 shows the weld build up required for the double beveled 2" thick plate. Don't forget to include the additional percentage for out-of-position conditions where the need applies.

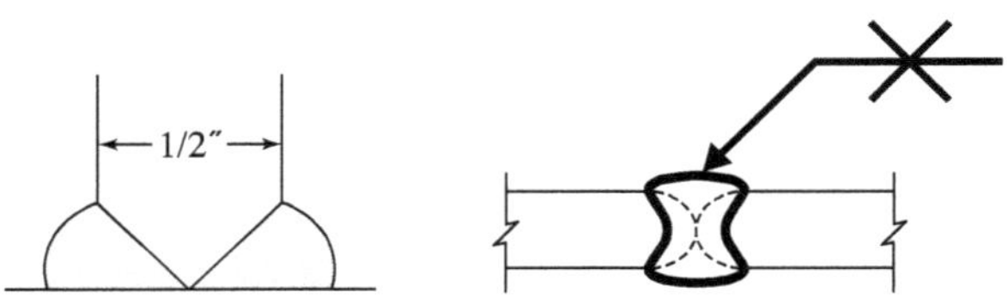

Figure 10.13
Double bevel weld 1/2"

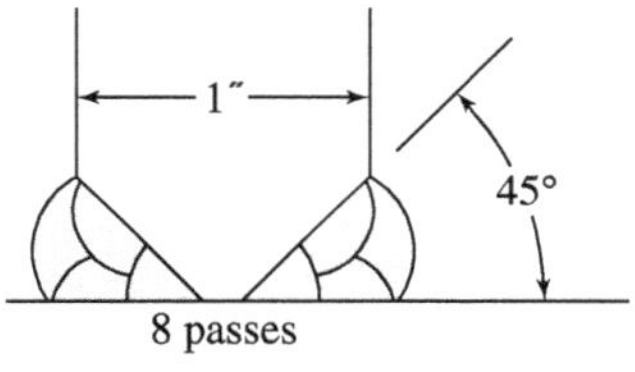

Figure 10.14
Double bevel weld 1"

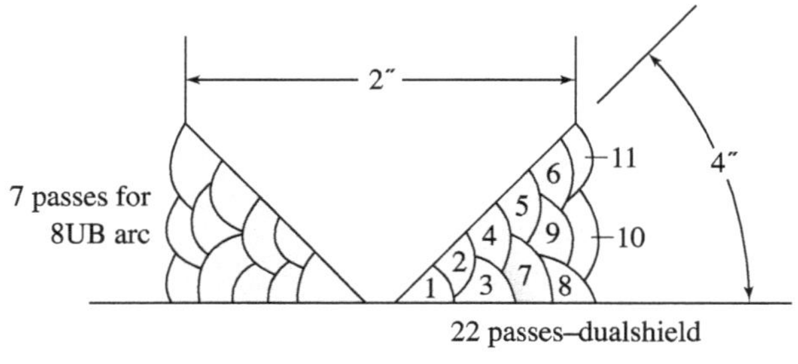

Figure 10.15 Double bevel weld 2"

STR—Straighten. Parts with lots of weld may bend from the isolated heat zone and will then need to be straightened (usually by heat, sometimes by use of a jig with hydraulic jacks) prior to completion of the part. This labor time requires a shop consult for accurate evaluation.

HAS—Headed Anchor Studs. When estimating these, include start up time of 30 minutes plus the minutes for each stud. Figure 10.16 summarizes the time per stud.

Size of studs	1/4 to 1/2″	5/8 to 3/4″	7/8 to 1″	Add for 45 degrees
Up to 10	2 min/ea.	2.5 min/ea.	3 min/ea.	Plus 1 min/ea.
11 to 20	1.5 min/ea.	2 min/ea.	2.5 min/ea.	Plus 3/4 min/ea.
21 to 50	1.25 min/ea.	1.5 min/ea.	2 min/ea.	Plus 1/2 min/ea.
51 to 100	1 min/ea.	1.25 min/ea.	1.5 min/ea.	Plus 1/2 min/ea.
101 plus	3/4 min/ea.	1 min/ea.	1.25 min/ea.	Plus 1/2 min/ea.

Figure 10.16 Estimated labor time for headed anchor studs

MISC—Miscellaneous. Anything else not covered in the functions already listed. Table 10.12 summarizes these estimates on the next page.

Fixture or Jig Time

Some fabrications require the making of a jig to hold parts and pieces in place while welding. Shop labor time for creating jigs or fixtures need to be added to the estimate; this time may be costed to this project as opposed to having a different job number for allocating this work.

The shop foreman usually is the one who might be making the decisions about creating jigs or fixtures to support quantity fabrications. Breaking points for making jigs will vary. The purpose behind building a fixture or jig is to reduce fabrication time.

If the quantity of the parts to be fabricated does not provide the benefit of a labor time reduction, then creating a jig to support the fabrication of the items does not make sense. It is helpful to the steel estimator to have a conversation about creating jigs or fixtures with a shop foreman prior to quoting a project that has quantity fabrications.

Oversize and special fabrications

It is a good idea to have a conversation within your organization about projects having special fabrication requirements, especially if you are concerned about out-of-the-ordinary fabrication labor and handling. For example, oversize fabrications may include large trusses, extra long and heavy beams, or frames with a finished size that is larger than your shop doors. Special fabrications are items that may require a high level of fabrication precision, a

Table 10.12 Miscellaneous Labor Estimates

Labor Category	Estimated Time
Bend	1 minute per foot plus 30 minutes for set up
Quantity bends	1/2 minute per foot plus set up time per worker
Bent anchors:	Flat bar and round bar including layout
Quantity of 5	14 minutes each
Quantity of 10	11 minutes each
Quantity of 20	10 minutes each
Quantity of 50	6 minutes each
Quantity of 100	5 minutes each
Stairs	1 hour per tread.6 hour per pre-manufactured tread
Stair Landings	.28 per square foot Otherwise, total of individual times per labor function per part
Handrails:	
1 1/2" standard	
pipe railing,3-line handrail	
Skewed railing	1.5 hours per post
Horizontal railing .	75 hours per post
Brackets	.3 hours
Return	1.5 hours each
Ladders	
Flat bar grab with 3/4"	
diameter rungs	1 hour per rung
Flat bar construction	
with 3/4" diameter rungs	
with cage	1.8 to 2 hours per foot
Standard time for layout	1 minute per foot
Standard time for cutting	1 minute per inch
Standard time for hole	1.5 minutes per hole to punch
Miscellaneous Labor Items	
Rough grind	4 minutes per foot
Sanding	2 minutes per foot
Heat camber	30 minutes each location
Thread	15 minutes set up plus 3 minutes each end
Rolling angle	1.5 hours each 20'-0 piece
Rolling plate	30–60 minutes each

large quantity of full penetration welds, a large quantity of plate work, or a large quantity of items to be beveled.

Direct Shop Labor

When you consider projects to bid, you must also anticipate the required shop schedule. Be certain that the required project schedule will fit your shop fabrication scheduling allowances and you can meet the project delivery requirements. You must understand both the direct shop labor hours that needed to complete particular project schedules as well as the number of people and hours that are available to work directly on those project schedules.

Understanding Labor Hours

Scheduling is always difficult for any shop to handle, as is maintaining a steady work flow for the staff you have available. After you have calculated the total labor hours on a project, you can judge the approximate time your shop will need to execute this project. Divide the total calculated shop hours for the job by the number of labor hours your shop has available per day.

Suppose your project requires 500 labor hours and you have 8 people who are available for directly applied labor hours. Then,

Labor hours = 500 total hours divided by 8 workers
= 62.5 hours/worker

In turn,
62.5 hours/worker divided by 7.5 hours per day
= 8.3 days/worker.

Another way to calculate this amont is to see that if we have 8 workers with 7.5 productive working hours per worker per day, then we can apply 60 hours per day for direct labor hours to the project. Divide the 500 total hours by the 60 hours per day—you will need 8.3 days for eight workers to complete the project.

Labor hours are generally estimated with consideration to best case scenario situations—with all materials being ordered and delivered on time, no one calling in sick, no equipment breaking down, no training for new hires, no shop clean up time, and no other unforeseen incidents that may interfere with shop schedules. Another assumption is that there are no other jobs that the shop hands need to work on at the same time. If you do need to estimate an allowance for other jobs, then try to determine how many workers you truly can apply fully to the project you are bidding in order to see how much longer you might need to run this project in your shop.

You may also have an option to hire additional workers for a short period of time. It is a good idea to know what special conditions you need to consider while bidding your work; then you avoid overselling your shop staff.

Understanding Cost Accounting

Cost accounting is necessary for fabrication shops. It enables shops to evaluate not only the costs of direct labor and direct materials, but also indirect labor and materials as well as overhead costs, and then assign those costs to individual jobs. Monitoring these costs will help estimators develop bids that more accurate reflect the complete costs of the project, not just the directly measurable ones. Expendable tools, equipment purchases, and shop labor for clean up and repair are some of the expenses that cost accounting methods help incorporate into the bids.

Costs for expendable tools such as grinding wheels, welding rods, welding tips, and brooms may be applied to the project. Estimators need to know about these expenses so they can factor in overhead expenses. Overhead expenses also include rent, electricity, phones, office (administrative) labor, and insurance. Different companies will have different policies regarding how to account for these expenses. Those policies will influence how bids are completed.

If these expenses are calculated into the labor cost per hour, then they should not be costed to the job again in another category. In this regard, labor for jigs and fixtures is somewhat of a gray area; make it a point to find out what the company policy is when you are estimating the work.

One of the ways to be more competitive in your estimating package is to eliminate the extremities of expendables or equipment purchases into the pricing. Expendables are already included in the pricing markups for overhead and profit. If the additional costs for purchased equipment were to be included in the steel estimate, the estimator would not be able to create a successful quote to make a low bid.

Applying labor hours to scheduling

In many cases, steel estimators are responsible not only for preparing estimates for bids, but also for overall negotiations and winning contracts. In other words, many are responsible for the company's sales. For them, a system needs to be developed by which they can effectively gauge the need for labor hours over time. Such a system would enable them to project the blocks of time when workers are available, and therefore work has to be generated (sold) for that shop. This system would help avoid peaks and valleys in projecting and scheduling the shop hours, in turn minimizing how erratically the company adds or removes workers.

The goal is to efficiently maintain the level of workers in the shop, adding as you grow. Effective management of the shop workers is a key element of creating a profitable fabrication business.

To determine how many hours of direct labor are available per week in the shop, multiply the number of shop workers you have by 7.5 hours per day and then by 5 days per week. In general, we use 7.5 hours per day, not 8.0, because we are looking for direct labor. Assume that, in general, the remaining 0.5 hour per day is indirect labor—the overhead labor that is not costed to a job. As a benchmark, for companies to be perceived as profitable, indirect labor hours should not exceed 6% of the total labor hours to be costed to projects.

Apart from lunch and breaks, many indirect hours are worked in a shop that do not go directly to the job—for example, meetings, cleaning, and repairing. It is unrealistic to think that all the workers in a shop will spend their entire 8 hours a day doing nothing but working.

Suppose you have a project with an estimate of 2800 hours over a 12-week schedule. Apply to that schedule (8) direct labor workers in the shop; they will be able to execute up to 300 hours per week to that project (5 days per week x 7.5 hours per day x 8 workers) Divide the total number of shop labor hours to be applied to the project (2800) by the total number of project weeks (12) to find out what the shop's actual application of labor needs to be. In this case, 2800 hours divided by 12 weeks is approximately 233 hours per week. This project schedule of 233 hours per week appears to be workable within your direct labor capacity of 300 hours. (Of course, this assumes that the project can be completed with the same number of hours applied from week to week.)

Additional considerations

When thinking about how to estimate direct labor hours for a particular project, you can think in terms of driving a car. At first you are stopped. Then it takes a bit before you are up to full speed, after which it takes a while to slow down again. A similar pattern happens in the fabrication shop with most projects that have been assigned to the workers. During the first week of a project, your workers may only be at 30–50% capacity while they stage materials and get set up. Then by Week Two they may be at full swing. This learning curve and ramp-up time are important to think about when you are trying to fit a project into a schedule.

Do NOT oversell your shop by thinking you can just bring in more people to get the job done. This approach is not a profitable way to conduct business. Overselling creates a labor bottle neck in your shop that will most likely result in overtime, as well as management issues directly related to the flush of personnel. You risk creating confusion and inefficiency.

Too often we think that workers can be hired to fill the shortfall in manpower and, in our estimating, simply increase the project labor hours. The truth is that new people require training and "settling in" time. At first their work will be slow while they are getting used to the new shop and required working procedures. This training time often adds more demand for labor hours on a project than you may think.

The result of rushing to add staff in order to keep up with an accelerated fabrication schedule will be an overrun in the labor hours actually applied to the job vs. the amount that was budgeted in the bid.

The additional labor hours spent training new employees is almost never accounted for by direct labor when quoting a project. A better way to manage new hires is to request that accounting assign a separate job number for new hire training. A certain percentage of new hire labor hours can then be costed to this separate job number rather than be applied to the project they are working on until they workers are fully trained.

Summary

This chapter has covered the basic theory on labor hours, followed by several applications of the theory. When estimating labor, the process is simplified by breaking down each function to increments of the hour. The individual time functions are added together to arrive at the total labor requirement per part. In turn, all those part totals are added together to arrive at the total labor hours for the project.

The actual labor for the project includes both direct and indirect labor. The amount of labor cost in the bid will vary among individual shops based on how indirect labor is costed by the company. Meanwhile, the steel estimator should check progress as the project moves along by comparing what was estimated against what was actually applied by the shop for the individual fabrications. This type of cross checking is necessary to help refine future bids and to understand the variance between what was bid and what the project cost.

After steel estimators arrive at the total labor hours for the project, they should step back and review this project as a whole, including how the project fits in with the overall company. The total scheduling time required for the shop to complete the work must be matched to the required time frame of the project as well as the shop's ability to meet that schedule based on its capacity and its commitment to other projects. By understanding the block of time required for fabrication, the estimators can determine how this new work will fit with the existing shop schedule.

11

Paint and Painting

Painting Structural and Miscellaneous Steel

This chapter on paint and painting will explain the prime painting of structural and miscellaneous steel fabrication. Prime paint is applied to fabricated steel parts and pieces to inhibit rusting. If the steel is not exposed to weather or caustic conditions and is not going to be finish painted, then a shop-applied standard paint may be acceptable.

Particular painting requirements for the steel will be indicated both in the Division 5 specifications for the fabricated steel and in the section for painting, 099000 (in Division 9).

Larger fabricators may have their own painting facility, including not only a paint booth, but also an area for wheelabrating or sandblasting as well. Most shops, however, perform only prime painting and will have to seek an outside source for sandblasting or wheelabrating, finish painting, and powder coating. Labor to stage, clean, and apply paint needs to be included in the steel fabrication estimate.

The amount of paint on the steel is measured by the "dry film thickness," which is a thickness of .001 of an inch and how paint is measured in the cured condition. A mil gauge is especially designed to measure dry film thickness; it is available in a wide range of types, from mechanical to digital.

Paint imperfections include runs, cracking, scratches, and ghosting—the last means you can see through to the base steel. These imperfections, which indicate improper coverage, are obvious paint errors that would be discovered upon visual inspection. Harder to see are paint holidays—pinholes in the paint that could incite rusting. These are best found with the use of a nanometer, a device that produces an electrical pulse. This current would be interrupted by pinholes in the paint, thereby alerting the user that there are imperfections.

To help you become more familiar with the painting requirements of steel, we will review a typical specification section on painting—Section 099000 Painting and Coating. The paint specifications are intended to cover all conditions of painting for the entire project. All interior and exterior surfaces that need to be painted or galvanized are referenced and explained; the types and colors of paints that are required are identified.

How to arrive at the amount of paint to be purchased and applied to achieve the specified dry film thickness is up to the expertise of the trade professionals. This chapter provides enough theoretical information to enable the reader to arrive at an acceptable amount of paint to allow for the cured coverage required.

To begin this process it is necessary to determine from Section 099000 the painting information to be applied to the fabricated steel. Once the steel specific information on the paint is found, we can calculate the amount of paint that needs to be procured.

Section 099000 gives us rules for painting, from approval submittal requirements on the paint we are using to the final inspection. The paint used on the steel, even if it is just prime paint, will have to undergo the same approval process as do all the other items supplied by the steel fabricator. Section 099000's specifications are shown in Figure 11.1:

1.1 SUMMARY

A. In general, the principle items of work include, but are not limited to, the following:
   1. Preparation of surfaces.
   2. Painting and finishing of all exposed-to-view interior and exterior surfaces, except as otherwise indicated or specified.
      a) Paint all surfaces unless otherwise indicated, scheduled, factory finished or indicated to receive a finish in other Sections of the specifications.
      b) Refer to the Finish Schedule and notes on the drawings and Part 4 - Paint Schedule at end of this Section for a general guide to the painting requirements.
   3. Field painting of all exposed-to-view mechanical and electrical items such as pipes, ducts, hangers, conduits, and like items in rooms or areas scheduled to be painted.
   4. Field painting of prime painted finished door hardware to match the door frame. Hardware includes, but is not limited to, coordinators' housing and associated door closer mounting brackets on door frames, astragals, and other items as required.
   5. Touch-up paint all blemished or otherwise disfigured paint on all surfaces which occur prior to acceptance of the building by the Owner.

B. Where items or surfaces are not scheduled or specifically mentioned, paint these items the same as adjacent similar materials or areas. If finish is not designated, the Architect will select these from the paint systems specified.

C. The following items do not require field painting:
   1. Composite panels.
   2. Exterior wall louvers.
   3. Toilet partitions.

Figure 11.1 Section 09900 specifications

The first section in Figure 11.1 summarizes the general information about the items to be painted. This summary explains that all items will be either primed or finish painted. It also identifies all the items to be field painted, how to paint items not identified, and refers us to the Finish Schedule and Notes on the drawings to find the general guide for painting.

1.2 RELATED SECTIONS

A. Section 05 12 00 - Structural Steel Framing: Shop priming of exposed steel shapes.

B. Section 05 50 00 - Metal Fabrications: Shop primed items.

C. Section 05 52 00 - Metal Handrails and Railings: Shop priming of exposed steel components.

D. Section 06 20 00 - Finish Carpentry: Shop finishing of all millwork items.

E. Section 06 42 16 - Wood Veneer Paneling: Shop finishing.

Figure 11.2 Related sections for items to be painted

The related sections in Figure 11.2 provide the specific locations for items to be painted. Note that there is no reference to fireproofing in this list. This tells us that the steel will not be fireproofed; therefore, it will need to be prime painted. From this information together with the individual sections on Structural Steel, Metal Fabrications, and Metal Handrails and Railings, we can tell that all the fabricated steel is to be prime painted.

The section on submittals shown in Figure 11.3 lets the steel fabricator

1.4 SUBMITTALS

A. General: Make submittals in accordance with Section 01 33 00.

B. Product Data: Submit complete list of products proposed for use, including technical data on each product to verify compliance; organize list to indicate painting systems to be used with each substrate.
   1. Submittal shall contain any proposed revisions to specifications (i.e. surface preparation, method of application, etc.) which contractor feels are necessary in their execution of the Contract.
   2. Any proposed revisions must be approved by the Architect prior to proceeding with the Work.

C. Submit paint manufacturer's product data sheets and Material Safety Data Sheets highlighting VOC limits for each paint or coating used in the building.

D. Samples: Using approved materials, prepare and submit samples of each type of finish, gloss, and color for approval. Label samples with color number, name and date. Provide three (3) samples each.
   1. Prepare paint color samples on 8-1/2 inch by 11 inch heavy, durable non porous paper.
   2. Furnish additional samples as required until colors and finishes are approved.

E. Contract Closeout Submittal: At time of Project Closeout coordinate submittal of extra maintenance materials with requirements of Section 01 77 00; refer to article entitled "Maintenance" in this Section for quantities and other requirements.

Figure 11.3 Submittals

know that a product data sheet will need to be submitted with the shop drawings to advise the designers of the type of paint we intend to use. The product data sheet may be obtained from the paint supplier or manufacturer.

With regard to samples, it is doubtful that the designers will ask for a finished sample of the prime painted items. If the fabricator were to be performing any finish painting, however, a sample of that finish painted project might be required.

If the fabricator are working with anodized aluminum or polished stainless steel, the requirements for samples may be found in this section as well.

1.5 QUALITY ASSURANCE

A. Single Source:
   1. To the maximum extent practicable, select a single manufacturer to provide all materials required by this Section, using additional manufacturers to provide systems not offered by the selected principal manufacturer.

   2. For each individual system, provide primer and other undercoat paint produced by same manufacturer as finish coat. Use only thinners approved by paint manufacturer and use only within recommended limits.

B. Visual Standards: Each distinct area of the finished Work shall be free of variations in color and sheen, runs, sags, holidays, blistering, checking, cracking, scratches and other signs of poor workmanship.

Figure 11.4 Quality assurance

The section on quality assurance in Figure 11.4 gives the guidelines for the workmanship expected with regard to painting. These visual standards are true for most shop painting standards.

1.7 PROJECT CONDITIONS

A. Environmental Conditions: Air temperature and substrate temperature and relative humidity shall be within the manufacturer's established limits. Do not apply exterior paint when the following conditions exist, unless requirements of paint manufacturers are more restrictive.
   1. Temperature: If surface and ambient temperature is above 90 degrees F, or below 50-degrees F.
   2. Relative Humidity: If relative humidity is above 85 percent.
   3. Weather: During damp and inclement weather or during excessively windy weather.

Figure 11.5 Project conditions

Environmental conditions affect paint application and quality; the requirements are usually outlined per Figure 11.5. Data sheets from paint manufacturers will include environmental condition requirements with regard to the ambient temperature and humidity. A quick check should be made to be sure that the project specification requirements are either met or exceeded with regard to the type of paint being used.

Extra costs could be incurred that may reflect on your quote if special conditions have to be created to meet these requirements. For example, the paint application requirements may require fabricators to paint their steel indoors as opposed to outside, or to send the steel out to a painting facility rather than painting at their own facility.

1.9 RIGHT OF REJECTION

A. Architect shall have the right to reject materials or work that does not comply with these specifications. Work so rejected shall be redone as directed. Work rejected and ordered to be redone shall be done at the Contractor's expense, and at no extra cost to the Owner.

1.10 MAINTENANCE

A. Extra Materials: Deliver to project site at a location designated by the Owner's Representative, extra paint materials in 5 gallon quantities for each color of finish coat material. Deliver paint materials in manufacturer's original unopened containers with each container clearly labeled.

Figure 11.6 Rejection and maintenance

Be aware that before leaving the shop, the quality of workmanship must be adequate per the requirements, as outlined by the project specifications in Figure 11.6. It is much more expensive to repair paint errors after the fabricated items have been shipped to the job site.

Primed or painted surfaces may get damaged during the installation. Because of the opportunity for damage during installation of the steel, most projects require a minimum of five gallons of repair or touch up prime paint be provided by the steel fabricator. Add this cost to the cost of the paint being used for the project.

PART 2 PRODUCTS

2.1 PAINT MATERIALS

A. Manufacturers:
1. Unless otherwise specified, the paint systems shown in the Paint Schedule at the end of this section are made up of products by Sherwin-Williams Co. trade names and numbers.
2. It is not the intent to limit products to that manufacturer, but rather to establish a quality which is required for this project.
3. Refer to PART 4 for Paint Schedule.

Figure 11.7 Paint materials

Figure 11.7 shows the section on paint materials and leaves the door open with regard to manufacturers. Item #2 gives fabricators the opportunity to use their own standard shop prime paint—every shop seems to have their preferred paint they like to use. It will be up to the steel fabricators to prove that the shop primer they want to use will be able to go through the approval process and also to prove that the paint product is compatible with the finish paint requirements for the painting contractor.

C. Substitution Requests: Submit for acceptance under provisions of Section 01 25 13. Indicate that manufacturer will conform to specified qualifications, including requirement for manufacturer to employ and maintain full time technical field representatives and testing equipment.

Figure 11.8 Substitution requests

If the prime paint that the steel fabrication shop wants to use is not provided for in the list of manufacturers, this section (Figure 11.8) of the 099000 specification refers to the substitution request form where the submittal for a substitute may be made. If there is a problem in the field with compatibility, it will be up to the steel fabricators to prove themselves in the right or provide the remedy for repair at their own expense.

F. Provide primer and finish coats which are compatible with each other and with prime coats provided under other Sections. Provide barrier coats over incompatible primers or remove and re-prime as required.

G. Tint each undercoat a lighter shade than finish coat so that numbers of coats can be easily discerned. No color mixing will be allowed at the job-site.

H. Thinner: Type as recommended by the paint manufacturer. Use thinner only when recommended by the paint manufacturer, and then only in a quantity as indicated on the label.

I. Primers: Primers, except metal primers, shall be white in color for inspection purposes.

Figure 11.9 Compatibility with other paints, thinner, and primer

In Figure 11.9, item F points out the need for compatibility and the solution for when prime and finish coats of paint are not compatible. Item I relieves the steel from any specific color requirement.

PART 3 EXECUTION

3.1 EXAMINATION

A. General: Examine surfaces to receive paint finish for conditions that will adversely affect execution, performance, or quality of work and which cannot be put into an acceptable condition through reasonable preparatory work as specified herein.

1. Surfaces which are unfit to receive the work of this section shall be repaired, replaced or re-finished such that they are acceptable and such that the work of this section may be done as specified. It shall be the responsibility of the General Contractor to ensure that these provisions are strictly enforced.
2. Commencement of Work constitutes acceptance of surfaces and conditions.

Figure 11.10 Examination

In Figure 11.10, the description on examination defines the painter's responsibility not to paint surfaces that would result in a poor quality product. This also extends to the steel fabricators' prime painting.

3.2 SURFACE PREPARATION (GENERAL)

A. General: Surface preparations and cleaning procedures shall be in strict accordance with the instructions and specifications of the paint manufacturer and with the requirements of this specification.

Figure 11.11 Surface preparation (general)

The information in Figure 11.11 directs the painter to refer to the paint manufacturer's recommendations for surface preparation. The same applies to the painter for the steel fabricator.

E. Uncoated Ferrous Metal: Thoroughly degrease surfaces using solvent (SSPC-SP 1) and remove rust and foreign matter by scraping, sanding, wire brushing, or other abrasion methods as necessary in accordance with SSPC-SP 2 and SSPC-SP 3. Remove pits and clean to bright metal before priming. Apply primer on the same day of surface preparation.

F. Shop-Coated Ferrous Metal: Thoroughly degrease surfaces and clean using solvent (SSPC-SP 1). Remove loose rust, blistered and peeling paint to bare metal by scraping, sanding, wire brushing, or other abrasion methods in accordance with SSPC-SP 2 or SP 3; feather edges of adjacent sound paint. Dull glossy surfaces by scuff-sanding and wipe down. Spot-prime all abraded portions, rust areas, and bare surfaces with specified primer on same day of surface preparation.

G. Galvanized Metal: Clean surfaces to remove factory films and oily residue as recommended by the paint manufacturer. Responsibility for insuring that the surface is properly prepared rests with the painting sub-contractor. Clean galvanized metal the same day to be painted.

Figure 11.12 Additional surface preparation requirements

In Figure 11.12, items *E, F,* and *G* indicate the surface preparation requirements for the steel. The references to SSPC or the Steel Structures Painting Council SP-1, SP-2, and SP-3 requirements are standards for "hand tool cleaning" of the steel surfaces. These standards are the least expensive to perform as opposed to the higher standard that reference the need for wheelabrating or sandblasting of the steel.

Painting over galvanized steel is usually performed in the field. Finish painting over galvanized steel tends to leave an "orange peel" or rough appearance to the finish.

The expectations of the designers sometimes cannot be met with regard to finish conditions on steel because the base material is not manufactured for aesthetic value. The reason for the application of galvanizing material is to extend the life of the product. Finish paint is applied to make the product look better. However, when finish paint is applied over a rough surface, it usually will exaggerate imperfections. This exaggerated rough appearance of finish paint over galvanized items will create an opportunity for rejection from the architect based just on the lack of aesthetic value. For this reason, finish painting over galvanized items is a specific exclusion in the steel fabricator's bid letter.

3.3 CLEANING PRIOR TO PAINTING

A. Remove dust and loose deleterious materials from all surfaces before beginning painting operations. Program the cleaning and painting so that dust and other contaminants from the cleaning process will not fall in wet, newly painted surfaces.

Figure 11.13 Cleaning prior to painting

The reference in Figure 11.13 to cleaning prior to painting is in reference to cleaning any surface. The paint manufacturers include the surface cleaning requirements with their product information. Still, it must be understood, when compared to the cleaning requirements included within the project specifications, the designers expect that the strictest requirements are to be followed.

Where surface preparation conditions conflict, the deciding factor will be what is contained in the document that was approved at the product information submittals.

The section shown in Figure 11.14 covering the application of the paint is general in presentation, but specific in the application. These are the rules that the painters must follow to achieve the level of quality that the designers are looking for.

Then the section in Figure 11.15 defines the requirements for paint repair; the section applies to the steel surfaces as well.

In Figure 11.16, the paint schedule section provides direction of the paint types to be used on specific items to be painted. Again the designers indicate they are not requiring the use of a particular type of paint, though they do direct that a product of equal or of increased quality is acceptable

In this section of the paint schedule (Figure 11.17), we know that the galvanized items will be finish painted. Although finish painting over galvanized may be performed in the field, it is best to specifically exclude it in the bid letter, just to be safe.

Applying finish paint over a galvanized finish creates an uneven or bumpy finish, often referred to as an "orange peel" effect. Many architects do

3.4 APPLICATION OF PAINT

A. Mixing: Mix paint materials in accordance with the manufacturer's instructions and directions. Mix often enough during application to keep the paint uniform and to ensure complete dispersion of pigment and a uniform composition.
   1. Prepare multiple component coatings using all of the contents of the container for each component as packaged by the manufacturer. Mixing of partial kits will not be permitted. Multiple component coatings that have been mixed shall not be used beyond their pot life. Only the components specified and furnished by the manufacturer, including thinner if required, shall be mixed.

B. Application: Apply paint in accordance with the manufacturer's directions. Use techniques best suited for substrate and type of material being applied. Brushes and rollers shall be of a type best suited for the type of material being applied.
   1. Apply intermediate and finish coats within the manufacturer's recommended top coating time periods.
   2. When applying paint to drywall, use a roller nap no greater than 3/8 inch so as to achieve a light stipple finish.
   3. Brush and level out paint applied to metal door frames to achieve a nearly sprayed-on appearance.
   4. If metal doors are not sprayed, finish may be applied with 1/4 inch nap roller.

C. Apply each coat of paint as a continuous film of uniform thickness, free from holidays, sags, crawls, pinholes, blisters, unevenness in color, or other evidence of poor workmanship. Repaint thin spots or areas missed in the application and allow to dry before applying next coat of paint.
   1. Give special attention to ensure that surfaces, such as edges, corners, crevices, welds and exposed fasteners receive a dry film thickness equivalent to that of flat surfaces.
   2. Each coat shall be free of dirt, dust, moisture, etc., prior to application of next coat.

D. Allow each coat of paint to thoroughly dry, full thickness of the film, before application of the succeeding coat. Paint is considered dry for recoating when the next coat can be applied without the development of any detrimental film irregularities such as wrinkling, lifting, or loss of adhesion of the previous coat.

E. Coverage for each paint material is specified as either the total minimum dry film thickness in mils, or the spreading rate in square feet per gallon over the surface designated. Actual coverage rate will vary depending upon the texture and porosity of the surface, climatic conditions, etc.
   1. The number of coats specified is the minimum required, irrespective of the coating thickness.
   2. In the event the required paint thickness is not achieved, apply additional coats until the required thickness is obtained.
   3. Do not exceed manufacturer's recommended maximum film build-up per coat (wet mils).

Figure 11.14 Application of paint

3.5 DAMAGED PAINT SURFACES

A. General: Before final acceptance of the work by the Architect , repair or re-finish painted surfaces which have been damaged at no additional cost. Refinish whole wall where portion of finish is not acceptable.

B. Areas of chipped, peeled, or abraded paint shall be hand or power sanded, feathering the edges. Prime and finish coat the areas using the same material as originally scheduled. Depending on the extent of repair and its appearance, an overall finish coat may be required by the Architect to achieve uniform appearance.

Figure 11.15 Damaged paint surfaces

PART 4 SCHEDULE

4.1 PAINT SCHEDULE

A. General: Provide the following paint systems for the various substrates wherever these materials occur in the Contract Documents, except as noted otherwise.
   1. Unless otherwise specified, the following paint systems are made up of products identified by Sherwin-Williams Company trade names and numbers.
   2. It is not the intent to limit products to that manufacturer, but rather to establish a quality which is required for this Project; other approved manufacturers are listed in Part 2 of this Section.

Figure 11.16 Paint schedule

4.2 EXTERIOR SURFACES

A. Galvanized Metal Doors and Frames: - Polyurethane, Semi-Gloss:
   1. Primer: "Pro-Cryl Universal Primer B66-310."
      a) VOC's : 110 g/L.
      b) Dry Mil Thickness: 1.5 minimum.
   2. Two Finish Coats: "Centurion Water Based Polyurethane (B65-700 Series)."
      a) VOC's : 66 g/L.
      b) Dry Mil Thickness: 4.0 to 6.0 minimum total.

   Note: Wash and etch surfaces using Sherwin Williams "Clean & Etch," or other material as recommended by paint manufacturer. Dilute material with water as recommended by manufacturer. Apply said pretreatment not more than 8 hours in advance of applying primer. Pretreatment, if required, shall be done at no additional cost.

B. Galvanized Metal Surfaces Including Structural Steel Members - Acrylic, Semi-Gloss:
   1. Primer: "Pro-Cryl Universal Primer B66-310."
      a) VOC's : 110 g/L.
      b) Dry Mil Thickness: 1.5 minimum.
   2. Two Finish Coats: "Sher-Cryl HPA High Performance Acrylic (B66-300 Series)."
      a) VOC's : 187 g/L.
      b) Dry Mil Thickness: 5.0 minimum total.

   Note: Wash and etch surfaces as specified above.

Figure 11.17 Exterior surfaces

E. Exposed Overhead Structure (Concrete and Steel Members, including pipes, ducts, hangers, conduits, and like items): Latex, Eggshell:
   1. One Finish Coat: "Waterborne Acrylic Dryfall (B42W1)"
      a) VOC's : 58 g/L.
      b) Coverage Rate: 130 to 225 square feet per gallon (approximate). An additional pass or separate coat may be required to achieve uniform color, hiding and appearance.
      c) Clean galvanized steel surfaces as specified above.
      d) Application: Apply paint using spray equipment, over-lap each pass 50 percent followed by cross-hatch pattern; adjust equipment as recommended by manufacturer to achieve dry-fall performance with overspray.

F. Ferrous Metal, all remaining metal items including metal doors and frames - Latex, Semi-Gloss:
   1. Primer: "Pro-Cryl Universal Primer B66-310."
      a) VOC's : 110 g/L.
      b) Dry Mil Thickness: 1.5 minimum.
   2. Two Finish Coats: "Centurion Water Based Polyurethane (B65-700 Series)."
      a) VOC's : 66 g/L.
      b) Dry Mil Thickness: 4.0 to 6.0 minimum total.

   Note: Omit full prime coat on shop primed steel surfaces. Spot prime as previously specified.

Figure 11.18 Finish painting for steel

not appreciate this e "orange peel" effect of painting over galvanizing, and often will reject products that have been finished this way because of the uneven surface. Concern for product rejection is the reason steel fabricators favor excluding finish painting over galvanizing.

In Figure 11.18, Sections *E* and *F* both identify the steel items to be finish painted. Section F gives the DFT requirement of the prime paint.

This paint specification has identified the items to be prime painted and the type of paint with which the primer needs to be compatible. We know that the acceptable prime paint is one that requires only the minimum of cleaning requirements and that the required minimum for primer is 1.5 DFT.

Because of its reference to finish painting of galvanized steel, this section indicates there will be some items to be galvanized. But these items are left for identification in the Division 5 specifications. As there is no reference to related sections regarding fireproofing, all steel items will be prime painted.

Calculate the Amount of Paint Required

To calculate the amount of prime paint to use on structural steel, the formula used assumes an industry standard for paint coverage of 1.5 mils dry film thickness. Theoretical paint use is based on the total weight of the fabricated steel to be painted.

The prime paint used will be a product that is compatible with the finish paint required at the project, if the steel is to be finish painted. Usually structural steel is hidden inside walls, thus, there is no need for any finish paint.

Sometimes the structural steel will be fireproofed and, as such, will not require any prime paint. The specifications for the structural steel in Division 5 will identify the painting requirements. Section 099000 Paint and Painting will identify the required prime and finished paints to be used.

When calculating the required paint to achieve the 1.5 DFT for primer based on the total weight of the steel to be painted, use the following information:

- Total weight divided by 2000 lbs equals total tons
- The theoretical square footage of the surface area to be primed is 377 square feet per ton of structural steel.
- A gallon of paint has an average 277 total square feet, including a 30% loss

Using a total weight of 200,000 lbs, the calculation is as follows:

- 200,000 lbs ÷ 2000 lbs/ton = 100 tons
- 100 tons x 377 ft^2/ton = 37,700 ft2

Next, divide 37,700 square feet by 277 square foot per gallon, the area of paint application per gallon providing 1.5 millimeters of dry film thickness, or dft of coverage.

- 37,700 ft2 ÷ 277 ft2/gallon = 136 gallons

136 gallons rounded up to the nearest 5-gallon purchase is 140 gallons. Paint comes in 5 gallon cans; suppose the cost is $25 per gallon. You would have to buy 140 gallons at $25.00 per gallon for a total of $3,500.00.

When calculating a dry film thickness requirement for more than 1.5 mils, you will need to adjust this formula accordingly.

Not figured into this paint cost are any applicable small tools or supplies. Additional costs include paint thinners or extenders, waste storage, and clean up rags.

Painting Miscellaneous Steel

Calculating paint for miscellaneous steel requires additional thought about paint loss. The theoretical paint coverage per gallon includes a 30% loss. With miscellaneous steel, that factor may be increased to a 40–60% loss depending on the nature of the fabrication.

Skinny framing with lots of bulk—like box trellis frames or sections of handrail—will require extra paint and handling because of the hanging or turning of the parts.

Theoretical Labor for Paint Application

The formula for calculating labor time is: multiply the number of gallons by 1.5 hours per worker per gallon. This amount does include set up and clean up time. If helpers/cleaners are only part of these hours, calculate their time as an additional percentage of the total time allowed. For example, for 10 gallons of paint, the time is:

10 gallons x 1.5 man-hours per gallon = 15 hours

If a helper is required for 10% of this time, then the time is:

15 man-hours x 1.10 = 16.5 hours

Steel estimators must also be aware of the cleaning requirements necessary to accommodate the adherence of the paint. Hand tool cleaning by solvents or grinding is typical for most prime paints. The quality level of the cleaning required for the paint is identified by the paint manufacturer with the product data sheets that are included with their product specifications.

Some prime paints require surface preparation by way of sandblasting or wheelabrating. Costs for sandblasting and wheelabrating will vary according to the level of surface preparation that needs to be accomplished.

Product Cleaning Prior to Painting

Cleaning requirements indicated by the paint product data sheets may be different than what is indicated in the project specifications. Product use is the governing factor and the product specific cleaning and painting requirements will be presented during the product approval review. Indicators for cleaning levels are set by the Steel Structures Painting Council requirements.

Shop Cleaning of Structural Steel—Specifications

In the following list of specifications, SSPC refers to Steel Structures Painting Council.

SSPC—SP1
Solvent Cleaning

The removal of oil, grease, dirt, soil, salts, and contaminants by cleaning with solvents, vapor, alkali, emulsion, or steam.

SSPC—SP2
Hand Tool Cleaning

Removal of loose rust, loose mill scale, and loose paint to degree specified, by hand chipping, de-scaling, sanding, scraping, and wire brushing.

SSPC—SP3
Power Tool Cleaning

Removal of loose rust, loose mill scale, and loose paint to degree specified, by power tool chipping, descaling sanding, wire brushing, and grinding.

SSPC—SP4
Flame Cleaning of New Steel

Dehydrating and removal of rust, loose mill scale, and some tight mill scale by use of flame, followed by wire brushing.

SSPC—SP5
White Metal Blast Cleaning

White metal blast cleaning is the removal of all visible rust, mill scale, paint, and foreign matter by blast cleaning by wheel or nozzle (dry or wet) using sand, grit, or shot. This is used in conditions where there is a very corrosive atmosphere, and the high cost of cleaning is warranted.

SSPC—SP10
Near White Blast Cleaning

Blast cleaning nearly to white metal cleanliness, until at least 99% of each element of surface area is free of all visible residues (for high humidity, chemical atmosphere, or a marine or other corrosive environment condition.)

SSPC—SP6
Commercial Blast Cleaning

Blast cleaning until at least two-thirds of each element of surface area is free of all visible residues (for rather severe conditions of exposure).

SSPC—SP7
Brush-off Blast cleaning

Blast cleaning of all except tightly adhering residues of mill scale, rust, and coatings, exposing numerous evenly distributed flecks of underlying metal.

Labor Allowances for Wheelabrating and Hand Blasting

This illustration shows calculations based on SSPC-SP5 White Metal Blast Cleaning. Assume the following:

- Wheelabrate = (tons x .6 hours per ton)
- Hand Blasting = (project square foot total divided by 800 square feet per day divided by 2 workers per day)
- Extra Handling = (total tons x 1.5 hours per ton)

Also assume 200 tons and 30,000 square feet.

- Wheelabrate = 200 tons x .6 hrs/ ton = 120 hrs

- Hand Blasting = 30,000 ft^2 ÷ 800 ft^2/per day ÷ 2 workers/day = 37.5 days x 8 hrs/day/worker x 2 workers = 600 hrs
- Extra handling = 200 tons x 1.5 hrs/ton = 300 hrs
- SSPC-SP5 Total = (120 + 600 + 300) hrs = 1020 hrs
  In turn, 1020 hrs ÷ 200 tons = 5.1 hrs per ton

Painting structural and miscellaneous steel is always a challenge to estimate accurately. Shop conditions, type of painting equipment, loss factors for overspray, and cleaning requirements are all variables that have a direct effect on estimating the correct amount of product.

The theoretical information within this chapter is deliberately generic. Adjusting it for specific shop conditions and painting needs will support the estimator in arriving at a reasonably close estimate.

Summary

This chapter on Paint and Painting reviewed the painting requirements as shown in Division 9, Section 09900 Painting and Coatings. Explanations were given on the specific points of information used by steel fabricators to correctly clean and coat the fabricated materials. Examples were provided that estimators can use to calculate the amount of product required to clean and prime paint the steel, as well as the required labor time to do so.

12

Shipping and Handling

In this chapter we will review the general requirements of shipping and handling, including considerations to special size or out-of-the-ordinary conditions.

Shipping

Structural and miscellaneous steel is loaded onto flatbed trailers, with a typical maximum weight of 40,000 lbs and usually within the standard shipping requirements of 8'-6" wide, under 14'-0" tall and 40'-0: long.

Items that ship first to the job site are generally the anchor bolts for the columns and any steel fabrications that need to be embedded into concrete. These small deliveries could be handled with a company truck or pickup for a minimal cost. Be aware that these items will ship separately as they are required prior to the concrete pour. Sometimes they will ship weeks ahead of the rest of the steel.

Steel columns and beams usually stack or nest easily onto flatbed trailers. Depending on the size of beams being used for the framing, it may be that the flatbed is full by weight long before the flatbeds are full by physical size. To calculate the shipping costs for steel beams and columns, the estimator can take the total weight and divide by 40,000 lbs to arrive at the number of shipments for pricing.

Large fabricated steel trusses may be over 8'-6" wide and over 40'-0" long. Special trucking for such conditions are available. Added costs for pilot cars and additional permits may have to be included in the pricing for the shipping of such items.

Over-width and over-height loads may require shipment during a special time of day or over a special route. The total number of shipments required with these special conditions along with physical size and total weight is required for the shipping to be priced accurately.

Height and width restrictions for bridges, power lines, narrow roads, and roads with limited lane access; heavy traffic; and weather conditions will not only have an impact on the pricing, but also an affect on delivery scheduling. Be sure you are aware of the time of year that these shipments occur. Include that information with your shipping inquiry.

Miscellaneous steel items—like steel stairs, stair landing frames, handrail, catwalks, mechanical frames, trellis frames, and ladders—

do not nest well. They will fill a flatbed with bulk rather than weight, resulting in more flatbed loads because of the volume. For loads like these, a maximum weight of 10,000–20,000 pounds may have to be allowed, even though we are using an 8' x 40' flatbed, because the load will have a lot of empty space. For this reason, pricing for shipping miscellaneous steel will have to be calculated separately from the structural steel.

Some odd-shaped fabrications will require special dunnage (wood underneath and in between steel items) or wood-framed supports that will have to be built to allow shipping by flat bed trucks. The cost of an additional shipment may be less than the shop labor for additional packaging or adjusting the load to save for one more truck.

Large fabrications may require the use of low boy or special trailers that have increased capacity, but also may be limited in their availability. The scheduling of special equipment may also affect cost and job site delivery schedules.

Additional shipments for items to be subcontracted will have to be considered. Shipments to the galvanizer, items to be formed, and items to be sandblasted and painted will have to be included both from the fabrication shop to the vender and then back to the shop or job site.

Some situations will not come to light or be resolved until the fabricator is actually working on the project. The steel estimator has to be aware that these conditions will exist. By adding the requirements of special shipping considerations to the qualifications in the bid letter, estimators can protect the steel fabricators from exposure to the backlash of such events.

It often happens that partial shipments that are not anticipated in the quote are requested by the customers. An added note in the bid letter that "special allowances for customer preference-driven shipments will not be accommodated without reimbursement" may offer some relief. Accommodating special requests and staggered shipments seems innocent until the accounting reveals that the shipping budget is in the red.

Steel estimators always have to be aware of the packaging or shipping requirements for the projects they are bidding because these additional costs for special trucking and handling may slip by unnoticed.

So much time is spent with finding, listing, and pricing materials, goods, and services by the estimators that additional shipping requirements may be overlooked while completing a quote. Taking a little extra time to give thought to shipping and handling will help prevent any negative financial impacts to the project.

In order to get accurate pricing from an outside source, steel estimators should have a total weight, the approximate shipping window of time, and the zip code where the job site is located. Any additional special shipments need to be identified by physical size, weight, and quantity of loads. Provide pictures if you have to. Make copies from the drawings of the special fabrications to send along with your quote request to get more accurate pricing. Freight travelling long distances at the wrong time of year may thoroughly undermine the quote for the shipping if the estimator hasn't considered all the aspects of the potential shipping requirements.

Handling

Labor hours to handle and stage materials are the most difficult and ambiguous factors to predict when creating an estimate for fabricated structural and miscellaneous steel. Steel estimators trust that the shop is aware of the need to be efficient with their materials and handling time. Therefore, labor hours are not usually added for material handling. If labor hours for handling are added to the pricing, however, they are generally the first place to cut money when working to be competitive on a quote.

When the market is good, handling may be factored ranging from .015–.07% of the total labor hours on a project, depending on the complexity. When the market is bad, this factor is simply eliminated. Consideration for handling tends to be largely ignored until the hours run over on a job and the search is on for the guilty party—which usually starts with the estimators.

When steel estimators think about the fabrication processes within the shop, the materials for the projects are in constant motion until delivered to the job site. From the time the truck arrives with delivery from the warehouse until the materials are offloaded at the final destination, the steel is in constant movement—from the saw to the worktable to the paint booth and out. Some of this time is allowed for in the estimated labor hours per function. But unless the shop is 100% efficient, there will always be exposure to handling time that cannot be accounted for.

Adding another factor into the mix for large structural jobs, materials are loaded onto trucks according to the zoning requirements determined between the steel fabricator and the erector prior to the start of fabrication. Who can factor in the sorting and staging time for the yard to be organized for this loading process? This additional handling can be avoided if zoning requirements are made known either during the bidding process or at least prior to the steel materials being ordered.

Steel estimators can do little to protect themselves from the unknown and unforeseen factors of extreme handling conditions. If everything was factored in, the opportunity to be low bidder would be eliminated. The only saving grace is in knowing that the competition you are bidding against may have the same exposure to unknown conditions as you do.

Steel estimator will have the first complete look at a project while creating the steel estimate. By maintaining good notes on discussions that occurred prior to bid time regarding special fabrications and difficult or extreme handling conditions, estimators will secure themselves their only protection from implication should they be low bidder on the job.

Summary

This chapter on Shipping and Handling is intended to give steel estimators some insight as to the specific aspects of shipping steel materials. Advice on obtaining accurate pricing from the shipping agency is provided, along with thoughts to special and extreme shipping requirements. Additional shipments to and from vendors, including extra handling and staging, have been presented for consideration. All of this information is designed to support estimators in thinking of shipping materials as more than just adding a price per pound.

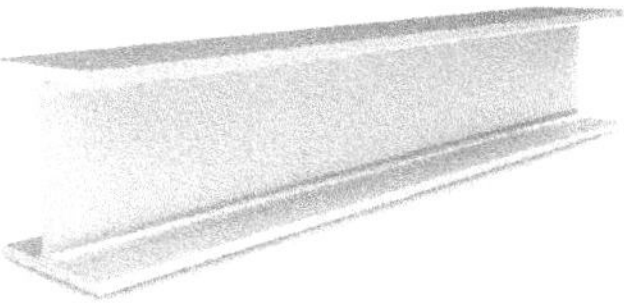

Notes and References

Reference Books and Manuals

A.I.S.C. Manual of Steel Construction
American Institute of Steel Construction, Inc.
400 North Michigan Avenue
Chicago, Illinois 60611

The American Institute of Steel Construction is the technical specifying and trade organization for the fabricated structural steel industry in the United States. This manual is the guide to the materials and practices for steel fabricators. The manual includes the AISC Code of Standard Practice, which outlines the commonly accepted standards of the Structural Steel Fabrication Industry.

Detailing For Steel Construction
American Institute of Steel Construction, Inc.
400 North Michigan Avenue
Chicago, Illinois 60611

This steel detailing manual will help you understand the entire process for steel detailing. It will also help you read and understand the shop detail drawings.

ASTM Standards for Structural Steel Fabrication
American Institute of Steel Construction, Inc.
400 North Michigan Avenue
Chicago, Illinois 60611

AWS D1.1: Structural Welding Code—Steel
American Welding Society
550 N.W. LeJeune Road
Miami, Florida 33126

The AWS D1.1 code covers the welding requirements for any type of welded structure made from commonly used carbon and low-alloy constructional steels.

These manuals provide very necessary information for compliance with trade practice and requirements. Be aware that these manuals are updated every few years and upgrade your library accordingly.

Guidelines

Pricing averages—based on project weight—used as a check only

These guidelines provide standards for where pricing should generally be in each category. They are intended for use as a back check of finished pricing, not for use in creating a fabricated steel price. If the price per pound for each category is significantly more or less than the averages shown, one might look for error of some pricing or weight conditions within the quote.

If the need to create a budget pricing for a project should arise, these average figures may be used for check value only based on the weight of the materials that are to be supplied. Because views on complexity of the project will vary with the level of practical steel fabrication experience, caution is advised regarding judgment in using these figures. This price range is based on the Northwest United States market, 4th quarter 2010.

Detailing (project weight x .16 to .30 per lb)

Detailing costs per pound will vary with the complexity of the steel detailing on a project. A simple frame project will be less per lb, whereas pricing for remodels and projects with lots of bracing, and miscellaneous framing, will be more per lb.

Labor (project weight x .20 to .30 per lb)

Structural Steel will be in the .20/lb range ;Miscellaneous Steel is in the .30/lb range.

Materials (project weight x .60 per lb – including waste)

This price range will change with the market. What we are looking for with this average is to make sure there are no huge errors in the materials purchase.

Fasteners (Material cost x .10% or .06 per lb)

Generally the costs of fasteners will run about 10% of the total material cost if using A307 anchor bolts and A325 grade steel to steel fasteners. This will vary if the fasteners are a special grade or are galvanized.

Buyouts (as required by project)

Prime Paint (Project Weight: .06–.10 lb or priced as buyout)

This cost per lb includes both labor and materials to hand tool clean and apply one shop coat at 1.5mils dft only. Sandblasting and wheelabrating are extra.

Galvanizing (Project Weight x .32–.48 lb or priced as buyout)

Handling (Total Labor Hours x .015–.07/lb)

Handling costs will vary by shop capacity and complexity of the project.

Shipping (Project Weight x .04–.06/lb)

This pricing is driven by fuel costs and special size loads. Large quantity loads may bring this pricing down (see Figure N.1).

This sample is based on the weight of 120,000 lbs of structural steel - simple frame, no specials

120000

Summary	Hours Weight	Cost lb/hr	Cost Extensions	Cost per lb		Sale Amount markup at .15%	Cost per lb
Detailing	274	$ 70	$ 19,200	$0.16		$ 22,080	$ 0.18
Engineering	0	$ 40	$ -	$0.00		$ -	
Labor	600	$ 40	$ 24,000	$0.20		$ 24,000	$ 0.20
Materials	120000	0.52	$ 62,400	$0.52		$ 71,760	$ 0.60
Hardware			$ 6,240	$0.05		$ 7,176	$ 0.06
Grating/Forming		$ 1.80	$ -			$ -	
Paint Material	82	25	$ 2,042	0.02	22620	$ 2,348	$ 0.02
Paint Labor	122	40	$ 4,900	$0.04	Sq Ft	$ 4,900	$ 0.04
Galvanizing	0	0.42	$ -			$ -	
Handling	42	40	$ 1,680	$0.014		$ 1,932	$ 0.02
Shipping	3.00	1250	$ 3,750	$0.0313		$ 4,313	$ 0.04
Totals			$ 124,211	$ 1.04 Per lb		$ 138,508	$ 1.15 Per Lb

Figure N.1 Shipping costs per lb

INDEX

www.ingramcontent.com/pod-product-compliance
Ingram Content Group UK Ltd.
Pitfield, Milton Keynes, MK11 3LW, UK
UKHW051912100926
8556IPUK00005B/854

9 780831 13439